NEUROMETHODS

Series Editor
Wolfgang Walz
University of Saskatchewan
Saskatoon, SK, Canada

For further volumes:
http://www.springer.com/series/7657

Stimulation and Inhibition of Neurons

Edited by

Paul M. Pilowsky, Melissa M.J. Farnham, and Angelina Y. Fong

Australian School of Advanced Medicine, Macquarie University, North Ryde, NSW, Australia

Humana Press

Editors
Paul M. Pilowsky
Australian School of Advanced Medicine
Macquarie University
North Ryde, NSW, Australia

Melissa M.J. Farnham
Australian School of Advanced Medicine
Macquarie University
North Ryde, NSW, Australia

Angelina Y. Fong
Australian School of Advanced Medicine
Macquarie University
North Ryde, NSW, Australia

ISSN 0893-2336 ISSN 1940-6045 (electronic)
ISBN 978-1-4939-6268-6 ISBN 978-1-62703-233-9 (eBook)
DOI 10.1007/978-1-62703-233-9
Springer New York Heidelberg Dordrecht London

Printed on acid-free paper

Humana Press is a brand of Springer
Springer is part of Springer Science+Business Media (www.springer.com)

Series Preface

Under the guidance of its founders Alan Boulton and Glen Baker, the Neuromethods series by Humana Press has been very successful since the first volume appeared in 1985. In about 17 years, 37 volumes have been published. In 2006, Springer Science + Business Media made a renewed commitment to this series. The new program will focus on methods either that are unique to the nervous system and excitable cells or which need special consideration to be applied to the neurosciences. The program will strike a balance between recent and exciting developments like those concerning new animal models of disease, imaging, in vivo methods, and more established techniques. These include immunocytochemistry and electrophysiological technologies. New trainees in neurosciences still need a sound footing in these older methods in order to apply a critical approach to their results. The careful application of methods is probably the most important step in the process of scientific inquiry. In the past, new methodologies led the way in developing new disciplines in the biological and medical sciences. For example, Physiology emerged out of Anatomy in the nineteenth century by harnessing new methods based on the newly discovered phenomenon of electricity. Nowadays, the relationships between disciplines and methods are more complex. Methods are now widely shared between disciplines and research areas. New developments in electronic publishing also make it possible for scientists to download chapters or protocols selectively within a very short time of encountering them. This new approach has been taken into account in the design of individual volumes and chapters in this series.

Wolfgang Walz

Preface

Many of our modern investigations into the nervous system are possible because of the pioneering work from Ludwig's Institute in the late 1800s (1, 2). From this single site came many of the techniques and discoveries that still provide information on the work being conducted today. Nevertheless, activation, inhibition or destruction of the nervous system or its component parts as a vital tool for the investigation of function has undergone remarkable development in the past 150 years. More recently, new approaches have been developed that allow for activation, inhibition or destruction of parts of the nervous system as therapeutic tools. Methods for activating or inhibiting parts of the nervous system have developed considerably since Borison and Wang's (3) initial experiments with Radon lesions in the medulla oblongata of dogs 60 years ago (Fig. 4 of (3)).

The objective of this book is to provide an overview of modern methods for generating lesions as well as for stimulating and inhibiting neural pathways. Most physiological studies require activation or inhibition of pathways in order to gain a mechanistic understanding of sensory inputs, integration and efferent pathways. Currently available approaches provide sensitivity and specificity that were unheard of at the time of Borison and Wang (3). An overview of the progress in the use of the neurotransmitter glutamate as a chemical excitant is provided. This method has its roots in the seminal work of many investigators in the 1970s (4). Since that time, investigations into the role of amino acid neurotransmitters increased dramatically as it became clear that they are responsible for the largest component of fast neurotransmission in the brain and spinal cord, as well as exerting effects on slowly acting metabotropic receptors and being required for other cellular activities.

A number of different approaches for activating and inhibiting neurons, nuclei and nerves are assayed. Many new techniques such as optogenetics and the use of the in situ perfused preparation are examined. In other sections, the use and validity of more well-known approaches are reassessed. The use of local anaesthesia to achieve reversible denervation in physiological studies—an approach that is common in clinical medicine—is examined in the context of baroreceptor denervation. Less invasive methods of examining baroreceptor function using heart rate variability and baroreflex sensitivity are also considered. At the intracellular level, cellular and integrative functions resulting from the activation or inhibition of neuronal G protein systems are assayed. Current approaches to antidromic activation are then assessed, as is the technique of dye labelling by electroporation of Neurobiotin. Activation and inhibition of single neurons during intracellular recording as well as using multibarrel electrodes extracellularly in the vicinity of an intracellular electrode are described. Analysis of sympathetic nerve activity in the periphery is a topic that has created considerable discussion for many years, and this is also examined. Another approach that is commonly used in neurophysiology is the compound action potential. Examination of the compound action potential in relation to vision and estimation of motor units in mice is described. Changes in peripheral nerve function following nerve injury and subsequent neuropathic pain is an important area of study and is discussed in relation to two models. Finally, a method for improving motor function in humans following stroke is described.

North Ryde, NSW, Australia

Paul M. Pilowsky
Melissa M.J. Farnham
Angelina Y. Fong

References

1. Fye WB (1986) Carl Ludwig and the Leipzig Physiological Institute: 'a factory of new knowledge'. Circulation 74(5): 920–928
2. Ludwig C, Cyon dE (1866) Die Reflexe eines der sensiblen Nerven des Herzens auf die motorischen Nerven der Blutgeffässe. Ber der Sächs Ges der Wissenschaften 18: 307–328
3. Borison HL, Wang SC (1951) Quantitative effects of radon implanted in the medulla oblongata: a technique for producing discrete lesions. J Comp Neurol 94: 35–55
4. Watkins JC, Curtis DR, Felix D (1971) Effect of intracellular injection of pyridine nucleotides on electrical responses of cat spinal motoneurones. Brain Res 35(2): 570–572

Contents

Contributors

STEPHEN ABBOTT • *University of Virginia, Charlottesville, VA, USA*
PAUL J. AUSTIN • *Discipline of Anatomy & Histology, School of Medical Sciences, University of Sydney, Sydney, NSW, Australia*
MARK C. BELLINGHAM • *School of Biomedical Sciences, The University of Queensland, Brisbane, QLD, Australia*
MELISSA M.J. FARNHAM • *Australian School of Advanced Medicine, Macquarie University, North Ryde, NSW, Australia*
ANGELINA Y. FONG • *Australian School of Advanced Medicine, Macquarie University, North Ryde, NSW, Australia*
ANDREA H. GAEDE • *Australian School of Advanced Medicine, Macquarie University, North Ryde, NSW, Australia*
ANN K. GOODCHILD • *Australian School of Advanced Medicine, Macquarie University, North Ryde, NSW, Australia*
STUART L. GRAHAM • *Australian School of Advanced Medicine, Macquarie University, North Ryde, NSW, Australia; Save Sight Institute, University of Sydney, Sydney, NSW, Australia*
CARA M. HILDRETH • *Australian School of Advanced Medicine, Macquarie University, North Ryde, NSW, Australia*
REFIK KANJHAN • *School of Biomedical Sciences, The University of Queensland, Brisbane, QLD, Australia*
ALEXANDER KLISTORNER • *Australian School of Advanced Medicine, Macquarie University, North Ryde, NSW, Australia; Save Sight Institute, University of Sydney, Sydney, NSW, Australia*
IDA J. LLEWELLYN-SMITH • *Cardiovascular Medicine, Physiology, and Centre for Neuroscience, Flinders University, Bedford Park, SA, Australia*
SIMON MCMULLAN • *Australian School of Advanced Medicine, Macquarie University, North Ryde, NSW, Australia*
PENELOPE A. MCNULTY • *Neuroscience Research Australia, Sydney, NSW, Australia; University of New South Wales, Sydney, NSW, Australia*
GILA MOALEM-TAYLOR • *School of Medical Sciences, University of New South Wales, Sydney, NSW, Australia*
SHYUAN T. NGO • *University of Queensland Centre for Clinical Research, The University of Queensland, Herston, QLD, Australia*
JACQUELINE K. PHILLIPS • *Australian School of Advanced Medicine, Macquarie University, North Ryde, NSW, Australia*
PAUL M. PILOWSKY • *Australian School of Advanced Medicine, Macquarie University, North Ryde, NSW, Australia*
ROHIT RAMCHANDRA • *Systems Neurophysiology Divison, Systems Neurophysiology Group, University of Melbourne, Parkville, VIC, Australia; Florey Institute of Neuroscience and Mental Health, Parkville, VIC, Australia*

Sarah E. Scheuer • *Neuroscience Research Australia, Sydney, NSW, Australia; University of New South Wales, Sydney, NSW, Australia*
Qi-Jian Sun • *Australian School of Advanced Medicine, Macquarie University, North Ryde, NSW, Australia*
V.J. Tallapragada • *Australian School of Advanced Medicine, Macquarie University, North Ryde, NSW, Australia*
Angelica G. Thompson-Butel • *Neuroscience Research Australia, Sydney, NSW, Australia; University of New South Wales, Sydney, NSW, Australia*
Anthony J.M. Verberne • *Department of Medicine, Clinical Pharmacology and Therapeutics Unit, Austin Health, University of Melbourne, Heidelberg, VIC, Australia*
Karin N. Westlund • *Department of Physiology, University of Kentucky, Lexington, KY, USA*
Yuyi You • *Australian School of Advanced Medicine, Macquarie University, North Ryde, NSW, Australia*
Liping Zhang • *Department of Physiology, University of Kentucky, Lexington, KY, USA*

Part I

Novel Experimental Techniques

Chapter 1

Optogenetic Stimulation of Neurons in the Brainstem of Freely Behaving Rats and Mice

Stephen Abbott

Abstract

Optogenetics is the combination of genetic and optical methods to achieve gain or loss of function of well-defined events in specific cells in living tissues. One of the strengths of this approach is that it can be used in conscious animals without impeding normal behavior. This method has been implemented to interrogate the neural determinants of behaviors including feeding, breathing, aggression, and arousal. Here I describe the procedures for stereotaxically guided delivery of opsin-encoding virus in deep brainstem structures of rodents, the implantation of guide cannulas and optical fibers to photostimulate neurons in vivo. The expected results and caveats are discussed.

Key words: Optogenetics, Brainstem, Channelrhodopsin 2, Lentivirus, Rodent, Method, Conscious

1. Introduction

One goal of neuroscience is to determine the neural basis of behavior. One approach to achieving this goal is the activation of groups of neurons while observing changes in animal behavior. Optogenetics combines genetic and optical methods to achieve gain or loss of function of well-defined events in specific cells in living tissues (1). One of the strengths of this approach is that it can be used in conscious animals without impeding normal behavior (2, 3). Generalized optogenetic methods are available elsewhere (4, 5), so this chapter will focus on some of the unique challenges associated with implementing optogenetic methods in the brainstem structures of rodents. Here I describe the procedures for stereotaxically guided delivery of opsin-encoding virus in deep brainstem structures of rodents, the implantation of guide cannulas and optical fibers to photostimulate neurons. The expected results and caveats are discussed.

Paul M. Pilowsky et al. (eds.), *Stimulation and Inhibition of Neurons*, Neuromethods, vol. 78,
DOI 10.1007/978-1-62703-233-9_1, © Springer Science+Business Media, LLC 2013

There are a growing number of light-sensitive proteins available, allowing the activation or inhibition of cells with varying degrees of temporal precision (6). Channelrhodopsin 2 (ChR2) is a rapidly gated light-sensitive cation channel derived from the algal species *Chlamydomonas reinhardtii* (7). ChR2 and variants of it are currently the most widely used opsins (6). Activation of ChR2 with blue light (peak activation 473 nM) causes channel opening (wild-type activation time ~0.2 ms; deactivation time constant ~10–12 ms) and cell depolarization enabling precise control over the rate of neuronal spiking up to a frequency of 50 Hz (8). Despite these impressive qualities, the performance of ChR2 depends on the particular variant of ChR2 used and the intrinsic properties of the targeted cell.

An important feature of optogenetics is the ability to limit the expression of ChR2 to a defined neuronal population using in utero electroporation (9), constitutive expression in transgenic mice (10, 11), or a viral expression strategy, which derives specificity from a cell-specific promoter or from the expression of Cre in transgenic mice (12, 13). The method described here is focused on the delivery of viral constructs targeting ChR2 expression to neurons using a Cre-loxP or promoter-driven strategy. The procedures for designing and producing ChR2 viral constructs are not included here but are available elsewhere (4). Several virus constructs are available through Addgene (http://www.addgene.org). Furthermore, useful information regarding advances in the field of optogenetics is available on the following Web sites: http://www.openoptogenetics.org/index.php?title=Main_Page (Openoptogenetics wiki) and http://www.stanford.edu/group/dlab/optogenetics/ (Optogenetics Resource Center).

2. Materials

2.1. Anesthetics, Analgesics, and Antibiotics

1. Rats: Ketamine (75 mg/kg) + xylazine (5 mg/kg) + acepromazine (optional sedative and mimetic: dose: 1 mg/kg) given as an intramuscular (i.m.) bolus.
2. Mice: Ketamine (dose: 100 mg/kg) + medetomidine (dose: 0.2 mg/kg) given i.m. Atipamezole (post-surgery reversal agent for medetomidine; dose: 2 mg/kg) given subcutaneously (s.c.).

   Caution: Ketamine is a controlled substance and should be handled according to the policies of the host institution.
3. Ketorolac (analgesic; 2–3 mg dose for rats and 0.2 mg dose for mice; s.c.).

4. Ampicillin (antibiotic; 10 mg dose for both rats and mice; i.m.).
5. Bupivacaine (concentration 0.75%, 3–4 infiltrations of <0.05 ml s.c.) (or other similar local anesthetic).

2.2. General Surgical Equipment, Instruments, and Supplies

1. Surgical tools including fine forceps, scalpel blades, fine bone rongeurs (Fine Science Tools, USA)
2. Hegenbarth Clip Applying Forceps 5″ and Michel Wound Clip 7 mm–5/16″ (Roboz, USA)
3. David Kopf stereotaxic frame assembly with ultraprecise micromanipulator (David Kopf Instruments, USA) or alternative stereotaxic setup with X–Y–Z micropositioner with 100-μm precision
4. Mouse stereotaxic adaptor (Narashige, Japan)
5. Dissection stereomicroscope with stand (Leica, USA)
6. Servo-controlled heating blanket (Harvard Apparatus, USA)
7. Picospritzer III (General Valve, USA)
8. Compressed 100% nitrogen tank for Picospritzer (local supplier)
9. Micropipette puller (Flaming/Brown model P-97, Sutter, USA)
10. Hair clippers (Oster, USA)
11. Handheld high-speed drill (Stoelting, USA)
12. 1-mm drill burrs (Stoelting, USA)
13. Sterile saline
14. 70% ethanol solution
15. Sterile cotton swabs
16. Iodine-based wash (Aplicare povidone-iodine, Henry Schein, USA)
17. Para lube eye ointment
18. Absorbable 5–0 sutures
19. Cyanoacrylate tissue adhesive (Henry Schein, USA)
20. Glass capillaries (O.D. 1.2, I.D. 0.68, length 76 mm, 1B120-3 WPI, USA)

2.3. Specialist Small Parts and Equipment for Guide Placement

1. Plastics One guide and accessories (For rats: guide cannula—C316G/SP with 5-mm guide projection; internal cannula—C316I/SPwith15-mmprojection;dummycannula—C316DC/SP with cannula to fit C316G/SP guide; dust cap—303 DC/1) (For mice: guide cannula—C316GS-4/SP with 3.5-mm guide projection; internal cannula—C316IS-4/SP with 10-mm projection; dummy cannula—C316DCS-4/SP with cannula to fit C316GS-4/SP guide; dust cap—303DC/1). Refer to Fig. 1
2. Small hand screwdriver (Plastics One, USA)

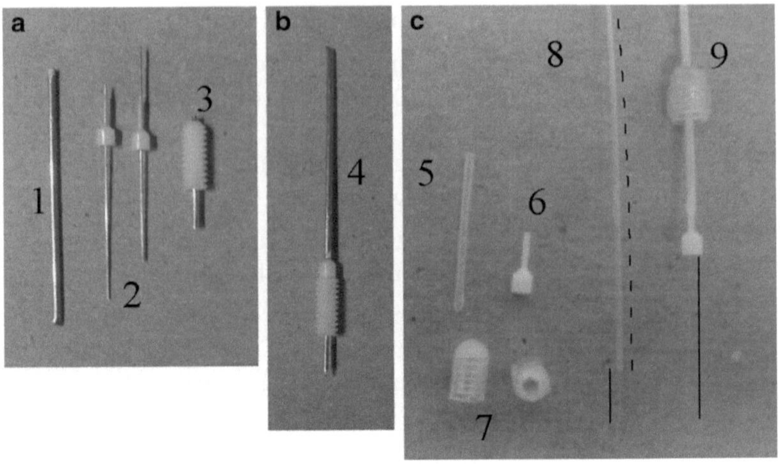

Fig. 1. Guide cannula and optical fiber for in vivo photostimulation. (**a**) Parts needed to implant Plastics One guide cannulas in rats. Materials for mice can be found in the materials section. A blunt 1½-inch 19-G steel needle with plastic fitting removed (1) is glued to an internal cannula (part no. C316I/SP (2)). To improve brain penetration, the internal cannula is cut to a sharp tip and retracted to fit the guide cannula (part no. C316G/SP with 5-mm guide projection (3)). (**b**) The guide cannula assembly ready for implantation (4). The internal cannula can be sharpened and left to protrude slightly from the tip of the guide cannula to improve tissue penetration. (**c**) Parts needed to fabricate the optical fiber assembly for in vivo photostimulation, including a 1-in. piece of PE50 tubing (5), plastic adaptor from an internal cannula (6), a dust cap (part no. 303DC/1) with a small hole drilled through the top (7), and a 200-μmb multimode fiber optic (8). The fiber optic (8) has a protective "cladding" (*dashed line*) that can be stripped to reveal a 65-μm core (*solid line*). The finished product (9) is fabricated by stripping 15 cm of cladding off the fiber optic and threading the dust cap, PE50 tubing, and plastic adaptor onto the fiber. The PE50 tubing and plastic adaptor are then glued to the fiber optic with cyanoacrylate. Once the glue is set, the fiber can then be inserted into a guide cannula and cut to the appropriate length using a diamond scribe.

3. Jeweler screws (Plastics One, USA)
4. Dental cement (Henry Schein, USA) and/or two part epoxy with hand applicator and mixing tubes (Plastics One, USA)

2.4. Specialist Equipment for In Vivo Photostimulation

1. 473-nm diode pumped solid state laser (CrystaLaser, catalog no. BCL-473-050; 50 mW, 473 nm, TEMoo CW laser with TTL on/off, analog output intensity modulation, PC connector at laser output).
2. Multimode optical fiber (ThorLabs, cat. no. BFL37-200 for 200-mm fiber with FC end to connect to PC adaptor of laser output; length can be specified, recommended length no shorter than 1 m).
3. Power meter (ThorLabs, sensor, cat no. S130A).
4. Diamond stylus (Thorlabs, cat no. S90W).

5. Fiber optic desheathing tool (Thorlabs, cat no. T12S25).
6. Stimulus generator with BNC adaptor (Astromed, s88x dual output square pulse stimulator). The minimum requirement for the stimulus generator is the capability generating square-wave TTL pulses within a range of 0–5 V, which is usually sufficient to control laser output.

Alternative configurations, tools, and current news concerning optogenetics are available at http://www.openoptogenetics.org.

3. Methods

3.1. Presurgery Preparation

1. Fabricate glass micropipettes with micropipette puller. Break the tip to an outside diameter of 30 μm under a stereomicroscope.
2. Prepare Plastics One guide cannula assembly. See technical notes (Sect. 4.1) and Fig. 1.
3. Determine the stereotaxic coordinates of the target region using a rodent brain atlas (14). For brainstem regions, the lambda (the point of intersection between the lambdoid and sagittal skull suture) is the proximal landmark.
4. Remove virus aliquot from −80°C freezer, thaw in 37 ° C water bath for 5 s, and immediately place on ice. See technical notes (Sect. 4.2).
5. Sterilize the prep and surgical areas with a 70% ethanol solution, diluted bleach, or cavicide. Sterilize surgical instruments with disinfectant and a dry bead sterilizer.

3.2. Animal Preparation

Animals should be housed and handled according to institutional guidelines. Recovery surgeries should be conducted under standard aseptic conditions.

1. Anesthetize animal with an intramuscular injection of the anesthetic solution provided in materials.Critical step: Assess the depth of anesthesia by lightly pinching the hind paw and lightly touching the eye of the animal. Adequate anesthesia is judged by the absence of a withdrawal reflex to a noxious pinch and the absence of the corneal reflex (i.e., protective closure of the eye). Administer additional doses of anesthetic (20% of initial volume) as necessary.
2. Remove the hair around surgical site using the hair trimmers and prep surgical site with 70% ethanol and iodine.
3. Place the animal on the servo-controlled heating pad and insert a rectal probe to monitor internal temperature and maintain body temperature between 36.5 and 37.0° Celsius.

4. Apply lubricant to the eyes of the animal to prevent drying.
5. Place 2–4 × 0.05 ml bupivacaine (or similar) injections subcutaneously at the site of the initial skin incision; the anesthetic will take effect over 5 min.
6. Place animal in the stereotaxic frame. First, fix the animal in the stereotaxic ear bars by positioning a fixed ear bar in the right ear canal; center the animal's nose to the stereotaxic frame and feed the opposite ear bar into the left ear canal while gently holding the animals head level; insert the bite bar between the jaws of the animal and lower the nose bar to secure the animal's head in place.

   Critical step: Correct positioning in the stereotaxic frame requires practice and is essential for accurate targeting of brain nuclei. Correct positioning will allow for swivel on the anterior–posterior axis while preventing lateral movement. Some ear bars will produce an audible sound as they pierce the ear drum indicating the ear bar is in the appropriate position. Atraumatic ear bars (Kopf Instruments, USA) are available if desired for recovery surgeries.
7. Make an anterior–posterior midline incision using a sterile scalpel blade to expose the surface of the skull ensuring that the incision provides enough exposure of the skull to identify stereotaxic landmarks. The connective tissue overlying the skull can be scraped away with a scalpel blade and cotton swabs. Stop any bleeding by applying light pressure to the site of the bleed with a cotton swab for 1 min. If small vessels in the skull bleed persistently, apply sterile bone wax to the bleed site with a cotton swab. See technical notes (Sect. 4.3).
8. Identify the stereotaxic landmarks, bregma and lambda, on the skull and adjust the bite bar so that these sites are level. This can be accomplished by measuring the relative vertical depth of each sight unpulled glass capillary mounted in a stereotaxic arm and adjusting the bite bar as necessary. Mark the site on the skull corresponding to the mediolateral and anterior–posterior coordinates of the target brain region. Note that species and age/weight differences will influence the accuracy of the coordinates provided by the brain atlas.
9. Make a small burr hole (diameter ≈ 2 mm) above the target brain region using a hand drill. Thin the skull and remove the final layers of bone with fine forceps.

   Caution: Applying excessive downward pressure while drilling will cause the drill to penetrate into the brain causing unrecoverable damage. Particular care is required when drilling near the large blood vessels on the surface of the brain.

10. The dura is a tough membrane that protects the brain; it must be manually pierced using a fine needle in rats. Make a small superficial incision to open the dura with a 28-gauge needle. Cerebrospinal fluid will be released once the dura is pierced. Remove excess fluid using sterile cotton swabs. For mice, the dura can be penetrated by the virus-carrying glass pipette without damaging the 30-μm tip.

3.3. Injection of Virus

1. Reverse fill the preprepared pipette with virus using PE50 tubing connected to a 10-ml syringe under a microscope and place the pipette in the micromanipulator arm holder. Note that the tip of the pipette will be sterilized during pulling. To improve accuracy, recalibrate the position of the pipette against bregma or lambda and move into position above the exposed brain. See technical notes (Sect. 4.4).
2. Connect PE tubing to the picospritzer with PE50 and silastic tubing. Lower the virus-filled pipette to the surface of the exposed brain and mark this point as vertical zero. Lower the pipette to the depth of the target brain region.
3. Set the pressure within the picospritzer at 50–60 psi and use 5–15-ms pulses to expel the virus from the glass capillary. Inject boluses of approximately 30–50 nl, waiting at least 30s before injecting another bolus. The final volume will be determined by the targeting requirements of the experiment. The injected volume can be measured by monitoring the movement of the meniscus against a calibrated microscope eyepiece, calibrated pipette, or grid paper glued to the pipette. Inject virus at a rate of not more than 100 nl/min. Once the full volume has been injected, allow the pipette to rest in place for a minimum of 5 min. See technical notes (Sects. 4.4 and 4.5).

For experiments that do not require the implantation of a cannula, for example, in animals that will be reanesthetized for an acute experiment, skip to step 25.

3.4. Placement of Guide Cannula

1. For rats, a minimum of two jeweler screws need to be implanted in the skull to secure the guide cannula and dental cement to the surface of the skull. To achieve this, use a hand drill to create two small burr holes roughly 3–4 mm away from the eventual position of the guide cannula. Screw in each screw using the hand screw tool while holding the screw with forceps. Each screw should protrude 0.5–1 mm from the surface of the skull to provide an anchor for the dental cement. Generally, mice do not require the use of screws to anchor the guide cannula in place.
2. Dust the modified guide cannula assembly (Fig. 1b) with dry ampicillin and secure in the micromanipulator. See technical notes (Sect. 4.6).

3. Position the guide cannula over the site of injection. For added accuracy, the cannula can be repositioned according to the location of bregma or lambda.
4. Lower the guide cannula into the brain until the bottom of the plastic portion of the guide cannula sits flush with the surface of the skull. See technical notes (Sect. 4.4).
5. Remove residual moisture, tissue, bone dust, or blood from around the guide cannula and jeweler screws with cotton swabs. See technical notes (Sect. 4.3).
6. Apply the cement/adhesive around the guide cannula and jeweler screws.

   For rats, we use C&B Metabond (Parkell, USA). Lower the guide cannula into the brain until the plastic pedestal touches the surface of the skull. Mix the three part cement according to supplier's instructions and apply to the dry skull around the guide cannula and over the jeweler screws with a disposable syringe. Build the cement up around the guide cannula. For mice, use Plastics One two-part epoxy. Apply the adhesive around the guide cannula with the hand applicator and mixing tubes. Wait 5 min for the epoxy to set. Once the cement or epoxy is set, release the guide assembly from the stereotaxic arm and retract the arm. Remove the modified internal cannula from the guide cannula. *Caution*: *The dummy cannula that maintains the patency of the guide cannula during recovery requires a minimum of three threads to screw into the guide*. See technical notes (Sect. 4.3).

3.5. Wound Closure

1. Irrigate the wound area with sterile saline and blot dry with sterile gauze. Close the wound with surgical sutures, wound clips, and/or small amounts of cyanoacrylate glue (i.e., "super glue" from any manufacturer). For rats, we use wound clips, which are applied with a hand tool, and cyanoacrylate glue. For mice, we use a combination of absorbable sutures and cyanoacrylate glue.
2. Once the wound is fully closed, insert the dummy cannula dusted with dry ampicillin to maintain the patency of the guide cannula during recovery.

3.6. Post-surgery

For rats, administer ketorolac and ampicillin along with 2–3 ml of saline or 2% glucose in saline (subcutaneous), to counteract dehydration, and then return the rat to a clean home cage. For mice, administer Antisedan, ketorolac, and ampicillin along with 1 ml of saline or 2% glucose in saline to counteract dehydration. Recover mice in a temperature-regulated recovery chamber to maintain core body temperature until consciousness is regained before returning them to a clean home cage. Additional doses of ketorolac and ampicillin may be administered on a daily basis depending on recovery. If an animal suffers serious discomfort, administer a stronger analgesic (e.g., buprenorphine). See technical notes (Sect. 4.7).

3.7. Inserting the Optical Fiber for In Vivo Photostimulation

1. Prepare Plastics One optical fiber assembly. See technical notes (Sects. 4.8 and 4.9) and Fig. 1.
2. Prior to behavioral experiments, it is necessary to habituate animal to the experimental conditions, preferably over consecutive days. If desired, a dummy fiber assembly created as per Zhang et al. (4), with a fiber cut to the length of the guide cannula, can be used to habituate the animal to the fiber assembly. On the day of the experiment, animals should be exposed to the experimental conditions for a minimum of 2 h.
3. Prepare the experimental setup. Set the power output of the optic fiber by connecting the FC adaptor to the PC output of the diode laser and adjust the power output to measure ~8–12 mW at the tip of the fiber using a power meter. Once the power output is set, connect a function generator to the BNC input on the diode laser.
4. To insert the optical fiber assembly, anesthetize the animal with isoflurane to sufficiently attenuate the righting reflex. Once the animal is unconscious, carefully insert the fiber into the guide cannula until the plastic adapter locks onto the cannula and then screw the modified dust cap to secure the fiber assembly in place. See technical notes (Sect. 4.10).
5. Return the animal to the experimental setup. The effects of light isoflurane anesthesia will dissipate over 5–10 min.
6. Conduct experiment. See technical notes (Sect. 4.11) for stimulation parameters and expected results.
7. Upon completing the experiment, deeply anesthetize and perfuse the animal. Process the brain to verify the placement of the guide cannula and optical fiber and validate the expression of ChR2. See technical notes (Sect. 4.12).

4. Technical Notes

4.1. Assembly for Implanting Guide Cannulas

There are several methods for implanting a guide cannula. Several companies sell accessories for stereotaxic equipment designed to hold Plastics One guide cannulas. We implant guide cannulas using the same micromanipulator arm and holder used for standard pipettes, which requires a custom-made guide cannula assembly (Fig. 1a, b). A long internal cannula is cut to a sharp tip and retracted so that the tip of the internal cannula protrudes (<1 mm) from the tip of guide cannula. It may be necessary to coarsely sand the edges of the cut tip of the internal cannula so that it moves freely inside the guide cannula. The sharp internal cannula improves penetration of the guide cannula into the brain and prevents tissue from clogging the guide cannula assembly during implantation. A 19-G 1½-inch blunt needle is then glued to the shoulder of the plastic adaptor on the internal cannula providing a rigid and stable area to clamp into the pipette holder. The major benefit of this assembly versus manufactured cannula holders is that it provides additional working room around the skull during surgery and is considerably cheaper than the premade product.

4.2. Storage and Handling of Virus

All viruses should be stored at −80°C. Lentivirus can be maintained for 6 h on ice without significant deterioration, whereas adeno-associated viruses are stable for days at 4°C. Viruses must be handled according to institutional guidelines.

4.3. Improving Adhesive of Cranial Cement/Adhesive to Skull

Failed adhesion is caused by either improper removal of the connective tissue that covers the skull, moisture remaining on the skull during the application of the cement/adhesive, or improper mixing of the cement/adhesive components. These issues can be addressed in the following ways. Connective tissue can be disrupted by applying a 30% hydrogen peroxide solution with a cotton swab. The reaction will dissolve connective tissue, presenting as white foam. This reaction requires only a few seconds and should be stopped by liberally irrigating the reactive area with PBS. Adhesion can be improved by scratching the surface of the skull with the tip of a scalpel blade before the screws and guide are implanted. After the guide cannula and jeweler screws have been implanted, evaporate any remaining moisture by applying a small amount of 70% ethanol with a cotton swab. Finally, most adhesives and cements come with mixing instructions that must be followed exactly to achieve the best results.

4.4. Making Injections and Implanting Guide Cannulae Below the Transverse Sinus

The transverse sinus presents a major challenge when implanting large diameter cannulas in the brainstem. Rupturing this vessel can cause excessive blood loss and death. If the target brain region is located below the transverse sinus, two approaches can be used to avoid damaging the vessel with only minimal reductions in accuracy. Firstly, the micropipette and guide cannula can be inserted into a region adjacent to the vessel, lowered 2–3 mm into the brain and then repositioned over the target region. This approach causes a modest amount of damage to the cerebellum, but the damage is not detrimental to the recovery of the animal. Alternatively, the guide cannula can be implanted on an angle so that the transverse sinus, which is a surface structure, is avoided (<15° in the posterior to anterior direction). This approach sacrifices accuracy for the preservation of the vessel and the integrity of the cerebellum and is recommended when it is not viable to implant the cannula as per the first option. The latter approach is also useful for targeting brainstem regions located caudal to the occipital suture.

4.5. Injecting the Virus

Rapidly injecting the virus will produce localized damage and may cause the virus to travel up the pipette tract. Removing the pipette rapidly will increase the movement of virus up the tract, causing undesired transduction of cells surrounding the tract.

4.6. Sterilizing Small Parts

Plastics One parts and jeweler screws can be sterilized by immersion in a 70% ethanol solution for 1 h prior to surgery.

4.7. Animal Recovery

A sufficient recovery period is required to allow adequate expression of ChR2 to achieve control of neuronal activity. Whole cell expression is achieved between 3 and 4 weeks after virus delivery and remains stable for at least 6 months, depending on the particular virus used. We have observed sufficient ChR2 expression in the proximal dendrites and soma after 7 days with high-titer virus, but expression in the distant axons and terminal fields requires at least 3 weeks of recovery time. Prior to the experiment, the patency of the guide cannula will be maintained as long as the dummy cannula remains in the guide cannula. We recommend that animals that have been implanted with guide cannulas be housed separately as the guide cannula can be displaced during social interactions with other animals.

4.8. Preparing the Optical Fiber Assembly for In Vivo Photostimulation

We use an optic fiber assembly modified from Zhang et al. (4). The modified design includes a 1-inch piece of PE 50 tubing threaded onto the optical fiber and glued to the top of the plastic adaptor taken from the internal cannula (Fig. 1c). The tubing provides additional structural support and improves the durability.

4.9. Determining the Length of the Guide Cannula and Optical Fiber

In the literature (2–4) it is recommended that the end of the guide cannula be targeted approximately 1 mm dorsal to the targeted brain region and that the optical fiber be cut to protrude less than 1 mm beyond the guide cannula. We have modified this approach to address unique issues associated with working the brainstem. We use a 5-mm (rats) or 3.5-mm (mice) guide cannula irrespective of the depth of the target brain region. This reduces tissue damage in brainstem structures that regulate critical homeostatic functions and improves animal recovery. The optical fiber on the other hand must be cut to a length that will reach the necessary depth required to stimulate the transduced neurons (<0.5 mm from the transduced neurons). To achieve this, the fiber assembly (Fig. 1c) is inserted into a guide cannula ex vivo and cut using a diamond stylus to the appropriate length based on the stereotaxic depth of the target brain region as measured from the surface of the skull.

Acutely inserting the optical fiber as described causes tissue trauma around the point of insertion and may reduce the number of experimental sessions per animal.

4.10. Inserting the Optical Fiber Assembly

On the day of the experiment, great care must be taken when inserting the fiber assembly into the implanted guide cannula to avoid excessive brain tissue trauma and importantly to avoid breaking the fiber inside the cannula. Inducing light anesthesia with a volatile anesthetic such as isoflurane alleviates much of risk associated with this procedure. To prevent fiber breakage during experiments, it is essential that the plastic adaptor is properly snapped onto the top of guide cannula and the modified dust cap is fully screwed into place.

Occasionally, the dummy cannula can be difficult to remove due to dried secretions from the healed wound; warmed saline and fine forceps are sufficient to break the dried seal.

4.11. Stimulation Parameters

Photostimulation parameters will depend on the opsin used, the neuronal population targeted, the level of expression of ChR2, and the desired magnitude of the stimulus. If the optical fiber is within 0.5 mm of the transduced cells, brief pulses of light (0.5–10-ms pulse duration) will produce one or more action potentials with a 15-mW light source. Longer light pulse may be necessary if ChR2-expression is low or if the fiber is misplaced, but caution is recommended as continuous illumination produces tissue damage appearing as a compressed, bleached area in post hoc histology. The maximal light-driven neuronal firing frequency is largely dependent on the opsin used and the targeted cell type. Ideally, neuronal recordings should be conducted to determine the optimal photostimulation parameters for each opsin within a given cell type (e.g., see Fig. 2).

Fig. 2. (continued) This is due in part to the characteristics of the ChR2 variant but also the intrinsic properties of the neuron itself and the level of transgene expression. Initially, the neuronal discharge is entrained at a 1:1 ratio to individual light pulses (far left expanded trace), but after this initial phase, individual light pulses occasionally fail to produce an action potential (middle expanded trace). As in (**a**), a brief silent period occurred following stimulus offset, before neuronal discharge returns to a basal discharge pattern. C1 An example of the effects of a single 20-ms light pulse delivered at low frequency (1 Hz). C2 A peristimulus histogram triggered from the onset of the stimulus. 20-ms pulses reliably produced two action potentials per pulse in this cell. D1 An example of the effects of a single 50-ms light pulse delivered at low frequency (1 Hz) in the same cell as C1. D2 Same as C2, except for a 50-ms pulse. 50-ms pulses produced 3–5 pulse per pulse in this cell. Note that the action potentials occur with a regular interspike interval.

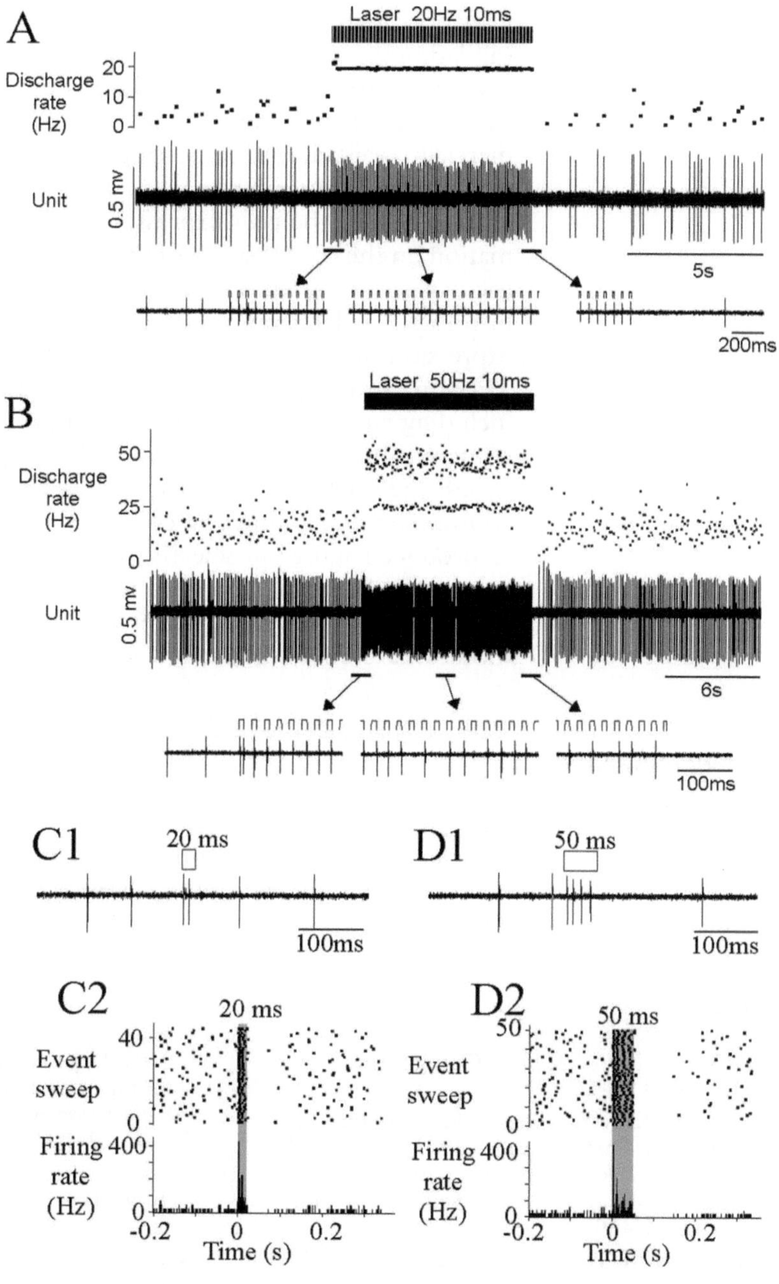

Fig. 2. Effects of photostimulation on ChR2-expressing neurons in the rostral ventrolateral medulla (RVLM). (**a**) An extracellular recording from a barosensitive presumptive C1neuron in the RVLM. Delivering a continuous train of light pulses (437 nm, 7–9 mW, 20 Hz, 10 ms) entrains neuronal discharge at a frequency equal to the stimulus frequency. This neuron fired in couplets during the first three stimulus pulses (far left expanded trace), before settling into a 1:1 ratio for the remainder of the stimulus (middle expanded trace). Neuronal discharge rapidly returned to a basal firing pattern after a brief period of inactivity following stimulus offset (far right expanded trace). (**b**) An extracellular recording from a different barosensitive RVLM neuron. Using ChR2 (H134R), C1 neuronal discharge cannot be entrained on a 1:1 basis reliably at stimulus frequencies greater than 30 Hz.

4.12. The Importance of Histology

Post-hoc analysis of the expression of ChR2 is a critical step in validating the selectivity of the viral construct. Most available ChR2 constructs are tagged with a fluorescent protein marker such as EFYP or mCherry, and these proteins can be amplified using commercially available antibodies and standard immunohistochemistry techniques. These markers allow the identification of the full cytoarchitecture of the cells expressing ChR2, providing critical information on the composition of the stimulated cell population. Once the experiment is completed, all animals should be processed for histology to determine the selectivity and extent of the ChR2-expression and the location of the fiber tip and guide cannula. There are several possible causes for undesired expression of ChR2 including viral tropism related to titer (15, 16) and poor selectivity conferred by promoters or ectopic Cre-expression. To avoid these issues, preliminary experiments must be conducted to verify the appropriate titer (diluted in cold PBS) and volume of virus needed to produce robust but selective ChR2 expression in the target cell group. Weak ChR2 expression reflects the opposite; ChR2 expression can be improved by increasing the viral titer or using a promoter that confers strong and selective expression.

Acknowledgments

Thanks to Dr. Natasha Kumar for helpful discussions and editing. This work is supported by a fellowship from the American Heart Association.

References

1. Deisseroth K (2011) Optogenetics. Nat Methods 8:26–29
2. Adamantidis AR, Zhang F, Aravanis AM, Deisseroth K, De Lecea L (2007) Neural substrates of awakening probed with optogenetic control of hypocretin neurons. Nature 450:420–U429
3. Aravanis AM, Wang LP, Zhang F, Meltzer LA, Mogri MZ, Schneider MB, Deisseroth K (2007) An optical neural interface: in vivo control of rodent motor cortex with integrated fiberoptic and optogenetic technology. J Neural Eng 4:S143–156
4. Zhang F, Gradinaru V, Adamantidis AR, Durand R, Airan RD, de Lecea L, Deisseroth K (2010) Optogenetic interrogation of neural circuits: technology for probing mammalian brain structures. Nat Protoc 5:439–456
5. Cardin JA, Carlen M, Meletis K, Knoblich U, Zhang F, Deisseroth K, Tsai LH, Moore CI (2010) Targeted optogenetic stimulation and recording of neurons in vivo using cell-type-specific expression of Channelrhodopsin-2. Nat Protoc 5:247–254
6. Yizhar O, Fenno LE, Davidson TJ, Mogri M, Deisseroth K (2011) Optogenetics in neural systems. Neuron 71:9–34
7. Nagel G, Szellas T, Huhn W, Kateriya S, Adeishvili N, Berthold P, Ollig D, Hegemann P, Bamberg E (2003) Channelrhodopsin-2, a directly light-gated cation-selective membrane channel. Proc Natl Acad Sci USA 100:13940–13945
8. Boyden ES, Zhang F, Bamberg E, Nagel G, Deisseroth K (2005) Millisecond-timescale, genetically targeted optical control of neural activity. Nat Neurosci 8:1263–1268
9. Huber D, Petreanu L, Ghitani N, Ranade S, Hromadka T, Mainen Z, Svoboda K (2008) Sparse optical microstimulation in barrel cortex

drives learned behaviour in freely moving mice. Nature 451:61–64

10. Petreanu L, Huber D, Sobczyk A, Svoboda K (2007) Channelrhodopsin-2-assisted circuit mapping of long-range callosal projections. Nat Neurosci 10:663–668
11. Arenkiel BR, Peca J, Davison IG, Feliciano C, Deisseroth K, Augustine GJ, Ehlers MD, Feng G (2007) In vivo light-induced activation of neural circuitry in transgenic mice expressing channelrhodopsin-2. Neuron 54: 205–218
12. Abbott SB, Stornetta RL, Fortuna MG, Depuy SD, West GH, Harris TE, Guyenet PG (2009) Photostimulation of retrotrapezoid nucleus phox2b-expressing neurons in vivo produces long-lasting activation of breathing in rats. J Neurosci 29:5806–5819
13. Depuy SD, Kanbar R, Coates MB, Stornetta RL, Guyenet PG (2011) Control of breathing by raphe obscurus serotonergic neurons in mice. J Neurosci 31:1981–1990
14. Paxinos, G.W., C, The Rat Brain in Stereotaxic Coordinates, Compact Sixth Edition 2008: Academic Press.
15. Nathanson JL, Yanagawa Y, Obata K, Callaway EM (2009) Preferential labeling of inhibitory and excitatory cortical neurons by endogenous tropism of adeno-associated virus and lentivirus vectors. Neuroscience 161:441–450
16. Nathanson JL, Jappelli R, Scheeff ED, Manning G, Obata K, Brenner S, Callaway EM (2009) Short promoters in viral vectors drive selective expression in mammalian inhibitory neurons, but do not restrict activity to specific inhibitory cell-types. Front Neural Circuits 3:19

Chapter 2

Stimulating Peripheral Afferents to Evoke Cardiorespiratory Reflex Responses in the In Situ Arterially Perfused Preparation

Angelina Y. Fong

Abstract

The in situ arterially perfused rodent preparation is an innovation that has allowed for significant progress in the study of cardiorespiratory reflex circuitry. This preparation provides a number of advantages over other preparations. The retention of peripheral cardiorespiratory afferents enables the study of reflex circuitry that is not possible in in vitro slice preparations. In addition, the in situ arterially perfused preparation provides unsurpassed mechanical stability of the brainstem compared with the in vivo preparation. This stability allows for better cellular recordings for prolonged periods.

Here, the basic technique for the in situ arterially perfused preparation including recording of a number of cardiovascular and respiratory parameters is described. In addition, some of the common techniques for stimulating peripheral afferent nerves that produce different cardiovascular and respiratory reflex responses are discussed.

Key words: Cardiorespiratory reflex, Baroreceptor, Chemoreceptor, Nociceptors, Somatic afferent, Locomotor-respiratory coupling, In situ arterially perfused preparation

1. Introduction

The study of neuronal mechanisms involved in cardiorespiratory control has been greatly aided by the development of reduced, in vitro preparations. In vitro brain slice and brainstem spinal cord preparations (1–9) have provided significant insights into the neurophysiological properties and synaptic mechanisms of cardiorespiratory networks. The major benefits of the reduced in vitro preparations include increased mechanical stability, visualisation of the neuronal network and the ability to precisely control the extracellular milieu. These advantages have allowed more direct

Paul M. Pilowsky et al. (eds.), *Stimulation and Inhibition of Neurons*, Neuromethods, vol. 78,
DOI 10.1007/978-1-62703-233-9_2,

interrogation of the synaptic properties of neurones within the cardiorespiratory network. However, with in vitro brain slice preparations, only a small portion of the brain and its neuronal networks are maintained, severely limiting the ability to study integration of afferent inputs and outputs. This limitation was overcome to some extent, by the development of in situ perfused preparations, including the arterially perfused rodent preparation (10–12).

Since its first description in mouse in 1996 (10), the in situ arterially perfused preparation, also known as the 'working heart-brainstem preparation', has been adapted for a variety of rodent species, including rats (12, 13), shrew (14, 15) and hamsters (16). Not only can the in situ arterially perfused preparation be adapted to many species, it also allows neurophysiological studies at ages that range from the day of birth (P0) through to adulthood (10, 16, 17), something that is not possible with most in vitro or in vivo preparations. However, the age range that can be utilised varies depending on the rodent species used. While adult mice can be used in this preparation, the age range for rats is typically no older than 6 weeks of age (equivalent to approximately ~150 g). In our experience, hamsters up to 7–8 weeks of age can also be used (16). Although the upper age range for use in this preparation is limited, it is generally considered that the sympathetic nervous system and the respiratory control systems are well developed before these time points.

The in situ arterially perfused preparation has a number of advantages compared with other approaches. First, autonomic reflex circuitry is intact, without the depressive and confounding effects of anaesthesia on autonomic reflex control. Second, while decerebration is possible in vivo, the procedures are often complex and time-consuming, and the potential for haemorrhage is high. These problems are negated in this preparation. Third, peristaltic pump perfusion enables greater mechanical stability of the brainstem, which provides a better environment for cellular recordings. This allows for cellular recording techniques, including extracellular, intracellular and whole-cell patch recording, which can be maintained during repeated stimulation (13, 18, 19). Repeated stimulation can be difficult in vivo, due to mechanical movement of the brainstem, resulting from both ventilation and cardiac pulsations. Fourth, a number of peripheral afferent nerves can easily be stimulated in the same preparation, with little additional dissection and equipment. These advantages mean that the in situ arterially perfused preparation is ideal for the study of central integration of peripheral afferents and autonomic reflex control. This chapter will outline the procedure for preparation of the in situ arterially perfused preparation and several common techniques for stimulating peripheral afferents that evoke cardiorespiratory reflex.

2. Equipment Setup and Solutions

This section will outline all the equipment and steps in setting up prior to performing the in situ arterially perfused preparation. This section will outline equipment required for setting up the perfusion circuit (Sect. 2.1), general laboratory equipment (Sect. 2.3) and solutions required.

2.1. Perfusion Circuit Setup

1. A low-pulsation peristaltic pump (e.g. Watson-Marlow 505S, 520S) with a minimum of six rollers on the pump head will improve the mechanical stability of the brainstem.
2. Circulating water bath.
3. Water jacket heat exchanger (for countercurrent heat exchange) to warm the perfusate inline to 31–32°C (Fig. 1b).
4. Two bubble traps (Fig. 1c) are essential for removing air bubbles and are highly recommended for dampening the pulsations of the heart and the pump rollers to aid the mechanical stability of the brainstem.
5. Polypropylene filter screens (1 × 25 μm + 1 × 40 μm, 25-mm diameter, Millipore PP4502500, PP2502500) and filter holder. Filtration is important as it prevents blood clots and other tissue debris from entering the perfusion.
6. Double-lumen catheter (e.g. Braintree Scientific, DLR 4) that will allow perfusion through one lumen and recording of the perfusion pressure through the second cannula.
7. Thick wall Tygon tubing (Masterflex 06409-14) that is relatively impermeable to gases.
8. Plastic chamber to hold the preparation.

The circuit, as illustrated and described in Fig. 1, is used to carry the perfusate in this setup (reserved volume between 100 and 200 ml, depending on the amount of fluid trapped within the circuit). The perfusate is passed through an inline heat exchanger (Fig. 1b), a bubble trap (Fig. 1c), polypropylene filter screens and a second bubble trap. The perfusate finally enters the preparation through a double-lumen catheter. The perfusate is allowed to overflow from the preparation, where it is collected and returned to the perfusate reservoir through the base of the chamber and reused to prevent loss of nutrients, such as essential amino acids. The perfusate is continuously bubbled with carbogen (95% O_2/5% CO_2).

2.2. Equipment Required for Dissection

1. Low-melting point wax (e.g. Paramat or Paraplast embedding medium) or Silgel® (Wacker-Chemie, Germany)
2. Surgical tools
   (a) Scissors (13–15 cm blade, straight or curved)

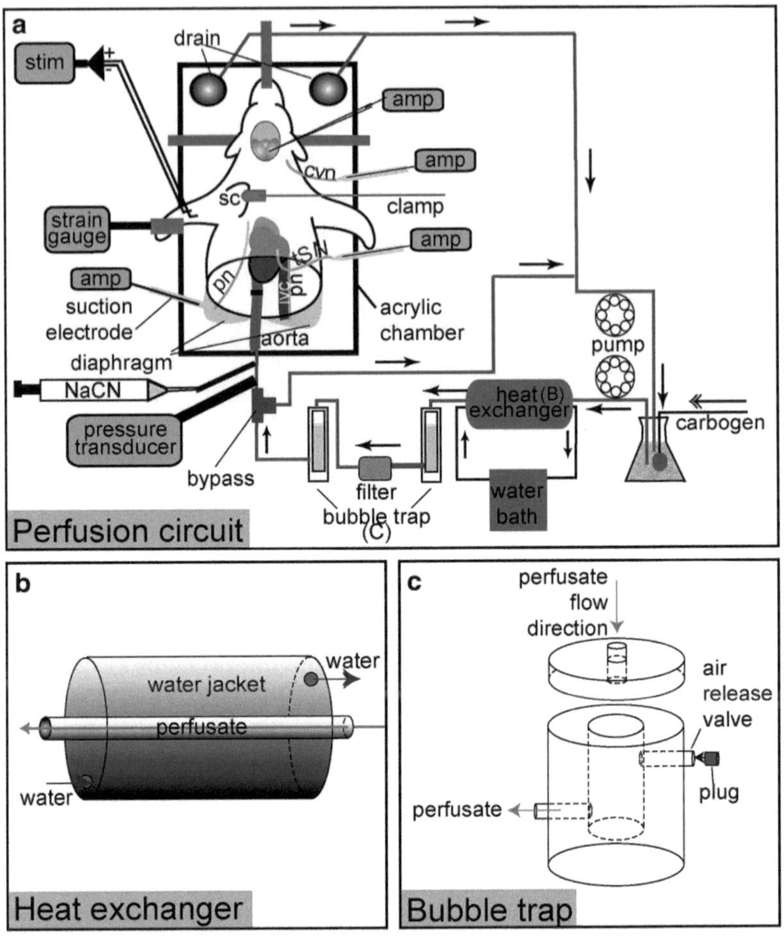

Fig. 1. Schematic diagram of the perfusion circuit and components used for the in situ arterially perfused preparation. (**a**) Schematic illustration of the decerebrated preparation, held in place by modified ear bars in an acrylic chamber. The preparation is perfused with a modified Ringer solution (Tables 1 and 2) through one lumen of the double-lumen catheter, retrogradely cannulated through the descending thoracic aorta. The second lumen is connected to a pressure transducer for recording of perfusion pressure. The perfusate is bubbled with carbogen (95% O_2 / 5% CO_2), warmed through an inline heat exchanger (**b**), through two bubble traps (**c**) and a nylon filter to trap debris. The overflow of the perfusate is collected from drains at the base of the acrylic chamber and recycled. The left phrenic nerves (pn) can be located by its insertion into the diaphragm and right pn can be found adjacent to the inferior vena cava (icv). Other efferent nerves that can be readily accessed in this preparation include the cervical vagus nerve (cvn) and thoracic sympathetic trunks (tSN). Peripheral somatic afferents can be activated by electrical stimulation of the forelimb, after stabilising the scapula (sc) and attaching the forepaw to a strain gauge. Peripheral chemoreceptors can be activated by direct inline injection of sodium cyanide (10–30 μl of 0.03% NaCN).

(b) Scalpel blades (#20 or #10)
(c) Scalpel blade handle (#3)
(d) Small dissection scissors
(e) Straight forceps
(f) Nerve hook

3. Decerebration aspirator (see Notes 6 and Fig. 4)
4. Dish with Silgel covering base

2.3. General Laboratory Equipment Required

1. Pressure transducer and bridge amplifier for perfusion pressure measurement
2. AC amplifiers for whole nerve and ECG recordings (e.g. Grass P511, Neurolog NL104A, or CWE BMA400)
3. Audio monitor
4. Data acquisition system, including an analogue to digital converter and computer for data acquisition (e.g. CED power1401 Mark II with Spike 2 version 7, or Powerlab Carbogen (95% O_2/5% CO_2)
5. Micromanipulator (e.g. Narashige MM-3)
6. Suture silk (4/0)
7. Pipette puller (e.g. Sutter P97, Sutter P2000)
8. Borosilicate glass capillary tubes (1.5 mm O.D., 0.86 mm I.D.)
9. Teflon-coated silver wire (0.003″–0.005″ bare)
10. Electrode holder
11. Silastic tubing
12. Needles: 18G (blunt ends), 23G
13. PE 50 tubing
14. Syringes: 1cc and 30cc
15. Induction chamber for isoflurane anaesthesia
16. 3-way taps
17. Superglue

2.4. Chemicals and Salts

1. Salts for Ringer solution (see Table 1, Notes 1).
2. High-molecular-weight oncotic agent (see Table 2, Notes 2) is added to the Ringer solution to maintain osmolarity between the perfusate and the tissue. Incorrect oncotic pressure will not provide adequate brainstem perfusion and will reduce the longevity of the preparation.
3. Neuromuscular blocking agent: vecuronium bromide, rocuronium or pancuronium. Neuromuscular blockade is not required for the preparation but is essential for ensuring mechanical stability during neuronal recordings.

Table 1
Stock solutions for easy dilution to prepare Ringer solution for the in situ arterially perfused preparation

	Salt	Concentration
1	NaCl	5 M
2	KCl	0.12 M
3	KH_2PO_4	50 mM
4	$MgSO_4 \cdot 7H_2O$	50 mM
5	$NaHCO_3$	1 M
6	$CaCl_2$	0.1 M

To make 2 L of Ringer solution for the in situ arterially perfused preparation, add 50 mL of stock solution 1–5 to 1.7 L of dH_2O and bubble with carbogen (95%O_2/5%CO_2) for 5 min before adding stock solution 6

Table 2
Oncotic agents used for maintaining osmotic pressure within the perfusate for the in situ arterially perfused preparation

Chemical	Concentration	Source	References
Ficoll type 70 (70 kDa)	1.25%	Sigma F2878	(16, 18, 20, 22)
Bovine albumin	2%	Sigma A7030	(11)
Polyethylene glycol 20,000	1–1.25%	Fluka 95172	(25, 26)
Dextran (260 kDa)	2.25–2.5%	Sigma D7265	(10)

4. Sodium cyanide (to activate chemoreceptors).
5. Arginine vasopressin—aids in maintaining perfusion pressure (see Notes 10).
6. Heparin.
7. Isoflurane.

2.4.1. Additional Equipment and Chemicals (Depending on the Peripheral Afferent Stimulation Desired)

1. Spinal/vertebral clamp
2. Force transducer (e.g. FT-100, CB Science)
3. Stimulator and isolation unit for control of electrical stimulus
4. Calibrated forceps for noxious pinch
5. Bradykinin, capsaicin for chemical stimuli

2.5. Solutions

The modified Ringer solution containing (in mM) 125 NaCl, 24 $NaHCO_3$, 3 KCl, 2.5 $CaCl_2$, 1.25 $MgSO_4$, 1.25 KH_2PO_4 and 1 D-glucose is made fresh on the day of use (see Notes 1). After bubbling with carbogen (95% O_2/ 5% CO_2), the pH should be ~7.4.

A high-molecular-weight compound, as listed in Table 2, is added to the 200 ml of Ringer solution as an oncotic agent (Notes 2) for the perfusate.

The neuromuscular blocking agent vecuronium bromide (3 μg/ml), rocuronium (10 μg/ml) or pancuronium (5 μg/ml) is usually added to the perfusate after the start of the perfusion, and a phrenic recording is successfully obtained.

2.6. Suction Electrodes for Whole Nerve Recording

The electrodes are borosilicate glass capillary tubes pulled on an electrode puller, with wide shoulders and a short taper. Strip the Teflon coating off the last 2–3 mm of a silver wire and place in the lumen. Seal the flat end of the pipette with silastic (silicone) tubing connected to a blunt needle attached to a 3-way tap and a 10-ml syringe (Fig. 2). Silastic tubing is preferred due to the flexibility of the material and its ability to form a tight seal around the glass capillary and silver wire. If an air leak develops around the silastic tubing, a small drop of superglue can be used to form a tight seal. The small amount of perfusate drawn up into the syringe when aspirating the nerve ending will provide the conductive bridge between the nerve and the silver wire. A second Teflon-coated silver wire is wrapped around the outside of the electrode and held in

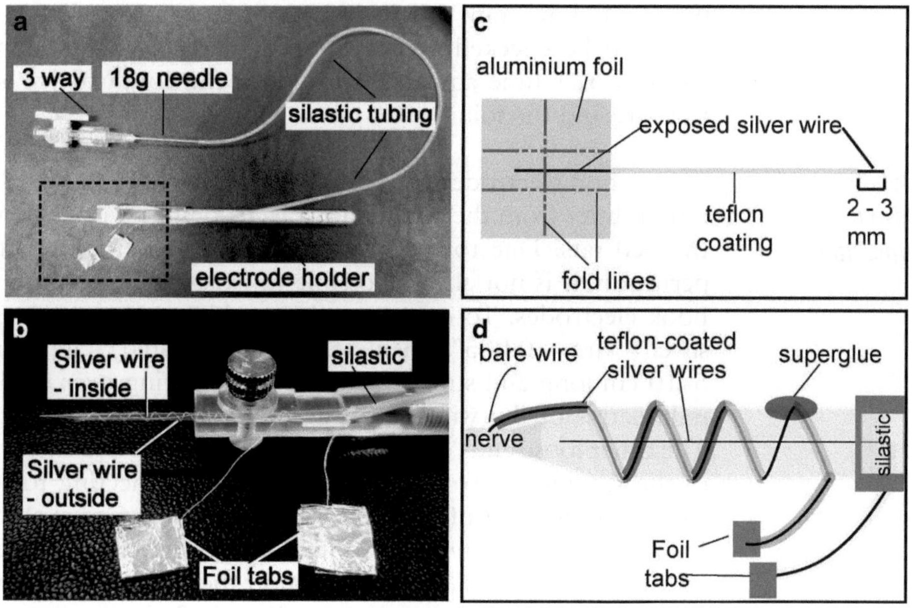

Fig. 2. Photos and schematic illustration of suction electrodes and stimulating electrodes commonly used in the in situ arterially perfused preparation. (**a**) Photograph of the suction electrode, connected to a silastic (silicone) tubing attached to an 18-G needle and 3-way tap, mounted on an electrode holder. The suction electrode (*dashed box*) is highlighted in (**b**), showing the Teflon-coated silver wire both inside and coiled around the outside of the borosilicate glass capillary. The ends of the silver wires are folded into 'foil tabs' (**c**) for easy connection to alligator clips. (**d**) Schematic representation of the construction of a suction electrode with the nerve aspirated inside the capillary and the correct placement of the reference wire (outside silver wire).

place with a drop of superglue (Fig. 2d). This is the reference wire for the bipolar nerve recording.

2.6.1. Potential Problems

1. If the nerve continuously slips out of the suction electrode:
   - The pipette tip is too large. Change to a different electrode with a smaller opening.
   - The seal between the silastic tubing at the end of the glass capillary or at the 3-way tap may be leaking. Check and reseal with superglue or trim the end of the silastic and reattach.
2. The nerve signal is weak:
   - Check to see if the nerve is drawn into the pipette properly. If only a very short piece of nerve is drawn into the electrode, the tip is too small. Either change to another electrode with a larger opening or gently file back the glass using emery paper.
   - The nerve may be damaged. Release the nerve from the electrode and trim to make fresh cut end.

Connect the loose ends of the wire to the preamplifier directly by clipping on alligator clips. The thin silver wires can be difficult to connect and this can be overcome by wrapping the wire in small pieces of aluminium foil to increase contact area (Fig. 2c). This needs to be checked and rewrapped regularly as the aluminium oxidises over time and this will reduce the conductivity between the wire and the foil.

2.7. Stimulation Electrodes for Whole Nerve Stimulation

Stimulation electrodes used in the in situ arterially perfused preparation differ from the stimulation electrodes used in in vivo anaesthetised rats. Due to the size of the preparation and continuous perfusion, it is not always possible to isolate the nerve on bipolar hook electrodes. To overcome this problem, take two lengths of silver wire (0.005″ diameter, Teflon-coated) approximately 5–10 cm long and strip away the Teflon coating over the last few millimetres of the wires at both ends to allow electrical contact. The other loose ends of the wire will be connected to the two poles of the stimulator. To improve the ease of connecting to stimulators, the loose ends of the silver wires can be wrapped in aluminium foil to increase the surface area for contact (Fig. 2c). Wrap one end of each of the two silver wires around the nerve that you wish to stimulate (e.g. vagus nerve) approximately 3–5 mm apart, making sure the wire is tightly wrapped around the nerve but without crushing the nerve. Excess lengths of exposed wires should be trimmed. The wires must then be secured in place and electrically isolated using either histology (low melting point) wax or Silgel. Low-melting point wax is ideal as it sets hard quickly but is fluid at relatively low temperature with a melting point of 55°C. This will prevent damage to the nerve. Silgel is another good option, although it will take longer for Silgel to set.

3. Methods

3.1. General Preparation of the In Situ Arterially Perfused Preparation

The following procedures and equipment are required for the in situ arterially perfused preparation:

1. Pretreat the animals with heparin sodium (<1,000 units) by intraperitoneal injection approximately 10–15 min prior to anaesthesia (see Notes 3).
2. Anaesthetise the animal using isoflurane (3–5%) in air. Wait until the animal is deeply anaesthetised, as determined by an absence of withdrawal to pinch and corneal reflex (see Notes 4).
3. Transect the animal just below the diaphragm (sub-diaphragmatically) using large scissors and immediately plunge the torso into cold Ringer solution bubbled with carbogen (see Notes 5).
4. Remove the scalp to expose the skull and slice off the top of the skull using the tip of a fresh scalpel blade, fitted onto a standard scalpel handle. Cut open the dura to expose the brain.
5. Remove or destroy the cortex of the rat, taking care not to damage the colliculi that lie directly below the cortex, just rostral of the cerebellum. Decerebration or decortication can be achieved either by aspiration or dissection (see Notes 6).
6. Once the decerebration/decortication is completed, remove the skin. Removing the skin is important to prevent hair and other debris that can clog the perfusion circuit.
7. Transfer the preparation to a dish containing Sylgard®/Silgel on the bottom on which you can securely pin down the preparation using pins. Cover the preparation with cold Ringer solution bubbled with carbogen (see Notes 7).
8. Pull down the diaphragm on the left to identify the insertion of the *left* phrenic nerve. Carefully lift the lobes of the left lung to find the phrenic nerve between the lobes. The lobes of the lung may be removed or left intact, as the lungs will remain deflated. However, removing the lungs will make it easier to find and record from the phrenic nerves. Gently rub the phrenic nerve with a pair of forceps, or nerve hook, to remove the connective tissue surrounding the nerve. This is important as excessive connective tissue will make it more difficult to obtain a good seal and hold the nerve in the suction electrode later. DO NOT cut the phrenic nerve at this point. The *right* phrenic nerve runs along the inferior vena cava and can be dissected

away at this point or later. The lobes of the right lung can be removed or left intact.

9. Separate the aorta from the vertebrae to facilitate cannulation (see Notes 8).
10. This is the end of the most basic components of the dissection for the in situ arterially perfused preparation. The dissection time up to this point should be kept to a minimum. Once you are proficient with this method, this dissection should be achievable within 10 min. If required, additional dissections (as discussed in Sect. 3.2) can be performed at this point.
11. Transfer the preparation to the perfusion chamber and retrogradely cannulate the descending aorta with the double-lumen cannula (see Notes 9). Advance the cannula 5–10 mm from the cut end and secure tightly with suture silk. The preparation can be placed onto modified ear bars (if required) or simply left in the bottom of the chamber. It is crucial that you make sure there are no points for leakage (e.g. torn arterial branches) between the suture and the tip of the cannula.
12. Arterial perfusion of the preparation can now begin, using carbogenated Ringer solution containing an oncotic agent. Perfusion to the preparation can slowly increase (over 1–2 min) until the flow rate is around 25–30 mL/min (for most preparations). This slow increase is important as it allows the vasculature to adapt to the change in perfusion pressure and for the preparation to rewarm progressively.
13. Cut the phrenic nerve at the insertion to the diaphragm. Using a suction electrode, tip diameter 200–300 μm, held on a micromanipulator, position the tip of the electrode next to the cut end of the phrenic nerve and draw back on the plunger of the attached syringe (Fig. 2d). The nerve should be drawn into the pipette and form a tight plug. Close the 3-way tap at this point to maintain the seal.
14. The heart will start beating soon after the perfusion begins and the preparation starts to rewarm (see Notes 10). Rhythmic respiratory muscle activity will usually begin within 5 min of the start of the perfusion, and the perfusion pressure has reached approximately 50 mmHg. To achieve this perfusion pressure, some preparation may require some 'fine-tuning' at this point, either by altering flow rate or by addition of AVP (see Notes 11).

3.2. Additional Dissections for Recording and Stimulation

Brainstem exposure: The dorsal surface of the brainstem can be exposed by cutting away the occipital bone and cerebellum. When removing the cerebellum, take care not to cut or damage the dorsal pontine respiratory group, located below the cerebellar peduncles. Damage to this region will produce apneustic bursting patterns. *This step is only required if direct brainstem microinjections or cellular recordings are to be performed.*

Sympathetic chain: The sympathetic chain can be found at the caudal levels of the thoracic vertebra running diagonally from the spinal processes towards the midline. Access to the sympathetic trunk can be gained by removing the last 3–5 ribs. However, take care not to cut too close to the vertebrae or you will damage the sympathetic chain.

Vagal nerve dissection: Turn the preparation over to expose the ventral side of the neck. This dissection can be made easier by placing a small cylinder, e.g. a piece of Tygon tubing used in the perfusion circuit (Masterflex 06409-14), under the neck to elevate and extend the neck. Cut away the *sternomastoid* muscle and blunt dissect through the *omohyoid* muscle to expose the common carotid artery and the adjacent vagus nerve. Blunt dissect the vagus nerve away from the common carotid artery. The recurrent laryngeal nerve can also be found running adjacent to the trachea.

For efferent vagal nerve recordings (16, 20): cut the vagus nerve as distal as possible and drape over the side of the neck for later recording with a suction electrode. Try to make the free section of nerve as long as possible to help placement of the electrode tip.

For vagal nerve stimulations (20): wrap two lengths of silver wires snugly around the vagus nerve, as described for stimulating electrodes.

4. Peripheral Afferent Stimulations to Evoke Cardiorespiratory Reflexes

As described above, a key advantage of this preparation is the ability to stimulate a variety of peripheral afferent nerves that can excite or inhibit central neural networks. The following describes a number of methods for stimulating and inhibiting these systems.

4.1. Baroreceptor Reflex Activation

Baroreceptors can be stimulated via graded increase in perfusion flow rate (21, 22), either computer or manually controlled. Perfusion pumps with serial input (e.g. Watson-Marlow 520DU) that allow for computer-controlled changes in perfusion pressure ramps will be more reproducible than manually controlled changes. In either method, the baroreceptors are stimulated by maximally increasing the pump flow rate for a few seconds, until the desired change in perfusion pressure is obtained. The rate of rise of perfusion pressure should be consistent between trials within the same preparation. This increase in perfusion pressure should lead to bradycardia, bradypnoea and sympathoinhibition.

4.2. Chemoreflex Activation

1. *Inline (intra-aortic) bolus injection of sodium cyanide* (11, 13, 21, 22): 10–30 μl of 0.03% solution of sodium cyanide. Connect a side port, constructed from a length of PE50 tubing and 23-G needle, to the inflow tube just distal of the

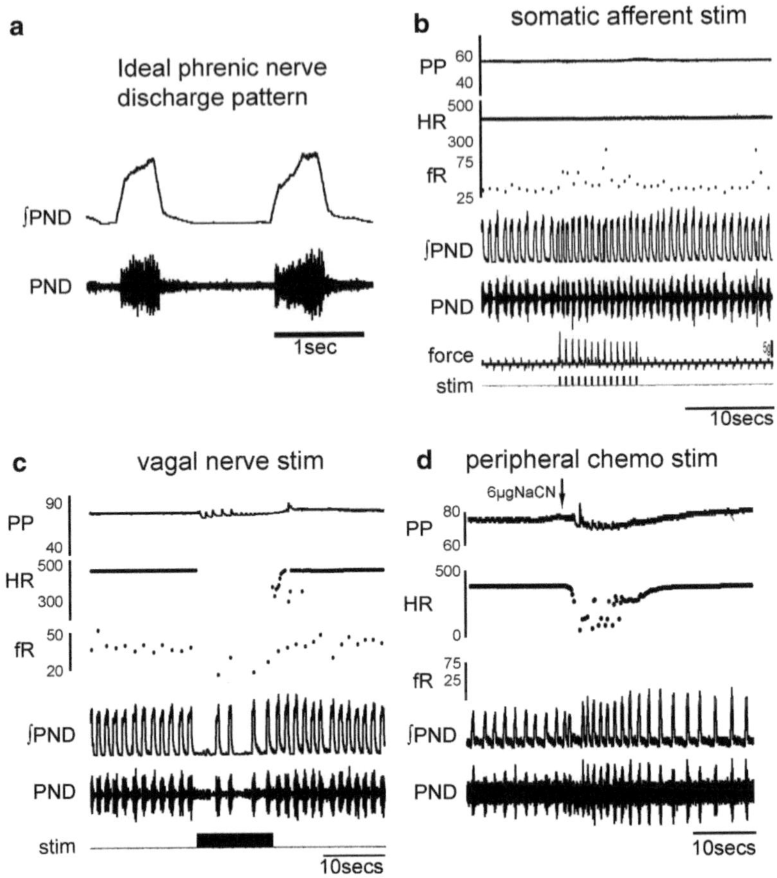

Fig. 3. Ideal phrenic nerve discharge (PND) and example effects of cardiorespiratory reflexes following activation of peripheral afferents. (**a**) Example of raw and rectified and smoothed PND in the working heart-brainstem preparation demonstrates a rapid onset with a ramping discharge pattern. (**b–d**) Example of peripheral afferent stimuli that excite and inhibit cardiorespiratory parameters measured in the in situ arterially perfused preparation. Train stimulation of somatic afferents (**b**) and peripheral chemoreceptor activation (**d**) produces tachypnoea, while vagal nerve stimulation (**c**) produces apnoea that mimics Hering-Breuer reflex activation. Vagal nerve stimulation (**c**) and chemoreceptor activation (**d**) also produce profound bradycardia and decrease in perfusion pressure.

connection to the double-lumen catheter. A bolus injection of between 10 and 30 μl of 0.03% sodium cyanide should elicit a rapid chemoreflex response characterised by bradycardia, tachypnoea and hyperpnoea (Fig. 3d).

2. Switching the perfusion mix to one bubbled with low O_2 (10% O_2, 5% CO_2, balanced N_2) will produce a more sustained peripheral chemoreceptor activation.
3. Switching the perfusion mix to one bubbled with high CO_2 (7–10% CO_2, balanced with O_2) will activate central chemoreceptors.

4.3. Somatic Afferent Stimulation

Somatic afferent nerves innervating the forelimbs can be activated by direct electrical stimulation of the forelimb flexor muscles in the limbs to produce isometric contraction (12, 18, 19, 21). The ipsilateral scapula is stabilised using a vertebral clamp and the distal end of the forelimb attached to a force displacement transducer connected to a recording amplifier to measure the tension generated by the electrically evoked forelimb contraction (see Notes 12). The electrical stimuli used to evoke muscular contractions are delivered using two silver wires (0.7 mm OD) introduced into the *biceps brachii* and *coracobrachialis* muscle groups running parallel to the *humerus*. The wires are then attached to a stimulation isolation unit. The following stimulation parameters are used to produce sufficient muscle contraction: pulse trains (0.4–1.3 Hz, 50- to 75-ms train duration, 1- to 3-ms pulse duration, 250 μA to 1 mA). Successful activation of somatic afferents is evidenced by a resetting of respiratory rhythm that results in tachypnoea, using train stimulation (Fig. 3b) occurring in late expiration, with pressor and tachycardic responses also apparent in sustained (tetanic) contractions (12, 18, 19, 21).

4.4. Nociceptors

Nociceptive afferents can be stimulated in a variety of ways:

Noxious pinch: The forelimb (23), hindlimb or tail, the latter two can only be done on the decerebrated arterially perfused rat preparation (11), can be stimulated with either a manually operated calibrated forceps or a pneumatic pincher. The design will need to provide accurate control of the distance between the ends of the forceps to allow for measurement and reproducibility of the stimulus intensity. The duration of the stimulation is between 4 and 10 s and will vary between preparations and is determined by the start of cardiorespiratory reflex responses (tachycardia, pressor response and bradypnoea).

Corneal stimulation: The cornea can be mechanically or chemically stimulated. For mechanical stimulation, the stimulus duration is between 1 and 2 s and the pressure ranges between 0.4 and 1.4 bar of air pressure (23). Chemical stimulation of the cornea can be achieved by applying a drop of bradykinin (10–20 μl of 1 μM) or capsaicin (10–20 μl of 0.12%, or 1 mM). The level of stimulation should be determined prior to paralysis (23).

Subcutaneous noxious stimulation: subcutaneous injections of bradykinin (5–10 μl of 1 μM) or capsaicin (5 μl of 0.12%, or 1 mM) can be used to stimulate subcutaneous chemoreceptors.

Thermal stimulation: Thermal sensors can be evoked by dripping heated saline (50–56°C) onto the skin of the forepaw (and hindpaw or tail in the decerebrated arterially perfused rat (11)), until a reflex cardiorespiratory response is obtained (11, 23).

Electrical stimulation: A branch of the brachial plexus, or the sciatic nerve in the decerebrated arterially perfused rat (11), can be electrically stimulated (10–40 V, 1–2 ms, 8–30 Hz). The nerves that form the brachial plexus emerge between the *scalene* muscles in the rat to form the brachial plexus in the axilla. The plexus can be accessed ventrally by cutting and reflecting the *clavicle* and the *pectoralis major* and *pectoralis minor* muscles. The *subclavian artery* runs within the axilla and care needs to be taken during the dissection not to damage the artery. Once the plexus is exposed, a branch can be isolated and a silver wire can be wrapped around the nerve for stimulation, as described above.

4.5. Pulmonary Stretch Receptor Activation (Hering-Breuer Reflex)

The pulmonary stretch receptors can be activated either via lung inflation (if left intact) or electrical stimulation of the vagus nerve (Fig. 3c) (20, 24). The parameters used for electrical stimulation of the vagus nerve are between 25 and 200 μA, at 50 Hz, with a pulse duration of 1–3 ms for 12 s. Electrical stimulation of the vagus slows phrenic bursting by increasing phrenic interburst interval (expiratory time). The threshold current required to produce a detectable increase in phrenic interburst interval should be determined in each preparation and the stimulus current chosen is normally 1.5–2 times threshold.

5. Notes

1. The Ringer solution can be readily made up using the following stock solutions listed in Table 1. To make 2 L of perfusate, measure out 1.7 L of distilled water and add 50 mL of stock solutions #1–5 and the D-glucose. Bubble with carbogen for at least 15 min before adding stock solution #6. It is important to bubble the Ringer solution with carbogen to adjust the pH to 7.3–7.4, prior to addition of stock solution #6 to prevent formation of precipitates. The Ringer solution can become cloudy after sitting on the bench for some time, but this can be reversed by bubbling the solution with carbogen again.
2. The high-molecular-weight compound, such as Ficoll® 70 (e.g. Sigma #F2878, Table 2), can be added to the perfusate only, rather than the entire volume of Ringer solution. The oncotic agent is not required for the Ringer solution used in the dissection stages. In general, 200 mL of Ringer solution is sufficient for one preparation. A fresh 200 mL of perfusate is prepared for each preparation. Once the oncotic agent is dissolved in the Ringer solution, the perfusate is filtered through a Whatman #1 filter to remove any particulates.
3. Heparin pretreatment is not necessary but is recommended by some researchers as it will improve the perfusion of the preparation by minimising blood clots in microvasculature. Heparin pretreatment can extend the amount of time spent

dissecting before the perfusion begins, allowing for more complex procedures to be carried out.

4. While all care should be taken to ensure that the animal is deeply anaesthetised before any surgical procedure begins, it is important that the heart continues to beat when the incisions are made. This is crucial as it will reduce the formation of clots in the microvasculature that may cause impairment in tissue perfusion.
5. Plunging the torso into cold Ringer solution will help rapidly reduce metabolism and aid in preserving the tissue. The Ringer solution is best chilled by placing the glass container in ice. The ideal temperature range for the Ringer solution is between 4 and 10°C.
6. An aspirator for decerebration can be made using a 5-mL syringe with a 200-μL pipette tip fitted over the syringe luer. Trim the opening of the 200-μL pipette tip to approximately 2–3 mm diameter to allow for easy suction of the cortex (Fig. 4). To aid suction, plunge the skull under the surface of the Ringer solution, then carefully and gently aspirate to remove the cortex. To assist in locating the colliculi, use the tip of the aspirator to gently separate the cortex from the cerebellum and push rostrally before aspirating.
7. Securely pinning down the preparation will help with the remaining fine dissection. The preparation can be pinned using any type of pins, from drawing pins, sewing pins, insect pins to syringe needles. The dish with Silgel/Silgard should be placed on ice to keep cold. The silicone compound in the dish will take longer to chill and can be stored in the refrigerator and placed into the ice just prior to when it is required.
8. In preparations where you do not wish to record thoracic sympathetic nerve activity, isolation of the thoracic aorta can be achieved by taking hold of the cut end of the aorta and tugging down sharply but firmly until most of the aorta is separated from the thoracic vertebra. Retrim the end of the aorta to get a nice, flush, cut end to make cannulation easier. However, if you wish to record from the thoracic sympathetic nerve trunk, then it is important to maintain the circulation to the spinal cord, as this is derived from dorsal penetrating arteries arising from the back of the aorta. Thus, you will need to maintain this circulation by cannulating the descending aorta distal to the level of T10, approximately at the level of the diaphragm.
9. The double-lumen cannula allows for perfusion and measurement of the perfusion pressure (Fig. 1). It is critical that a slow but continuous flow through the cannula tip is maintained to prevent the development of air bubble in the perfusion line.
10. On occasion, the heartbeat will be irregular or arrhythmia will develop. In most cases, the rhythm will settle down and become regular as the preparation is 'tuned'. On other occasions, a bolus

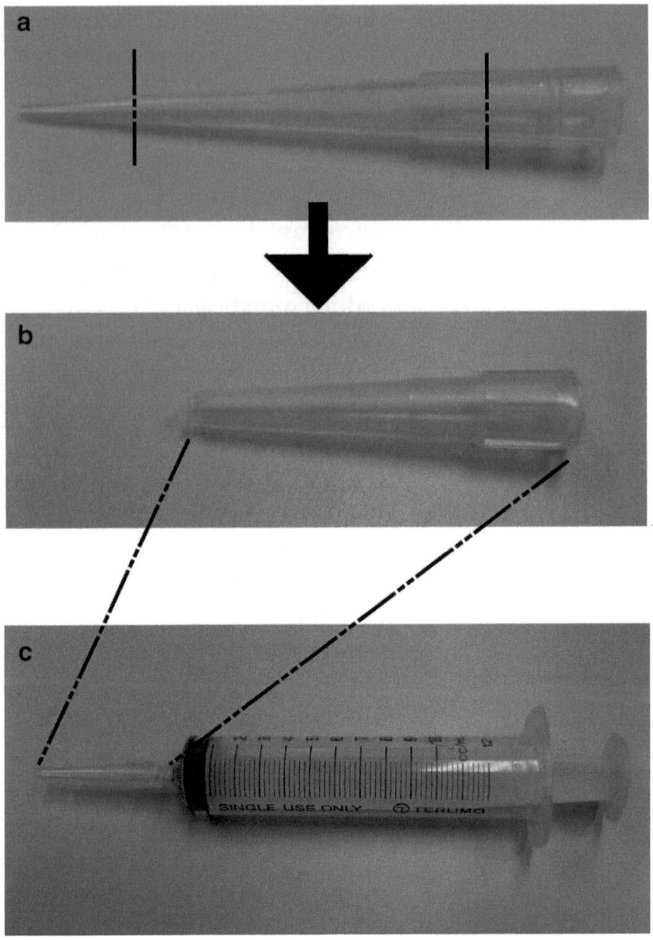

Fig. 4. Construction of aspirator used for decerebration. Trim a 200-μL pipette tip as illustrated by the *dashed lines* (**a**) to increase the opening. Trim the large end of the pipette tip to allow for a tight fit onto the luer end of a 10-mL syringe (**b**). Finally, firmly attach the trimmed pipette tip to a 10-mL syringe (**c**).

injection of sodium cyanide to evoke a severe but transient bradycardia may allow the rhythm to reset to a regular one.

11. You will need a perfusion pressure of approximately 50–70 mmHg to obtain sufficient perfusion to the brainstem. In some preparations, this is achieved when an adequate flow rate is reached (~30 ml/min). However, in some preparations, despite an adequate flow rate, the perfusion pressure remains low. Adequate perfusion is required for the establishment of the eupneic bursting pattern of the respiratory discharge (Fig. 3a) and will increase sympathetic activity, which will assist in producing vasomotor tone to help maintain the perfusion pressure. There are two ways to increase perfusion pressure. The simplest option is to increase flow rate until a ramping phrenic discharge pattern is established. Once the desired bursting pattern is established, slowly decrease the flow rate

while observing the bursting pattern. If the ramping discharge begins to revert to an apneustic or flat discharge pattern, slightly increase the perfusion rate again until the ramping discharge pattern is re-established (Fig. 3a). The second option is to induce pharmacological vasoconstriction by adding, for example, 200–400 pM arginine vasopressin (AVP) to the perfusate (10).

12. Typically, a force production of ~5–10 g of isometric contraction is required for activation of somatic afferents. Thus, it is vital that the forelimb is mechanically stable. The scapula can be secured using a small vertebral clamp, and the forepaw held on a strain gauge approximately parallel to the table. The forelimb will also need to be pre-tensioned to >10 g of force prior to any electrical stimulation of the forelimb muscles.

References

1. Smith JC, Ellenberger HH, Ballanyi K, Richter DW, Feldman JL (1991) Pre-Botzinger complex: a brainstem region that may generate respiratory rhythm in mammals. Science 254: 726–729
2. Wang X, Huang ZG, Dergacheva O, Bouairi E, Gorini C, Stephens C, Andresen MC, Mendelowitz D (2005) Ketamine inhibits inspiratory-evoked gamma-aminobutyric acid and glycine neurotransmission to cardiac vagal neurons in the nucleus ambiguus. Anesthesiology 103:353–359
3. Del Negro CA, Morgado-Valle C, Hayes JA, Mackay DD, Pace RW, Crowder EA, Feldman JL (2005) Sodium and calcium current-mediated pacemaker neurons and respiratory rhythm generation. J Neurosci 25:446–453
4. Kline DD, King TL, Austgen JR, Heesch CM, Hasser EM (2010) Sensory afferent and hypoxia-mediated activation of nucleus tractus solitarius neurons that project to the rostral ventrolateral medulla. Neuroscience 167:510–527
5. McDougall SJ, Peters JH, LaBrant L, Wang X, Koop DR, Andresen MC (2008) Paired assessment of volatile anesthetic concentrations with synaptic actions recorded in vitro. PLoS One 3:e3372
6. Chen Q-H, Andrade MA, Calderon AS, Toney GM (2010) Hypertension induced by angiotensin II and a high salt diet involves reduced SK current and increased excitability of RVLM projecting PVN neurons. J Neurophysiol 104:2329–2337
7. Bailey TW, Appleyard SM, Jin Y-H, Andresen MC (2008) Organization and Properties of GABAergic Neurons in Solitary Tract Nucleus (NTS). J Neurophysiol 99:1712–1722
8. Ballanyi K, Onimaru H, Homma I (1999) Respiratory network function in the isolated brainstem-spinal cord of newborn rats. Prog Neurobiol 59:583–634
9. Kron M, Reuter J, Gerhardt E, Manzke T, Zhang W, Dutschmann M (2008) Emergence of brain-derived neurotrophic factor-induced postsynaptic potentiation of NMDA currents during the postnatal maturation of the Kolliker-Fuse nucleus of rat. J Physiol 586:2331–2343
10. Paton JF (1996) A working heart-brainstem preparation of the mouse. J Neurosci Methods 65:63–68
11. Pickering AE, Paton JF (2006) A decerebrate, artificially-perfused in situ preparation of rat: utility for the study of autonomic and nociceptive processing. J Neurosci Methods 155:260–271
12. Potts JT, Spyer KM, Paton JF (2000) Somatosympathetic reflex in a working heart-brainstem preparation of the rat. Brain Res Bull 53:59–67
13. Paton JF, Li YW, Kasparov S (1999) Reflex response and convergence of pharyngoesophageal and peripheral chemoreceptors in the nucleus of the solitary tract. Neuroscience 93:143–154
14. Smith JE, Paton JF, Andrews PL (2002) An arterially perfused decerebrate preparation of Suncus murinus (house musk shrew) for the study of emesis and swallowing. Exp Physiol 87:563–574
15. Smith JE, Paton JF, Andrews PL (2001) Cardiorespiratory reflexes in a working heart-brainstem preparation of the house musk shrew, Suncus murinus. Auton Neurosci 89:54–59
16. Gajda BM, Fong AY, Milsom WK (2010) Species differences in respiratory rhythm

generation in rodents. Am J Physiol Regul Integr Comp Physiol 298:R887–R898
17. Dutschmann M, Wilson RJ, Paton JF (2000) Respiratory activity in neonatal rats. Auton Neurosci 84:19–29
18. Fong AY, Potts JT (2008) Neurokinin-1 receptors modulate the excitability of expiratory neurons in the ventral respiratory group. J Neurophysiol 99:900–914
19. Potts JT, Rybak IA, Paton JF (2005) Respiratory rhythm entrainment by somatic afferent stimulation. J Neurosci 25:1965–1978
20. Fong AY, Potts JT (2006) Neurokinin-1 receptor activation in Botzinger complex evokes bradypnoea. J Physiol 575:869–885
21. Potts JT, Fong AY, Anguelov PI, Lee S, McGovern D, Grias I (2007) Targeted deletion of neurokinin-1 receptor expressing nucleus tractus solitarii neurons precludes somatosensory depression of arterial baroreceptor-heart rate reflex. Neuroscience 145:1168–1181
22. Boscan P, Paton JF (2002) Integration of cornea and cardiorespiratory afferents in the nucleus of the solitary tract of the rat. Am J Physiol Heart Circ Physiol 282: H1278–H1287
23. Boscan P, Paton JF (2001) Role of the solitary tract nucleus in mediating nociceptive evoked cardiorespiratory responses. Auton Neurosci 86:170–182
24. Harris MB, St-John WM (2005) Phasic pulmonary stretch receptor feedback modulates both eupnea and gasping in an in situ rat preparation. Am J Physiol Regul Integr Comp Physiol 289:R450–R455
25. Molkov YI, Zoccal DB, Moraes DJA, Paton JFR, Machado BH, Rybak IA (2011) Intermittent hypoxia-induced sensitization of central chemoreceptors contributes to sympathetic nerve activity during late expiration in rats. J Neurophysiol 105:3080–3091
26. Abdala AP, Rybak IA, Smith JC, Paton JF (2009) Abdominal expiratory activity in the rat brainstem-spinal cord in situ: patterns, origins and implications for respiratory rhythm generation. J Physiol 587:3539–3559

Chapter 3

Excitatory Responses to Microinjection of Glutamate Depend on Dose Not Volume: A Meta-Analysis of Studies in Rat RVLM

Andrea H. Gaede and Paul M. Pilowsky

Abstract

Microinjection of the excitatory amino acid glutamate is commonly used to stimulate neuronal cell bodies in brainstem nuclei that are crucial for cardiovascular regulation, respiratory control, and other functions. One such nucleus, the rostral ventrolateral medulla (RVLM), integrates afferent information to coordinate cardiovascular responses to changes in the environment. In the RVLM, an increase in mean blood pressure of ≥25 mmHg following glutamate microinjection is widely accepted as evidence of accurate localisation of this nucleus. However, the dose of glutamate, and injection volume, varies considerably between investigators, and the optimal parameters are controversial. Here we examined data from 34 publications over the past 20 years in which glutamate doses, ranging from 0.0051 to 8.5 μg, were injected to identify the rostral ventrolateral medulla (RVLM) in rat. The aim of this chapter is to describe, in a pragmatic way, the ideal parameters for this method. Our meta-analysis reveals that there is a dose–response relationship between glutamate and blood pressure at low doses, but once the dose is sufficient to elicit a ≥35 mmHg rise in blood pressure, the response plateaus (~0.5 μg). Interestingly, the injection volumes used in the studies examined do not show any correlation with the observed blood pressure responses. Neither strain nor weight of rat had any influence on the blood pressure response induced by glutamate. Glutamate microinjection is a stable and reproducible method for activating cell bodies, but not fibres of passage, in most brain regions, and may also be a useful tool for normalising other drug responses.

Key words: Brainstem, Cardiovascular control, Blood pressure, Excitatory amino acid, Medulla oblongata

1. Introduction

The use of chemical microinjection to excite or inhibit neurons is widespread, yet there is still uncertainty about the pharmacokinetics and pharmacodynamics of this drug application method in vivo. Work by Nicholson and colleagues (1, 2) has carefully modelled the likely volume of distribution of drugs microinjected into neural

Paul M. Pilowsky et al. (eds.), *Stimulation and Inhibition of Neurons*, Neuromethods, vol. 78,
DOI 10.1007/978-1-62703-233-9_3, © Springer Science+Business Media, LLC 2013

tissue. Their findings suggest that many factors determine how far drugs spread following a bolus microinjection into the brain. These factors include rate of injection and injection volume and content. In addition, the concentration or amount of substance injected may be of central importance.

Glutamate microinjection was introduced in the early 1980s as a neuronal excitant for investigating the roles of activating cell bodies in different parts of the central nervous system (3, 4) and has since become a widely used method. Initial studies employed relatively high concentrations of glutamate, often in the molar range. It quickly became apparent that much lower concentrations of glutamate could be used to achieve equally robust responses. Currently, most studies report concentrations of glutamate that are 100 mM or less, in volumes of 200 nL or less. Microinjection of glutamate, and other receptor agonists and antagonists, into the brainstem is extremely valuable in elucidating the role of different nuclei in cardiovascular and respiratory regulation. Most studies rely on the information provided by the comprehensive work from Nicholson's laboratory to determine the likely spread of injected agents. Such estimations are of crucial importance because they provide some idea as to whether or not an agent will spread from the area of interest to other areas that may subserve completely different functions. Clearly, the factors that may affect injection spread are dependent upon many variables.

Here we provide a meta-analysis of studies that have injected glutamate into the RVLM to determine the importance of drug concentration compared with the importance of volume injected, for the blood pressure responses obtained. We restricted our survey to studies where glutamate was microinjected into the RVLM of adult rats. Several variables were recorded, including the weight of rats, the strain of rats, the anaesthetic used, and any associated experimental pathophysiological states.

2. Methods

A comprehensive review of the literature was conducted to select reports published between 1982 and 2009, where glutamate microinjections were made into the RVLM, and blood pressure recorded. Studies were only considered for analysis if all of the following criteria were available: the weight range of rats used, the type of anaesthetic, the method of glutamate microinjection, the dose and volume of glutamate, the strain and sex of rat, and the presence or absence of any pathological condition.

Data were analysed using GraphPad Prism version 5.03. The data are presented as *xy* plots with blood pressure on the *y*-axis and glutamate concentration (M), dose (g), or moles on the *x*-axis.

The log values of the glutamate doses were taken to allow for statistical comparison between doses and to fit a first-order polynomial line (straight line). Grouped data are expressed as the mean ± SEM. Significance was determined as $P < 0.05$. We determined whether there was a significant relationship between the glutamate concentration, moles injected, or grams injected and the resultant blood pressure response. Clearly, these values are simply different ways of expressing the same information, but we have included all three units commonly presented in publication. In addition, we have plotted the volume injected against the blood pressure response since authors use volumes ranging from 10 nL to 200 nL in the studies canvassed: a 20-fold variation. In one study by Bergamaschi et al. (5), a dose of 8.46 μg in 200 nL was administered to Wistar, and Wistar two-kidney one clip (2k1c), rats (5). On average, this dose elicited 43 mmHg and 75 mmHg rises in MAP, respectively. This study was excluded from determining the effects of volume because the dose used is five times the highest dose of any other study included, and the ΔMAP effects may be confounded by dose, strain, presence of induced hypertension, and volume.

Sixty-one data points from 34 publications fulfilled the criteria for inclusion in this study. 23.5% of papers were published in the journal *Hypertension*, 23.5% were published in *Brain Research*, 17.6% were published in the *American Journal of Physiology*, and 14.7% were published in *Autonomic Neuroscience: Basic and Clinical*, or its predecessor *Journal of the Autonomic Nervous System*. The remaining articles are listed in the references and did not exceed 10% for any one journal.

We used the equation from Nicholson's laboratory to determine the estimated spread of injected volume and compared this with the blood pressure response observed (1). Formula (1) below, adapted from Lipski et al. (4), calculates the radius in microns from the given volume in nanolitres according to Fig. 1 in the Nicholson 1985 paper:

$$r = \left(\frac{3u \times 10^6}{4\pi\alpha} \right)^{1/3} \quad (1)$$

where r radius in μm, U injection volume in nL, and α volume fraction available to the injectate, normally given as ~0.21.

Hence, an injection of 100 nL and $\alpha = \sim 0.21$ will have a radius of ~484 μm. The volume fraction (α) can vary depending on the lipid solubility of the drug, uptake mechanisms available, and molecular shape and size of the drug:

$$484\mu\text{m} = \left(\frac{3 \times 100 \times 10^6}{4 \times \pi \times 0.21} \right)^{1/3}$$

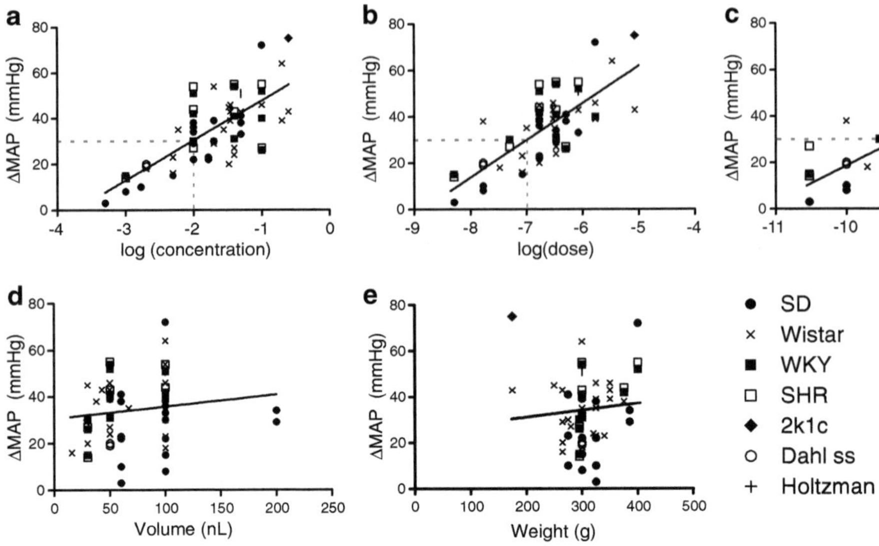

Fig. 1. Correlations between the (**a**) concentration (log (M); moles per litre), (**b**) dose (log (g)), and (**c**) mol (log (mol); moles) injected and changes in mean arterial pressure (ΔMAP). In each case, a clear dose–response effect is evident. The R^2 values for (**a**), (**b**), and (**c**) are 0.3 (P<.0001), 0.5 (P<.0001), and 0.5 (P<.0001), respectively. (**a–c**) *Dashed line* shows concentration, dose, and moles, respectively, at MAP = 30 mmHg; the typical change in blood pressure used to identify the RVLM. (**d**) A plot of the relationship between the volume (nanolitres; nL) and (ΔMAP); R^2 is 0.02 (P=0.3). (**e**) Plot of weight of the rats compared with (ΔMAP); R^2 is 0.01 (P=0.5). *SD* Sprague-Dawley, *WKY* Wistar-Kyoto rat, *SHR* spontaneously hypertensive rat, *2K1C* Wistar 2-kidney 1-clip hypertension model, *Dahl SS* Dahl salt sensitive.

3. Results

Observed changes in mean arterial blood pressure (ΔMAP) following glutamate microinjection into the RVLM were collected from 34 publications (5–38). There was a clear dose–response relationship between the amount of glutamate microinjected and the resultant ΔMAP (Fig. 1a–c). Conversely, there was no clear relationship between the volume of injectate and the blood pressure response for glutamate doses between 0.0051 and 1.69 μg (Fig. 1d). Glutamate dose values were log transformed and plotted against MAP response; the results were analysed by linear regression. The slope of the fitted line did not deviate significantly from zero. The ΔMAP responses, weights, and glutamate dose observed in the different strains of rats are presented in Table 1. All studies included in this meta-analysis were carried out in male rat.

We also examined whether a relationship exists between the strain or weight of rat and the ΔMAP response to a given dose of glutamate. The ΔMAP was not significantly different between strains of rat (Table 1). Furthermore, the weights of rats were not significantly different between strains, and the magnitude of blood pressure effects does not show any correlation with weight (Fig. 1e). The slope of the line fitted to these data did not significantly deviate from zero.

Table 1
ΔMAP responses, weights, and glutamate dose observed in the different strains of rats

	Sprague-Dawley	Wistar	Wistar-Kyoto	Spontaneously hypertensive	Total	Dahl $n=2$	Holtzman $n=1$
Mean ΔMAP	28 ± 4	35 ± 2	38 ± 4	42 ± 5	34 ± 2	19.5 ± 0.5	51
ΔMAP range	3.0–72[a]	16–64	15–54	14–55	3.0–72[a]	19–20	51
Mean weight	320 ± 10	301 ± 9	318 ± 13	320 ± 15	312 ± 6	300	300
Weight range	275–400	175–375	295–400	295–400	175–400	300	300
Mean glutamate dose	0.32 ± 0.11	0.85 ± 0.36	0.45 ± 0.16	0.31 ± 0.086	0.55 ± 0.16	0.017	0.85
Glutamate dose range	0.0051–1.7	0.017–8.5	0.0051–1.7	0.0051–0.85	0.0051–8.5	0.017	0.85

Blood pressure values are in mmHg; weights are in grams, and doses are in μg
[a]Includes dose–response curve

4. Notes

The principal novel findings of this analysis are first that glutamate clearly elicits dose-dependent blood pressure responses when injected into the RVLM within a range of doses from 0.03 to 50 nmol. In this range of doses, ΔMAP responses extend from 3 mmHg (0.03 nmol) to 75 mmHg (50 nmol) and that this finding is consistent across many strains of rat and many weights. Secondly, we observed no correlation between the volume of glutamate microinjection and the resultant MAP responses. The caveat here is that the volumes used ranged from only 10–200 nL. It is possible that if higher volumes were used, then different findings may have resulted. This issue is discussed in more detail below. Finally, we noted that for concentrations of glutamate between 10 and 100 mM, reliable responses of between 20 and 50 mmHg can be obtained, regardless of rat strain, size, or pathophysiological condition. For example, in spontaneously hypertensive rats, although the baseline arterial pressure was higher than their normotensive controls, the ΔMAP to a given dose of glutamate was the same.

Concentrations smaller than 10 mM would make it difficult to accurately locate the RVLM, whilst concentrations in excess of 100 mM appear to provide no additional benefit. Qualitatively, on reviewing the papers, we note that, as in our own experience, glutamate has a rather short-time course of action and that generally most variables return to control within 2–5 min. Furthermore, responses to glutamate do not display tachyphylaxis, and reinjection of the same dose at the same site elicits reproducible changes in MAP. In addition, dose–response curves do not reveal tachyphylaxis to glutamate as there is a linear relationship between increasing dose and increasing response (6, 27).

The Nicholson equation includes a volume fraction that is highly dependent on factors such as the lipid solubility of the agent injected and mechanisms available to remove the injected fluid. It should also be noted that this calculation provides the radius immediately after injection and assumes that the injectate diffuses completely into the pore space. According to the work of Nicholson (1), this modelling means that less than half of the available drug is present immediately following injection. Over the next time period, and depending on uptake and degradative mechanisms affecting the injected agent, its sphere of influence may be greater or smaller.

Many other factors apart from the volume fraction (or porosity) may affect the results. In the case of glutamate, rapid uptake by glial cells and neurons means that the area affected may be smaller than in situations where rapid uptake mechanisms are not present. Furthermore, unlike other agonists and antagonists, glutamate is quickly removed from the synaptic cleft so that its effects are likely to be more limited. A detailed explanation of the factors affecting diffusion of substances in the brain can be found in the recent review by Syková and Nicholson (2).

We consider it unlikely, albeit possible, that increasing the volume above 200 nL would have a substantial effect on the MAP responses reported by the authors examined here. There are two reasons for this. First, the RVLM is a longitudinal nucleus. In rat, it is approximately 200 μm in diameter and 600 μm in length. Increasing the volume to even 400 or 500 nL would only have the effect of causing a distribution of fluid outside the domain of the RVLM and yet would not reach the caudal ventrolateral medulla or any other cardiovascular nucleus since the rate of decay increases exponentially. There is, however, the risk that such large injections may affect nearby nuclei that have different functions, including the Bötzinger nucleus (respiratory rhythm) and the laryngeal motorneurons that are located more dorsally. The retrotrapezoid nucleus, which is implicated in chemosensitivity, may also be affected by extremely large injections. As such, it is essential that care be taken in planning injections to account for the specific neuroanatomy of the region being investigated.

The speed with which injections are made is one possible source of so-called volume effects. Injections made too rapidly may produce pressure-related responses. Changes in MAP and sympathetic nerve activity in response to pressing on the ventral lateral medulla have been reported several times previously. In one case, the effects were abrogated by blocking glutamate receptors (39). Therefore, it is likely that, in addition to dose, the pressure of the injection may nominally alter the magnitude of the blood pressure response.

5. Recommendations

The findings of this study allow us to make a number of recommendations in relation to glutamate microinjection studies:

1. Doses of glutamate between 50 and 100 mM should be used because this range allows for a consistently reliable MAP response. Higher doses appear to confer no additional advantage.
2. The volume used in microinjection studies is extremely variable, as is the injection method. We believe that the inside, and outside, diameter of the injection device should also be provided and that the volume being injected should be tailored to the specific nuclei being studied.
3. As reported by Seyedabadi et al. (27), for example, glutamate injections can be repeated, and if the electrode is kept in the

same location, reliably similar responses can be obtained. This suggests that glutamate microinjection may be a useful tool for normalising the responses observed after the injection of other agents. Such an approach, if successful, would markedly reduce variability in studies investigating the effects of pharmacological agents throughout the central nervous system, thereby reducing the number of animals required for such studies. In other words, if the microinjection of glutamate elicits a smaller response in the RVLM due to injection location or condition of the animal, then one might expect a test substance to also elicit a smaller response. Where such an approach may be particularly valuable is in situations where the test substance produces an extremely prolonged response and only one dose per animal is possible.

6. Conclusions and Perspectives

The novel features that we report here are that the effects obtained by microinjection of glutamate are essentially entirely dependent upon the dose used. The volume of injectate, the age of the rat, the size of the rat, and the anaesthetic used appear to have little impact on the blood pressure response. One important caveat to this statement is that if a sufficiently large volume is injected, then spread to distant sites with different effects is possible. Interestingly, despite a range of volumes between 10 and 200 nL, this problem was not apparent in the papers reviewed here. Differences in volume may become an issue with substances other than glutamate, or if volumes are increased dramatically above 200 nL. The results of this study support previous studies demonstrating that glutamate microinjection is a simple and reliable way to activate cell bodies but not fibres of passage in the RVLM and other brain sites (3, 40–43) and that the effect of glutamate microinjection depends principally on dose.

Acknowledgements

AHG is supported by a Macquarie University Research Excellence Scholarship. PMP is supported by funding from the National Health and Medical Research Council of Australia (457080, 604002, 1024489, 1030297), the Australian Research Council (DP110102110, LP120100463), the Heart Foundation of Australia (G 11S 5957), and Macquarie University.

References

1. Nicholson C (1985) Diffusion from an injected volume of a substance in brain tissue with arbitrary volume fraction and tortuosity. Brain Res 333:325–329
2. Sykova E, Nicholson C (2008) Diffusion in brain extracellular space. Physiol Rev 88:1277–1340
3. Goodchild AK, Dampney RA, Bandler R (1982) A method for evoking physiological responses by stimulation of cell bodies, but not axons of passage, within localized regions of the central nervous system. J Neurosci Methods 6:351–363
4. Lipski J, Bellingham MC, West MJ, Pilowsky P (1988) Limitations of the technique of pressure microinjection of excitatory amino acids for evoking responses from localized regions of the CNS. J Neurosci Methods 26:169–179
5. Bergamaschi C, Campos RR, Schor N, Lopes OU (1995) Role of the rostral ventrolateral medulla in maintenance of blood pressure in rats with Goldblatt hypertension. Hypertension 26:1117–1120
6. Adams JM, Madden CJ, Sved AF, Stocker SD (2007) Increased dietary salt enhances sympathoexcitatory and sympathoinhibitory responses from the rostral ventrolateral medulla. Hypertension 50:354–359
7. Adams JM, Bardgett ME, Stocker SD (2009) Ventral lamina terminalis mediates enhanced cardiovascular responses of rostral ventrolateral medulla neurons during increased dietary salt. Hypertension 54:308–314
8. Araujo GC, Lopes OU, Campos RR (1999) Importance of glycinergic and glutamatergic synapses within the rostral ventrolateral medulla for blood pressure regulation in conscious rats. Hypertension 34:752–755
9. Bergamaschi CT, Campos RR, Lopes OU (1999) Rostral ventrolateral medulla: a source of sympathetic activation in rats subjected to long-term treatment with L-NAME. Hypertension 34:744–747
10. Brooks VL, Freeman KL, Clow KA (2004) Excitatory amino acids in rostral ventrolateral medulla support blood pressure during water deprivation in rats. Am J Physiol (Heart C) 286:H1642–H1648
11. De Paula PM, Machado BH (2001) Changes in regional vascular resistance in response to microinjection of L-glutamate into different anteroposterior coordinates of the RVLM in awake rats. Auton Neurosci (Basic) 87:301–309
12. Ito S, Komatsu K, Tsukamoto K, Sved AF (2000) Excitatory amino acids in the rostral ventrolateral medulla support blood pressure in spontaneously hypertensive rats. Hypertension 35:413–417
13. Gordon FJ, McCann LA (1988) Pressor responses evoked by microinjections of L-glutamate into the caudal ventrolateral medulla of the rat. Brain Res 457:251–258
14. Kagiyama S, Tsuchihashi T, Phillips MI, Abe I, Matsumura K, Fujishima M (2001) Magnesium decreases arterial pressure and inhibits cardiovascular responses induced by N-methyl-D-aspartate and metabotropic glutamate receptors stimulation in rostral ventrolateral medulla. J Hypertens 19:2213–2219
15. Kasamatsu K, Sapru HN (2005) Attenuation of aortic baroreflex responses by microinjections of endomorphin-2 into the rostral ventrolateral medullary pressor area of the rat. Am J Physiol (Reg I) 289:R59–R67
16. Kawabe T, Chitravanshi VC, Kawabe K, Sapru HN (2006) Cardiovascular effects of adrenocorticotropin microinjections into the rostral ventrolateral medullary pressor area of the rat. Brain Res 1102:117–126
17. Kido H, Sasaki S, Oguni A, Harada S, Morimoto S, Takeda K, Nakagawa M (2004) Effects of angiotensin II type 1 receptor antagonist on pressor responses to pulsatile compression of the rostral ventrolateral medulla in rats. Hypertens Res 27:427–432
18. Machado BH, Bonagamba LGH, Dun SL, Kwok EH, Dun NJ (2002) Pressor response to microinjection of orexin/hypocretin into rostral ventrolateral medulla of awake rats. Regul Peptides 104:75–81
19. Madden CJ, Satoru I, Rinaman L, Wiley RG, Sved AF (1999) Lesions of the C1 catecholaminergic neurons of the ventrolateral medulla in rats using anti-DÎ^{2}H-saporin. Am J Physiol (Reg I) 277:R1063–R1075
20. Marchenko V, Sapru HN (2003) Cardiovascular responses to chemical stimulation of the lateral tegmental field and adjacent medullary reticular formation in the rat. Brain Res 977:247–260
21. Martins-Pinge MC, Becker LK, Luccizano Garcia MR, Zoccal DB, Neto RV, Basso LS, De Souza HCD, Lopes OU (2005) Attenuated pressor responses to amino acids in the rostral ventrolateral medulla after swimming training in conscious rats. Auton Neurosci (Basic) 122:21–28
22. Minson JB, Chalmers JP, Caon AC, Renaud B (1987) Separate areas of rat medulla oblongata with populations of serotonin- and adrenaline-containing neurons alter blood pressure after L-glutamate stimulation. J Auton Nerv Syst 19:39–50

23. Muratani H, Averill DB, Ferrario CM (1991) Effect of angiotensin II in ventrolateral medulla of spontaneously hypertensive rats. Am J Physiol (Reg I) 260:R977–R984
24. Ross CA, Ruggiero DA, Park DH (1984) Tonic vasomotor control by the rostral ventrolateral medulla: Effects of electrical or chemical stimulation of the area containing C1 adrenaline neurons on arterial pressure, heart rate, and plasma catecholamines and vasopressin. J Neurosci 4:474–494
25. Sakima A, Yamazato M, Sesoko S, Muratani H, Fukiyama K (2000) Cardiovascular and sympathetic Effects of L-glutamate and glycine injected into the rostral ventrolateral medulla of conscious rats. Hypertens Res 23:633–641
26. Schreihofer AM, Ito S, Sved AF (2005) Brain stem control of arterial pressure in chronic arterial baroreceptor-denervated rats. Am J Physiol 289:R1746–R1755
27. Seyedabadi M, Goodchild AK, Pilowsky PM (2001) Differential role of kinases in brain stem of hypertensive and normotensive rats. Hypertension 38:1087–1092
28. Smith JK, Barron KW (1990) The rostral and caudal ventrolateral medulla in young spontaneously hypertensive rats. Brain Res 506:153–158
29. Smith JK, Barron KW (1990) Cardiovascular effects of L-glutamate and tetrodotoxin microinjected into the rostral and caudal ventrolateral medulla in normotensive and spontaneously hypertensive rats. Brain Res 506:1–8
30. Tsuchihashi T, Kagiyama S, Onaka U, Abe I, Fujishima M (1997) Pressor and sympathetic responses to excitatory amino acids are not augmented in the ventrolateral medulla of Dahl salt-sensitive rats. Brain Res 750:195–200
31. Tsuchihashi T, Kagiyama S, Ohya Y, Abe I, Fujishima M (1998) Antihypertensive treatment and the responsiveness to glutamate in ventrolateral medulla. Hypertension 31:73–76
32. Tsuchihashi T, Kagiyama S, Matsumura K, Lin Y, Abe I, Fujishima M (2000) Cardiovascular responses to glutamate and angiotensin II in ventrolateral medulla of hypertension induced by chronic inhibition of nitric oxide. Hypertens Res 23:359–364
33. Urbanski RW, Sapru HN (1988) Evidence for a sympathoexcitatory pathway from the nucleus tractus solitarii to the ventrolateral medullary pressor area. J Auton Nerv Syst 23:161–174
34. Vieira AA, Colombari E, De Luca Jr LA, de Almeida Colombari DS, Menani JV (2007) Central cholinergic blockade reduces the pressor response to L-glutamate into the rostral ventrolateral medullary pressor area. Brain Res 1155:100–107
35. Willette RN, Krieger AJ, Barcas PP, Sapru HN (1983) Medullary gamma-aminobutyric acid (GABA) receptors and the regulation of blood pressure in the rat. J Pharmacol Exp Ther 226:893–899
36. Willette RN, Punnen-Grandy S, Krieger AJ, Sapru HN (1987) Differential regulation of regional vascular resistance by the rostral and caudal ventrolateral medulla in the rat. J Auton Nerv Syst 18:143–151
37. Xu Y, Krukoff TL (2004) Adrenomedullin in the rostral ventrolateral medulla increases arterial pressure and heart rate: roles of glutamate and nitric oxide. Am J Physiol Regul Integr Comp Physiol 287:R729–R734
38. Yajima Y, Ito S, Komatsu K, Tsukamoto K, Matsumoto K, Hirayama A (2008) Enhanced response from the caudal pressor area in spontaneously hypertensive rats. Brain Res 1227: 89–95
39. Morimoto S, Sasaki S, Miki S, Kawa T, Nakamura K, Ichida T, Itoh H, Nakata T, Takeda K, Nakagawa M, Yamada H (1999) Pressor response to compression of the ventrolateral medulla mediated by glutamate receptors. Hypertension 33:1207–1213
40. Curtis DR, Lodge D, McLennan H (1979) The excitation and depression of spinal neurones by ibotenic acid. J Physiol 291:19–28
41. Flatman JA, Lambert JD (1979) Sustained extracellular potentials in the cat spinal cord during the microiontophoretic application of excitatory amino acids. J Neurosci Methods 1:205–218
42. Talman WT, Perrone MH, Reis DJ (1980) Evidence for L-glutamate as the neurotransmitter of baroreceptor afferent nerve fibers. Science 209:813–815
43. Zieglgansberger W, Puil EA (1973) Actions of glutamic acid on spinal neurones. Exp Brain Res 17:35–49

Chapter 4

Acute Activation and Inhibition of the Sympathetic Baroreceptor Reflex

Melissa M.J. Farnham

Abstract

Many laboratories around the world use baroreceptor denervation in their investigations of the cardiovascular system. The most common method, performed in the rat, is a permanent denervation where the carotid sinus and aortic depressor nerves are transected. Here, we describe detailed step-by-step methods for acute reversible baroreceptor denervation in the rat. In this method, local anaesthetic agents are superfused around the region of the carotid sinus. With this method, complete baroreceptor denervation can be achieved for longer (bupivacaine, 30–60 min) or shorter (lignocaine, 10–30 min) periods, without harming the physiological state of the rat.

Key words: Barodenervation, Bupivacaine, Lignocaine, Carotid sinus

1. Introduction

Homeostasis is necessary to provide adequate perfusion pressure to vital organs such as the heart and brain and for the delivery of oxygen to, and removal of waste from, tissues. The baroreceptor reflex is the primary sympathetic reflex involved in maintaining blood pressure homeostasis. Baroreceptors are crucial for defending the organism against high arterial blood pressure. The baroreceptor reflex is activated by increased blood pressure which stretches baroreceptor fibres in the wall of the carotid sinus and the aortic arch, generating action potentials. The afferent baroreceptor neurons synapse in the nucleus of the solitary tract where they excite second-order excitatory neurons that project to different parts of the brain resulting in reflex responses including bradycardia and sympathoinhibition (for a review, see Pilowsky and Goodchild (1)). This chapter will only deal with the sympathetic baroreflex (Fig. 1).

Paul M. Pilowsky et al. (eds.), *Stimulation and Inhibition of Neurons*, Neuromethods, vol. 78,
DOI 10.1007/978-1-62703-233-9_4, © Springer Science+Business Media, LLC 2013

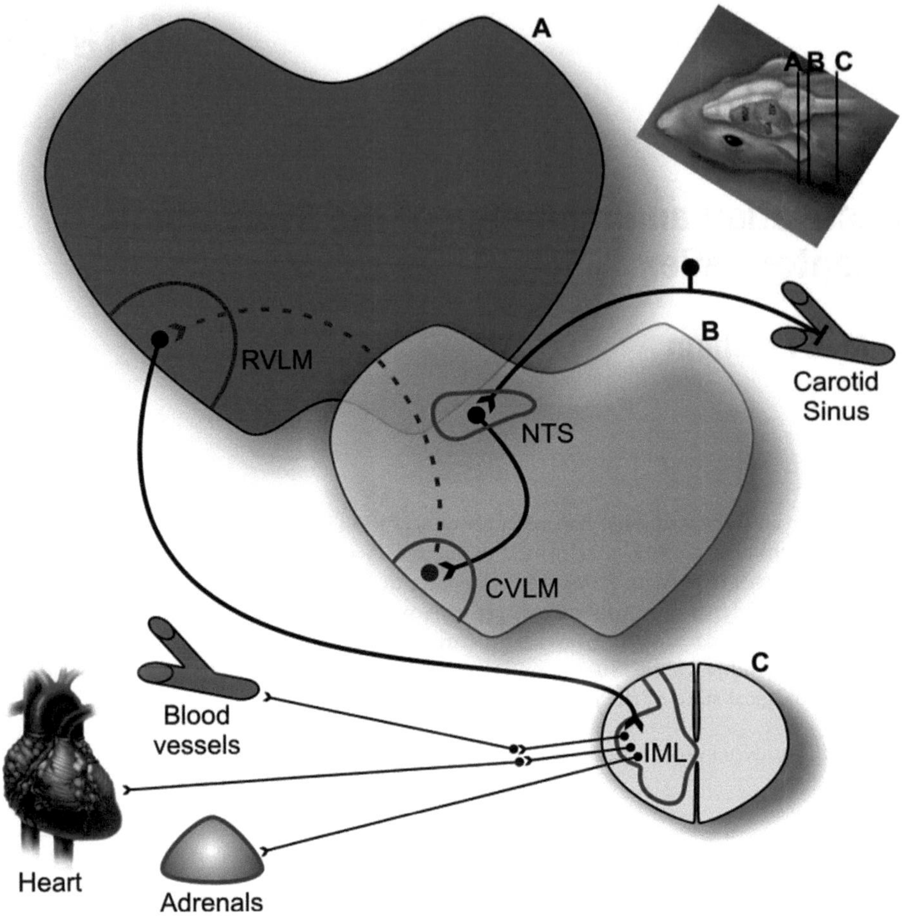

Fig. 1. Schematic cartoon of the sympathetic baroreceptor pathway. Note the three coronal levels on the *inset*. The *dashed line* represents inhibitory GABAergic neurons that convert excitatory baroreceptor signals to inhibition of bulbospinal sympathoexcitatory neurons. The *solid lines* are glutamatergic projections. Increased blood pressure causes activation of stretch-sensitive sodium channels, in stretch receptors (baroreceptors) in the carotid sinus, and the subsequent generation of action potentials that project to the nucleus of the solitary tract (NTS). Activated neurons in the NTS project to and excite GABAergic neurons in the CVLM. These inhibitory CVLM neurons release GABA onto the tonically active presympathetic neurons in the rostral ventrolateral medulla (RVLM) thereby reducing sympathetic drive to the intermediolateral cell column (IML) of the spinal cord which in turn causes a reduction in sympathetic drive to the heart, adrenal gland and resistance blood vessels.

Many laboratories, including our own, investigate the extent to which the baroreflex is altered by pharmacological manipulations such as exogenous administration of neuropeptides, drugs, hormones or their antagonists (2–6). In most studies, the heart rate baroreflex is measured (7–11), which is mediated primarily through the parasympathetic nervous system. It is less common to find studies that quantitate the sympathetic baroreflex; however, this is routinely done in our lab.

Transection of the carotid sinus and aortic depressor nerves are the most commonly performed procedures for chronic barodenervation in rat (9, 12, 13). Some scientists will attempt to, in addition to cutting the nerves, strip the adventitia around the carotid bifurcation and paint the artery with a 10% phenol solution to ensure complete destruction of baroreceptor afferents (14). This phenol approach is not straightforward and can result in significant morbidity and mortality, even in skilled hands. Other methods that cause interruption of the central baroreceptor pathway involve removing a part of the skull to expose the brainstem. Electrodes are then inserted into the NTS, or other sites such as the caudal ventrolateral medulla (CVLM), to silence the neurons in the region with injections of muscimol or electrolytic lesioning (9, 15).

More recently, we reported a novel method that is completely effective in blocking sympathetic components of the baroreflex but has two major advantages of being acute and reversible (16). Only two other groups have performed a method similar to this one, but neither measured sympathetic nerve activity (17, 18). A similar approach to the one described in this article is used clinically during carotid surgery if sudden bradycardia and hypotension occurs (19). This chapter will detail all the equipment, reagents and surgical techniques necessary to perform this acute, reversible baroreceptor denervation. It is important to note that this method does not address the heart rate baroreflex since the animals are vagotomised and the aortic depressor nerve transected.

2. Materials

2.1. Anaesthesia and Surgery

1. Urethane (ethyl carbamate; 1.3–1.5 mg/kg, 10% solution in saline, i.p., e.g. Sigma–Aldrich): make fresh on experiment day. Urethane should be stored in an airtight container, in the dark. Urethane is highly deliquescent and is therefore best purchased in small amounts and stored with desiccant. Urethane also causes chemical burns at concentrations exceeding 10%.
2. 0.9% sodium chloride (NaCl, saline).
3. Heparinised 0.9% saline (~40 U/ml) made fresh on the day of experiment. We achieve this by adding 3 drops of 5,000 U/ml heparin (e.g. DBL Heparin Sodium, Hospira) to 100 ml of 0.9% saline. Use for the arterial cannula to prevent clotting.
4. 26-gauge needles.
5. Syringes (1, 2, 10 ml).
6. Atropine sulphate (100 μg/kg; e.g. Pfizer): administer in the same injection as urethane. This reduces bronchial secretions and assists ventilation. If normal vagal function is needed, this should be omitted along with the vagotomy.

7. Surgical tools—e.g. scalpel, curved haemostats, iris scissors, fine forceps (straight and curved), a microvascular clip, small scissors and large-toothed forceps.
8. Silk sutures (e.g. Pearsalls, UK): 5/0 for tracheotomy and 2/0 for ligating nerves and blood vessels.
9. Cyanoacrylate adhesive (e.g. Super Glue, Krazy Glue, etc.).
10. Polyethylene tubing (PE9658; o.d. 0.96 mm; i.d. 0.58 mm): cut into 5-cm lengths for barodenervation cannulae and 10-cm lengths for cannulation of blood vessels.
11. Silastic tubing (o.d. 1.65 mm; i.d. 0.76 mm): used to connect to the PE9658 tubing to the 18-G needle.
12. 18-gauge drawing up needle for all cannulas—used to connect the cannula to the 3-way tap (vascular) or a 1-ml syringe (barodenervation).
13. 3-way taps—for construction of vascular cannula.
14. 14-gauge cannula (e.g. Optiva) for tracheotomy, needle removed and sheath shortened to 1 cm.
15. Neuromuscular blocker—e.g. pancuronium bromide (0.8 mg/kg i.v. initially, then 0.8 mg/kg/h, i.v., e.g. Astra Pharmaceuticals).
16. Silver bipolar electrode—either purchased or self-made.
17. SilGel (604, 2 part A: 1 part B, Wacker Chemie, Germany) or liquid paraffin oil.

2.2. Baroreflexes

1. Phenylephrine (PE, 10 μg/kg; i.v., e.g. Sigma–Aldrich): made fresh on the experiment day in 0.9% saline
2. Lignocaine (0.5 ml, Xylocaine 1%, e.g. AstraZeneca, Australia)
3. Bupivacaine (0.5 ml, Marcaine® 0.5%, e.g. AstraZeneca, Australia)
4. 1-ml syringes

2.3. Euthanasia

Potassium chloride (KCl, 3 mol/L, 0.5 ml)

3. Methods

3.1. Ethics

All procedures and protocols detailed here were approved by the Animal Care and Ethics Committee of Macquarie University (16). Experiments were conducted in accordance with the Australian code of practice for the care and use of animals for scientific purposes.

3.2. Anaesthesia

In these experiments anaesthesia was induced and maintained by administering an intraperitoneal injection (using a 26-gauge needle) of urethane (see above). Atropine sulphate (100 μg/kg) is included in the injection to reduce bronchial secretions prior to vagotomy. Absence of a withdrawal reflex to a hind paw pinch is used to ensure an adequate surgical plane of anaesthesia. Additional doses of 30–40 mg of urethane should be administered intravenously (i.v.), as required, to suppress nociceptive reflexes. Once the rat is adequately anaesthetised, the fur is removed with small animal clippers from sites where incisions are to be made (ventral neck, medial leg and back). Core body temperature is measured and maintained between 36.5 and 37.5°C using a rectal probe connected to a homeothermic-heating blanket (e.g. Harvard Apparatus, USA).

3.3. Cannulations

The cannulae are constructed by using a small piece of silastic to join one end of the PE9658 tubing with the 18-G needle, which is then connected to a 3-way tap and filled with either saline (venous) or heparinised saline (arterial).

With the rat in a supine position, a midline incision is made from the mandible to the sternum. The fat pad directly under the skin is separated to reveal the muscles underneath. The right external jugular vein is located lying superficial and lateral to the sternum, proximal to the clavicle. The right common carotid artery is located between the sternomastoid and sternohyoid muscles and directly underneath the omohyoid muscle. Better access is obtained by transecting the omohyoid muscle with a cautery (e.g. JC750 Heavy duty cautery set, RB Medical UK) to minimise any bleeding, but taking care not to bring the cautery close to underlying nerves or major blood vessels. The right common carotid artery and jugular vein are cannulated (see Sect. 3.4) to enable recording of arterial blood pressure and administration of drugs, respectively.

To access the femoral veins, the rat is placed in a supine position, and a longitudinal incision is made over the femur. The fat pad is retracted towards the midline to reveal the deep femoral neurovascular bundle. The femoral vein is carefully dissected away from the artery and nerve and cannulated for administration of PE only. A femoral vein is used to reduce volume-loading effects on the heart that are common following injections through the jugular vein cannula. The wound can then be closed around the cannula with sutures, staples or cyanoacrylate adhesive (e.g. Super Glue).

3.4. Insertion of the Intravascular Cannulae

To insert a cannula, approximately 1–2 cm of the vessel is cleared of connective tissue and fascia. This section is suspended across a pair of fine, blunt, curved forceps (for stability) and permanently ligated at the distal end with a silk suture. The proximal end is occluded with a microvascular clip (in the case of an artery). A pair of iris scissors is used to cut an opening at the distal end (but proximal to the ligation) perpendicular to length of the vessel and no more than 1/3 of the circumference. The cannula is then inserted into this slit, the microvascular clip is removed, and the cannula is advanced ~2 cm proximally. The cannula is secured in place with several silk sutures and the curved forceps removed. It is important not to advance the carotid artery cannula so far that it enters the heart. This will result in a blood pressure reading of 250/0 mmHg, and the condition of the animal will deteriorate rapidly.

3.5. Tracheotomy

The trachea is cannulated to permit artificial ventilation. After the cannulation of the carotid artery and jugular vein, the sternohyoid muscle is separated to reveal the trachea. The surrounding connective tissue and fascia is cleared away with blunt dissection. A small hole is made in the trachea, just caudal to the larynx, and the shortened 14-gauge cannula (without the needle) is inserted. The cannula is secured with silk sutures.

3.6. Vagotomy

The vagi are located alongside the common carotid arteries. The aortic depressor nerves are in the same neurovascular bundles as the vagi and common carotid arteries, and both are transected bilaterally to eliminate parasympathetic afferent and efferent pathways to the cardiopulmonary viscera. An important consequence of this part of the procedure is the elimination of the heart rate baroreceptor reflex.

3.7. Placement of the Barodenervation Cannulae

The common carotid bifurcations are located lateral to the caudal part of the digastric muscles. The tissue surrounding both bifurcations is gently cleared away, using blunt dissection, from both the vessels and the local nerves, being careful not to damage surrounding nerves or the carotid body. Prepare 2 × 5-cm lengths of PE9658 tubing by ensuring the ends are flat (so as not to damage the artery or tissue) and adding a 1-cm piece of silastic tubing to one end. An 18-gauge drawing up needle is attached to the other end of the silastic. The cannulae are positioned so that the tips lie at the carotid bifurcations (Fig. 2). The positioning of the tip in the bifurcation is critical to concentrating the local anaesthetic to the area containing the baroreceptors and afferent nerves. A small drop of cyanoacrylate adhesive is very effective for securing the cannula to the muscle wall and keeping it in place. The neck skin wound can then be sutured closed around all the cannulae.

3.8. Greater Splanchnic Nerve Dissection

A dorsal midline incision is made from the bottom of the ribcage to the pelvis. The lumbodorsal fascia on the left side is incised with gentle cautery to minimise bleeding and to access the retroperitoneal cavity. The cavity is held open, and the adrenal glands are

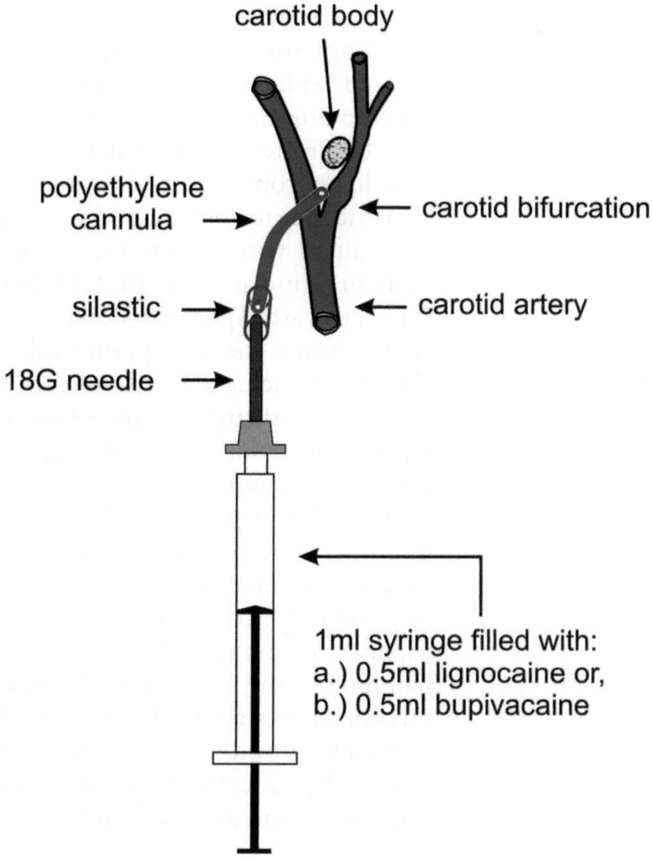

Fig. 2. Schematic diagram of barodenervation cannula placement. The barodenervation cannula is comprised of a 5-cm PE9658 tube with a 1-cm piece of silastic tubing connected to one end. An 18-G drawing up needle is attached to the other end of the silastic which can then be connected to a syringe for barodenervation. The cannula is positioned so that the tip is at the carotid bifurcation, which has been cleared of surrounding tissue and membranes. The cannula is secured to the muscle wall with cyanoacrylate adhesive. Barodenervation is achieved by administering 0.5 ml of either lignocaine (1%) or bupivacaine (0.5%) to the area of the carotid bifurcation (Figure and legend reprinted from Farnham and Pilowsky (16). Copyright (2009) with permission from Elsevier).

located as a landmark. Saline-soaked cotton tips are used to gently clear away the fascia to reveal the greater splanchnic nerve and celiac ganglion, located towards the midline. The celiac ganglion is identifiable as a 'bulge' between the adrenal and the midline, and the large nerve running into it from the midline is the greater splanchnic nerve. The splanchnic nerve is gently cleared of fascia, tied distally with a silk suture and transected at the celiac ganglion (distal to the silk suture).

3.9. Stereotaxic Step-Up Prior to Experimentation

The rat is placed in ear bars in a stereotaxic frame. The arterial cannula is connected to a blood pressure transducer (e.g. Edwards Lifesciences) via the 3-way tap, which feeds into a bridge amplifier (e.g. Scientific Concepts). The tracheal cannula is connected to the ventilator, and the rat is ventilated with room air supplemented with 100% O_2. Neuromuscular blockade is achieved with pancuronium bromide (0.4 mg given as a 0.2-ml bolus injection i.v.) administered through the jugular vein cannula. All skeletal muscle neuromuscular-blocking drugs (*-onium class drugs) can also affect autonomic ganglia if the dose is excessive. Paralysis is maintained with a slow i.v. infusion of 10% pancuronium bromide in saline at a rate of 2 ml/h using a syringe pump (e.g. 11 Plus, Harvard Apparatus) connected to the jugular vein cannula. At this stage the retroperitoneal cavity needs to be held open. This can be achieved with silk ties to the surrounding skin edges. The dissected greater splanchnic nerve is placed across a silver bipolar electrode for recording, which is connected to a headstage (e.g. Super Z, CWE) and then a bio-amplifier (e.g. BMA-400 AC/DC, CWE: sample rate 2 kHz, 10–20 k × gain, 0.1–3 Hz filtering). The retroperitoneal cavity is filled with an electrically isolated material such as liquid paraffin oil or a hard setting SilGel (e.g. 604, 2 part A: part B, Wacker Chemie, Germany). In our experience, nerve activity is generally reduced by surgery and handling of the nerve. It is therefore necessary to allow the nerve recording to stabilise for no less than 60 min. Data is acquired using a Cambridge Electronic Design ADC system (model 1401; Cambridge Electronic Design, Cambridge, UK) and Spike 2 acquisition and analysis software.

3.10. Baroreflex and Barodenervation

Before performing the barodenervation a control baroreflex test should be conducted to ensure that the carotid sinus nerve was not damaged during placement of the barodenervation cannulae. Slowly fill the femoral vein cannula with PE (10 µg/kg). Filling is deemed complete when a sharp increase in arterial blood pressure is recorded. A control set of baroreflexes is obtained by administering 0.1 ml of PE (10 µg/kg) through the femoral line, three times. This dose of PE is sufficient to increase mean arterial pressure by ~100 mmHg. Allow a recovery period of 10–15 min between doses. A sharp increase in arterial pressure coupled with a sharp decrease in the splanchnic sympathetic nerve activity is normally observed (Fig. 3).

After blood pressure and splanchnic sympathetic nerve activity return to control levels, barodenervation is performed with local anaesthetic. The choice of agent depends on the duration of anaesthesia required. Either lignocaine (0.5 ml, Xylocaine 1%, AstraZeneca, Australia) or bupivacaine (0.5 ml, Marcaine® 0.5%, AstraZeneca, Australia) is slowly infused into the area surrounding each of the carotid bifurcations via a 1-ml syringe connected to the cannulae (Fig. 2). Allow the anaesthetic agent 10 mins to act before retesting the baroreflex. Test the baroreflex in the same way as

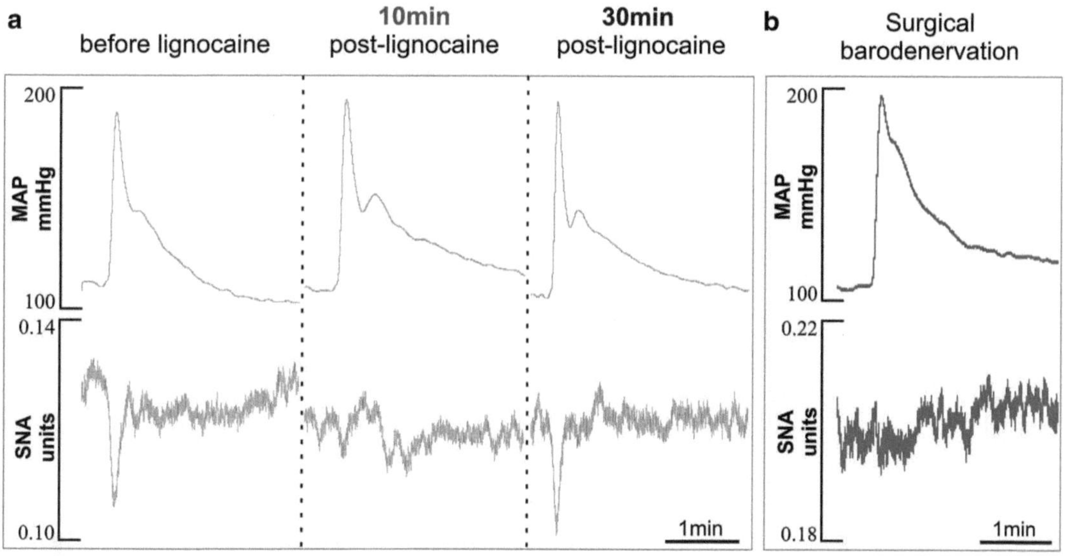

Fig. 3. Individual example of lignocaine barodenervation. (**a**) Smoothed traces of mean arterial pressure (MAP) and splanchnic sympathetic nerve activity (SNA) of an individual rat during a PE challenge before lignocaine, 10-min post-lignocaine and 30-min post-lignocaine. At 10-min post-lignocaine, there is an absence of any discernable sympathetic baroreflex, and at 30-min post-lignocaine, the baroreflex is returned. (**b**) A representative example of surgical barodenervation where a PE challenge caused a large increase in MAP, but no change in SNA without prior administration of local anaesthetics to the carotid bifurcation. Scale bar = 1 min (Figure and legend reprinted from Farnham and Pilowsky (16). Copyright (2009) with permission from Elsevier).

described above by administering 0.1 ml of PE (10 μg/kg, i.v.). You should observe a sharp increase in arterial pressure (similar to the intact condition), but with little or no change in splanchnic nerve activity (Figs. 3 and 4).

At the conclusion of the experiment, the residual noise in the splanchnic nerve can be obtained by killing the rat with an i.v. injection of 0.5 ml of 3 mol/L potassium chloride or by crushing the proximal end of the nerve with fine forceps. This residual (zero level) noise permits normalisation of nerve activities between rats.

4. Notes

4.1. Choice of Local Anaesthetic

- Neither lignocaine nor bupivacaine affects resting mean arterial pressure (16).
- Bupivacaine is a more complete and longer lasting barodenervation agent compared to lignocaine, which can be seen by comparing Figs. 3 and 4. Bupivacaine is effective for twice as long as lignocaine, which is expected as the half-life is twice as long.
- The viscosity of the anaesthetic might also improve the model. The anaesthetic agents used in Farnham and Pilowsky (16)

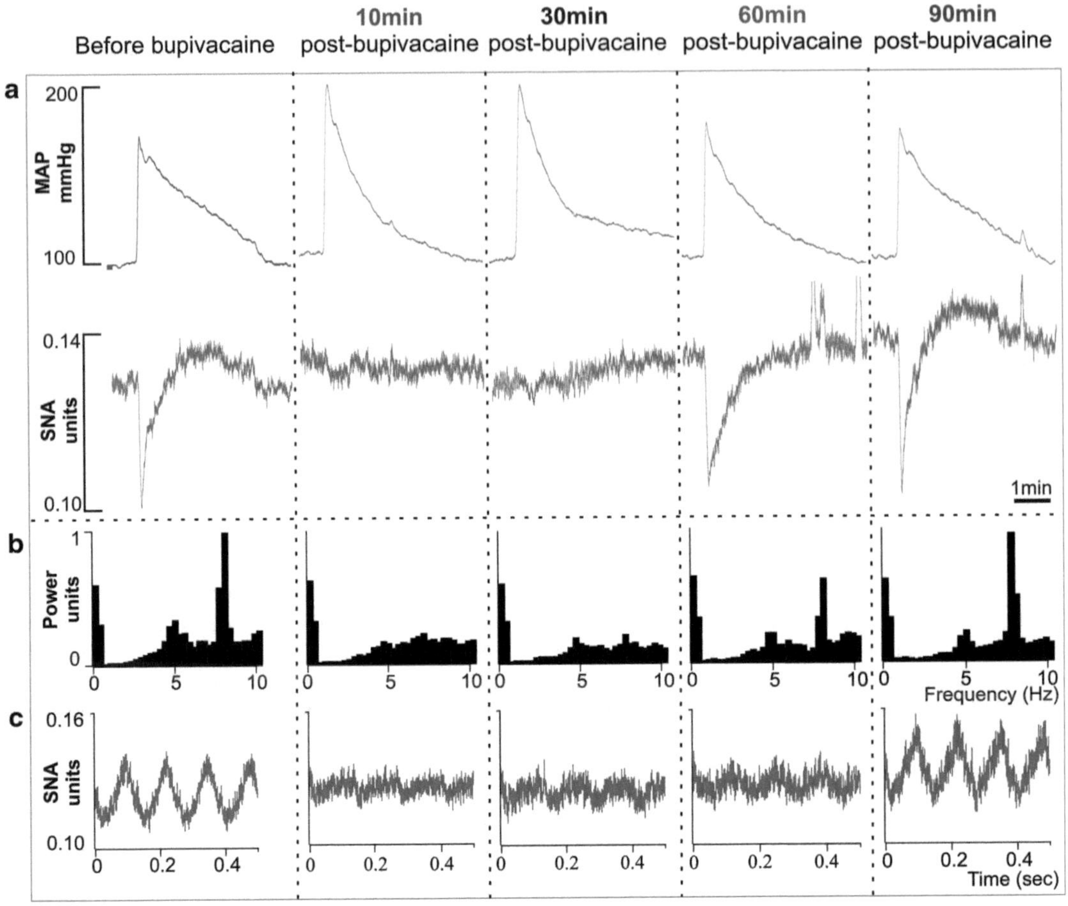

Fig. 4. Individual example of bupivacaine barodenervation. (**a**) Smoothed traces of mean arterial pressure (MAP) and sympathetic nerve activity (SNA) of a rat during a PE challenge before bupivacaine, 10-min post-bupivacaine, 30-min post- bupivacaine, 60-min post-bupivacaine and 90-min post-bupivacaine. It can be seen that at 10-min and 30-min post-bupivacaine, there is an absence of the sympathetic baroreflex, and at 60-min and 90-min post-bupivacaine, the baroreflex is returned. (**b**) Power spectra of the 2-min periods preceding the PE challenges in (**a**). The 8-Hz peak (corresponding to heart rate) present before barodenervation is abolished at 10-min and 30-min post-bupivacaine and reappeared 60-min and 90-min post-bupivacaine. (**c**) Ensemble averages of SNA triggered from heart rate show the pulse modulation over the same time periods as in (**b**). Bupivacaine abolished the pulse modulation at 10-min and 30-min post-bupivacaine. At 60-min post-bupivacaine the pulse modulation is evident, but does not reach control levels until 90-min post-bupivacaine. Scale bar = 1 min (Figure and legend reprinted from Farnham and Pilowsky PM (16). Copyright (2009) with permission from Elsevier).

paper were in a liquid form. Since there is no containment around the barodenervation cannula and carotid bifurcation, the liquid will simply obey the laws of gravity and run out of the area and out of the neck wound. More viscous liquids or gels would theoretically stay in the area longer, providing longer lasting anaesthesia. It may be possible therefore, by testing other local anaesthetic compositions, to manipulate the barodenervation to range between 10 min and several hours.

- Local anaesthetics can have toxic effects at high doses which can manifest as cardiovascular changes or events, such as reduced heart contraction, arrhythmias and hypotension (20, 21). Repeated doses at short intervals are not advised.

4.2. Conscious State

This method is restricted to anaesthetised preparations as the anaesthetics have access to the vagi and the laryngeal nerves. Anaesthetising the vagi and laryngeal nerves in a conscious animal may affect the animal's ability to breathe.

References

1. Pilowsky PM, Goodchild AK (2002) Baroreceptor reflex pathways and neurotransmitters: 10 years on. J Hypertens 20: 675–1688
2. Krieger MH, Moreira ED, Oliveira EM, Oliveira VLL, Krieger EM, Krieger JE (2006) Dissociation of blood pressure and sympathetic activation of renin release in sinoaortic-denervated rats. Clin Exp Pharmacol Physiol 33:471–476
3. Miyawaki T, Goodchild AK, Pilowsky PM (2002) Activation of mu-opioid receptors in rat ventrolateral medulla selectively blocks baroreceptor reflexes while activation of delta opioid receptors blocks somato-sympathetic reflexes. Neuroscience 109:133–144
4. Goodchild AK, Van Deurzen BTM, Sun Q-J, Chalmers J, Pilowsky PM (2000) Spinal GABA(A) receptors do not mediate the sympathetic baroreceptor reflex in the rat. Am J Physiol 279:R320–R331
5. Farnham MMJ, Li Q, Goodchild AK, Pilowsky PM (2008) PACAP is expressed in sympathoexcitatory bulbospinal C1 neurons of the brain stem and increases sympathetic nerve activity in vivo. Am J Physiol 294:R1304–1311
6. McMullan S, Dick TE, Farnham MMJ, Pilowsky PM (2009) Effects of baroreceptor activation on respiratory variability in rat. Respir Physiol Neurobiol 166:80–86
7. Lancien F, Mimassi N, Conlon JM, Mével J-CL (2011) Central pituitary adenylate cyclase-activating polypeptide (PACAP) and vasoactive intestinal peptide (VIP) decrease the baroreflex sensitivity in trout. Gen Comp Endocrinol 171:245–251
8. Chang KS, Morrow DR, Kuzume K, Andresen MC (2000) Bupivacaine inhibits baroreflex control of heart rate in conscious rats. Anesthesiology 92:197–207
9. Schreihofer AM, Ito S, Sved AF (2005) Brain stem control of arterial pressure in chronic arterial baroreceptor-denervated rats. Am J Physiol 289:R1746–1755
10. Gao X, Zhang F, Han Y, Wang H, Zhang Y, Guo R, Zhu G (2004) AT1 receptor in rostral ventrolateral medulla mediating blunted baroreceptor reflex in spontaneously hypertensive rats. Acta Pharmacol Sin 25:1433–1438
11. Schreihofer AM, Mandel DA, Mobley SC, Stepp DW (2007) Impairment of sympathetic baroreceptor reflexes in obese Zucker rats. Am J Physiol 293:H2543–2549
12. Kawabe T, Kawabe K, Sapru HN (2007) Cardiovascular responses to somatosensory stimulation and their modulation by baroreflex mechanisms. Clin Exp Hypertens 29: 403–418
13. El-Mas MM, Abdel-Galil A-GA, El-Gowelli HM, Daabees TT (1997) Short-term aortic barodenervation diminishes (alpha)1-adrenoceptor reactivity in rat aortic smooth muscle. Eur J Pharmacol 322:201–210
14. Krieger EM (1964) Neurogenic hypertension in the rat. Circ Res 15:511–521
15. Reis D, Ross C, Ruggiero D, Granata A, Joh T (1984) Role of adrenaline neurons of ventrolateral medulla (the C1 group) in the tonic and phasic control of arterial pressure. Clin Exp Hypertens 6:221–241
16. Farnham MMJ, Pilowsky PM (2009) Local anaesthetics for acute reversible blockade of the sympathetic baroreceptor reflex in the rat. J Neurosci Methods 179:58–62
17. Behnia R, Koushanpour E, Sinclair DM (1996) Pressure-heart rate relationship in intact and after stepwise elimination of three major baroreflex loops in dogs. Anesth Analg 83:965–974
18. Seagard JL, Hopp FA, Drummond HA, Van Wynsberghe DM (1993) Selective contribution of two types of carotid sinus baroreceptors to the control of blood pressure. Circ Res 72:1011–1022

19. Al-Rawi PG, Sigaudo-Roussel D, Gaunt ME (2004) Effect of lignocaine injection in carotid sinus on baroreceptor sensitivity during carotid endarterectomy. J Vasc Surg 39:1288–1294
20. Ford DJ, Singh P, Walters C, Raj PP (1984) Duration and toxicity of bupivacaine for topical anesthesia of the airway in the cat. Anesth Analg 63:1001–1004
21. Brosh-Nissimov T, Ingbir M, Weintal I, Fried M, Porat R (2004) Central nervous system toxicity following topical skin application of lidocaine. Eur J Clin Pharmacol 60:683–684

Part II

Intracellular Signalling

Chapter 5

Pharmacological Dissection of G Protein-Mediated Second Messenger Cascades in Neurons

Mark C. Bellingham

Abstract

Despite the advent of transgenic gene manipulation in vivo and in vitro and the rapid rise of RNA interference, in order to target protein function by ablation or suppression of specific gene products, pharmacological manipulation of G protein-coupled signalling pathways remains a highly useful tool for understanding receptor coupling to effector mechanisms in neurons. The key to drawing valid conclusions from pharmacological manipulations is a solid knowledge of the signalling pathway molecules and on- and off-target effects of the pharmacological tools to be used. This chapter outlines the actions, application and pitfalls of pharmacological ligands commonly used to dissect the major second messenger signalling cascades activated by G proteins in neurons. Despite the limitations of many currently used pharmacological tools, careful and judicious use of these tools allows the exploration and definition of these pathways at all levels.

Key words: G protein, Second messenger, Cyclic AMP, Cyclic GMP, Phospholipase C, Phospholipase A2, Arachidonic acid, Protein kinase, Protein phosphatase

1. Introduction

GTP-binding proteins (G proteins) are one of the most widely distributed means of signal transduction across the neuron membrane, coupling extracellular membrane receptors for a multitude of neurotransmitters, hormones and other bioactive molecules to a variety of signalling pathways within the neuron (1, 2). G proteins are membrane-associated heterotrimeric proteins, composed of α (36–52 kDa), β (35–36 kDa) and γ (8–10 kDa) subunits (3). The α subunit is unique to each G protein (Table 1) and is thought to be the primary determinant of the functional properties of the G protein. In contrast, the βγ subunit complex, assembled from a repertoire of 5β and 12γ subunits, does not bind preferentially to

Paul M. Pilowsky et al. (eds.), *Stimulation and Inhibition of Neurons*, Neuromethods, vol. 78,
DOI 10.1007/978-1-62703-233-9_5, © Springer Science+Business Media, LLC 2013

Table 1
G protein α and βγ subunits are associated with specific effector enzymes and second messengers

Subunit	Common effector enzyme	Cell types	# of clones
$G\alpha_o$	AC↑ PLC↑	Neurons	1
$G\alpha_i$	AC↓ PLC↑	Ubiquitous	3
$G\alpha_s$	AC↑	Ubiquitous	2
$G\alpha_T$	GC↑ PDE↑	Photoreceptors	2
$G\alpha_z$	AC↓	Neurons	1
$G\alpha_{q/11}$	PLC↑	Ubiquitous	4
Gβγ	AC, VDCC↓ PLC, GIRK, PI3K↑	Ubiquitous	5β, 13γ

AC adenylyl cyclase, *PLC* phospholipase C, *GC* guanylyl cyclase, *PDE* phosphodiesterases, *VDCC* voltage-dependent Ca^{2+} ion channels, *GIRK* G protein-gated inwardly rectifying K^+ channels, *PI3K* phosphoinositide 3-kinase. Ubiquitous expression found in neurons and glia

particular α subunits and has a number of distinct neuronal functions, including direct modulation of ion channels (4) (Fig. 1).

In the inactive state, the α subunit contains a guanine binding site which is occupied by guanosine diphosphate (GDP) (5). When a ligand occupies a receptor, the receptor protein changes conformation to allow association of this site with a G protein (Fig. 1). This association catalyses the exchange of GDP for guanosine triphosphate (GTP) (6). In the presence of Mg^{2+}, this complex then dissociates in a temperature-dependent manner to form an active α subunit and βγ subunit (5). These active subunits are able to bind to and regulate specific effector enzymes, such as adenylyl and guanylyl cyclases, phospholipases, or other proteins, such as ion channels (4, 7) (Fig. 1). The α subunit has usually been assigned a greater regulatory role, but βγ subunits have also been shown to directly regulate protein kinases and ionic channel proteins (4). After hydrolysis of GTP by the slow, intrinsic GTPase activity of α subunits, α and βγ subunits reassociate, ending further activation of effector proteins (5).

The obligatory step of G protein activation imposes a significant delay of >50 ms in producing a response following binding to the receptor site, and subsequent steps in more complex signalling cascades can increase this response delay to more than 200 ms (8–10). In addition, responses can persist and undergo significant amplification due to the successive production of second messenger

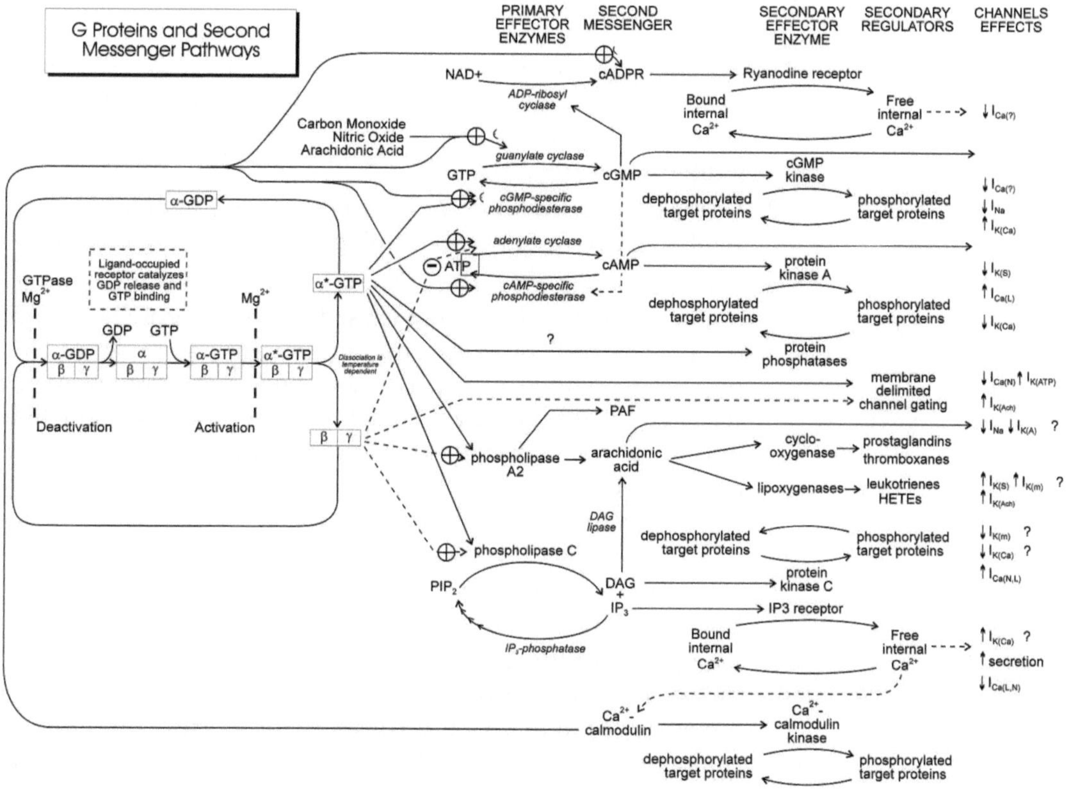

Fig. 1. The G protein life cycle and downstream signalling cascades. See text for details and abbreviations. *Circled plus signs* indicate activation of the target, *circled minus signs* indicate inhibition of the target.

molecules which outlive the original G protein activation (1, 9, 10). Delayed onset and long persistence are thus hallmarks of neuronal responses involving G proteins (1).

1.1. Principles Underlying Pharmacological Dissection of G Protein-Mediated Responses in Neurons

The fundamentals of determining whether a neuronal response is mediated by a specific G protein-initiated signal cascade appear straightforward. The main signalling cascades are well characterised (see Fig. 1 for an overview of the main signal cascades and their molecular components) and pharmacological tools for inhibiting or activating most molecular components of these are readily available. Ideally, one would like to use a single specific inhibitor of a key cascade molecule or enzyme to demonstrate that this inhibitor blocks the test response and move on to other work. Downfalls of this simplistic approach include significant crosstalk between signalling cascades, especially those involving Ca^{2+}-dependent processes and the non-specificity of the pharmacological tool chosen. Even more problematic is how to interpret a negative result—if an inhibitor did not block the test response, does this rule out a role for the hypothesised signalling cascade or is it merely a failure to adequately present the drug to the test system?

In general, it is wise to:

1. *Test your hypothesised signalling cascade at two levels.* The closer to the ultimate effector protein, the better. Consistent results at two levels tend to rule out "off-target" effects of a single inhibitor and the possibility of signal cascade crosstalk.
2. *Carry out positive control experiments.* For compounds which yield negative results before ruling out a signalling cascade.
3. *Use positive test procedures.* These are designed to produce a response which mimics that caused by the G protein-coupled receptor ligand, by regulating an effector molecule. Such a response should at least partially occlude the response to the receptor ligand. General caveats to these procedures include difficulties in establishing whether a maximal response has been produced by the procedure, and the possible presence of receptor reserve in the system being tested. In general, mimicry of a receptor-mediated response by manipulation of effector molecules should be interpreted conservatively and confirmed by a negative test procedure.
4. *Carry out negative test procedures.* These are designed to block the response to the receptor ligand by regulating an effector molecule thought to be involved in producing the ultimate response. The two chief difficulties with negative tests are adequate delivery of the regulating drug into the system being tested and possible crossover of the drug to other effector molecules. Negative tests which fail to block receptor-mediated responses are commonly due to the former; demonstration that the drug, given in the same or lower concentration and by the same method of delivery, produces an effect on another response can be used to rule this out. The possibility of crossover effects is best guarded against by the knowledge and use of suitable drug concentrations and demonstration of similar effects by other drugs targeting the same effector molecule.

In the materials and methods below, I have indicated sources of the drugs/molecules (if a Google search for the substance common name revealed three or more sources, the substance is indicated as widely available), the range of concentrations used when applied to neurons (most generally as part of the external solution bathing brain slices but also often acutely isolated or cultured neurons; the concentration is that applied in the external solution unless stated otherwise) and alternate/chemical names for compounds. I have highlighted potential known pitfalls in using substances by using the note of *Drawbacks.*

This formulary does not attempt to be comprehensive; I have limited it to compounds which have been commonly (or less commonly) used for pharmacological manipulation of intracellular signalling pathways in neurons. I acknowledge that there are a multitude of references which I have not cited about the use of these substances to test neuronal functions, due to space: I encourage researchers to find the most appropriate references for substances used in their experimental work and cite these, rather than follow my choices.

2. Materials, Methods and Notes

2.1. Determining Whether a G Protein Is Involved

It is possible to determine whether a G protein is necessary for production of a response, and the specific Gα subunit type involved can also be identified in some instances. The latter information can then inform choices about the downstream signal cascade to be tested as per Table 1.

Positive tests for involvement of a G protein in transduction of the receptor signal to the functional response generally rely on persistent activation of a response in the intracellular presence of non-hydrolysable GTP analogues, either alone or after withdrawal of the agonist (4, 5):

(a) GTP analogues or reduction of GTPase activity produce a similar response.

 (i) GTPγS (widely available, 0.1–0.5 mM in pipette solution or direct application to inner membrane side (11–14))—a GTP analogue which irreversibly activates G proteins and is not susceptible to GTPase activity (15, 16). It is non-membrane permeant and thus needs to be applied intracellularly, for example, by inclusion as a component of the whole-cell patch pipette internal solution (17) or by direct application to the inner side of isolated membrane patches (4). *Drawback*: GTPγS may also directly stimulate formation of inositol(1,4,5)-triphosphate (IP_3) in membranes.

 (ii) GMP-PNP (widely available, 250 μM in pipette solution (18), 5′-guanylyl-imidodiphosphate)—non-hydrolysable GTP analogue which binds tightly to GTP binding sites, causing persistent activation (18) and is the preferred analogue for activation of ADP-ribosylation factor (15). *Drawbacks* of (i) and (ii):

 - Effectiveness of these analogues depends on endogenous GTP levels; thus, it may be difficult to achieve appropriate concentrations and negative results are inconclusive.

 - Effects are often slow to appear and to reach equilibrium.

(iii) Cholera toxin (widely available, 500 ng/ml in culture media (14))—the cholera toxin A subunit persistently activates the α subunit of $G\alpha_S$ by catalysis of its ADP-ribosylation, reducing intrinsic GTPase activity (15); a similar effect can occur at $G\alpha_t$ and $G\alpha_{i/o}$ but additionally requires presence of a receptor ligand (15).

(iv) *Pasteurella multocida* toxin (Sigma-Aldrich, 5 μg/ml (19))—selective for $G\alpha_{q/11}$, activating PLC and IP_3 signalling (20).

(v) Aluminium fluoride (AlF^{4-})—persistently activates α subunits by inhibiting their GTPase activity (15). Can also use sodium or potassium fluoride (10 mM) in recording patch pipette to combine with endogenous Al^{3+} or with added $AlCl_3$ (100 μM) to enhance formation of AlF^{4-} (21). *Drawbacks*: high intracellular levels of fluoride may alter Ca^{2+} and Mg^{2+} concentrations (15) and may also stimulate hydrolysis of PIP_2 (22).

(vi) Mastoparan (Sigma-Aldrich, 1 μM (23))—a cationic amphiphilic peptide from the venom of the wasp *Vespula lewisii* which can directly activate G proteins, preferentially G_i and G_o, rather than G_s (2). It is membrane permeant. MAS 7 is a synthetic analogue with ~fivefold more potency, while MP-S activates $G\alpha_s$ more markedly than $G\alpha_{i/o}$ (15). *Drawbacks*: has a variety of effects, including stimulation of exocytosis (15), phosphoinositide breakdown and PLA_2 activity, inhibition of calmodulin (15) and increasing free intracellular Ca^{2+} by inhibition of the SERCA pump (IC_{50} 1 μM) (24).

(vii) Melittin (Sigma-Aldrich, 1 μM (25, 26))—a cationic amphiphilic bee venom polypeptide that directly activates $G\alpha_{11}$ and $G\alpha_i$ proteins by a mechanism similar to mastoparan (27). *Drawbacks*: also binds calmodulin, activates PLA_2 and inhibits PKC.

(viii) Compound 40/80—cationic amphiphilic compound which directly activates G proteins (most probably $G\alpha_i$) (15). *Drawbacks*: also inhibits PLC, calmodulin and ADP-ribosylation; releases arachidonic acid and histamine and modulates PLA_2.

(ix) Substance P (widely available)—directly activates G proteins and induces Ca^{2+} mobilisation (15). *Drawback*: also activates neurokinin receptors.

(x) Ruthenium red (widely available, 30 μM (28))—a polyvalent cation which can activate $G\alpha_i$ and $G\alpha_o$ GTPases. *Drawbacks*: also a non-specific blocker of several transient receptor potential channels (29) and inhibits ryanodine-sensitive and voltage-gated Ca^{2+} channels and mitochondria Ca^{2+} uptake (30, 31).

(b) Absolute requirement for intracellular GTP + agonist to get a response (e.g. (6, 32)), suitable only for membrane patches as complete removal of endogenous GTP is required. *Drawback*: difficult to achieve zero GTP levels.

(c) Mimicking response with direct application of pre-activated G protein.

Negative tests

(a) Pertussis toxin (List, PTX, 500 ng/ml in culture media (14, 33))—selective for α subunits of $G\alpha_t$, $G\alpha_o$ and three forms of $G\alpha_i$. PTX catalyses the hydrolysis of NAD^+ and the transfer of the resulting ADP moiety to a cysteine residue near the C terminus of these Gα subunits, thus preventing functional interaction of the transmembrane receptor with the G protein (15). *Drawbacks*: 1–2 days required to produce ribosylation, thus cannot be used in acutely prepared brain slices (can use pre-activated PTX + NAD), and can produce a false negative if βγ subunit is active unit.

(b) Blockade of response by GDPβS (Boehringer Mannheim, 250–1,000 μM (12–14, 18, 34, 35), guanosine-5′-*O*-2-thiophosphate)—a membrane-impermeant blocker of GTP responses applied by inclusion in the pipette solution. It is a non-hydrolysable GDP analogue which competes with GTP for the α subunit binding site and locks the GTPase cycle into an inactive state when present at higher concentrations than cellular GTP (~25 μM) (18).

(c) Suramin (widely available, 20 μM (36))—suramin and suramin analogues at sub-micromolar to micromolar concentrations suppress the rate limiting step of GDP release from Gα subunits, thus preventing dissociation of the α and βγ subunits (27). Two suramin analogues, NF449 and NF503, are selective for $G\alpha_s$ at low micromolar concentrations but also inhibit $G\alpha_i$ or $G\alpha_q$ at >30-fold higher concentrations (27). *Drawbacks*: Suramin also inhibits phospholipase D activity (IC_{50} = 15 μM), is a purinergic P_2 receptor agonist (37) and inhibits ryanodine-sensitive Ca^{2+} release channels (31).

(d) *N*-ethylmaleimide (widely available, NEM, 10–300 μM (38–40))—a sulfhydryl alkylating agent, which mimics the action of PTX by alkylating the PTX-sensitive cysteine residue of $G\alpha_o$; in addition, a cysteine residue near the N terminus is also alkylated, decreasing $G\alpha_o$ affinity to βγ complexes (15, 33, 41). NEM (10–300 μM) is membrane-permeant and acts rapidly in cultured neurons and brain tissue slices (38–40), but its indiscriminant chemical modification of sulfhydryl bonds in proteins (e.g. (42)) may limit its biological utility, particularly at higher concentrations which disrupt agonist-receptor interactions (33).

(e) YM-254890 (Sigma-Aldrich, 1 μM (43))—a cyclic depsipeptide derived from broth cultures of *Chromobacterium* sp. QS3666 which selectively blocks effects of Gα_q on PLC (IC_{50} ~ 30 nM (44)).

(f) Gallein (widely available, 10 μM (36))—a small molecule inhibitor of βγ subunit effects, with an IC_{50} of 240 nM (45); M119 is a structurally related molecule, acting at similar concentrations (IC_{50} 200 nM (46), used at 10 μM (36)).

2.2. What Is the Primary Effector Enzyme?

(A) Adenylate cyclase (AC) [cAMP-dependent phosphodiesterase (cAMP-PDE)].

Most likely if the G proteins Gα_s (activating AC) or Gα_i (inhibiting AC) are involved and cAMP levels are altered (47). The action of AC is to generate cAMP by hydrolysis of ATP, while cAMP-PDE utilises cAMP (47). There are 10 AC isoforms in mammals, with varying degrees of Ca^{2+} sensitivity (47). βγ subunits can independently stimulate or inhibit some AC isoforms (47).

Positive tests

(a) Activators of AC mimic (and occlude) response.

(i) Forskolin (widely available, 1–10 μM (48–52))—cell-permeant, rapid, reversible activator of AC (EC_{50} = 5–10 μM (48–50) to as low as 8 nM (51)). Does not affect activity of cAMP-PDE or guanylyl cyclase. Several derivative compounds are available which show enhanced water solubility and increased potency or oral activity. 1,9-dideoxy-forskolin is an analogue which does not activate AC (EC_{50} > 1 mM) and can be used as a negative control compound (49, 52).

Negative tests

(a) Inhibitors of AC reduce response.

(i) SQ22,536 (Biomol, 100–200 μM (36, 53), 9-tetrahydro-2-furyladenine)—cell-permeable AC inhibitor (IC_{50} = 200 μM) (49).

(ii) MDL-12,330A hydrochloride (widely available, 300 μM (36), also called RMI 12330)—cell-permeable, irreversible inhibitor of AC (IC_{50} = 250 μM).

(iii) 2′,5′-dideoxyadenosine (widely available, 100 μM (53))—cell-permeable AC inhibitor which acts to block the activation of the PKA catalytic subunit (IC_{50} = 250 μM) (49, 53). *Drawback*: also acts as an agonist at the intracellular adenosine binding site (P-site) (54).

(iv) NKY80 (Sigma-Aldrich, 10–100 μM (55), 2-amino-7-(2-furanyl)-7,8-dihydro-5(6H)-quinazolinone—a non-nucleoside compound which selectively inhibits type V adenylyl cyclase (IC_{50} 8.3 μM) in comparison to type II (IC_{50} 1.3 mM) and type III (IC_{50} 132 μM)) (54).

(b) AMP-PNP (widely available, 1–5 mM in pipette solution (53), 5′-Adenylimidodiphosphate) is a non-hydrolysable form of ATP which competes with ATP for its binding site on AC; as it is non-membrane permeable, it has to be applied intracellularly in the pipette solution.

(B) Phospholipase C (PLC).

Most likely if the G protein involved is PTX-insensitive (most likely $G\alpha_{11/q}$) and IP_3 + DAG are produced (56). PLC cleaves phosphatidylinositol 4,5-bisphosphate (PIP_2) into DAG and IP_3 (56). DAG remains bound to the membrane, and IP_3 is released as a diffusible messenger into the cytosol to bind to IP_3 receptors of the endoplasmic reticulum, causing Ca^{2+} release from intracellular stores (56). In addition, free Ca^{2+} and DAG together work to activate PKC (56). There are three PLC subtypes, β, γ and δ, each of which has several isoforms; the three subtypes are thought to act in concert, each contributing to the neuronal response to PLC (56).

Positive tests

(a) Activators of PLC increase the response.

(i) *m*-3M3FBS (Calbiochem, 50 μM (57, 58) 2,4,6-trimethyl-*N*-[*m*-3-trifluoromethylphenyl) benzenesulphonamide)—*o*-3M3FBS is an inactive analogue used as a control (57). Whether this compound activates PLC has been questioned, as it fails to generate IP_3 in some test systems and releases Ca^{2+} from both endoplasmic reticulum and mitochondria (59).

Negative tests

(a) Inhibitors of PLC reduce the response.

(i) U-73122 (Sigma-Aldrich, 1–10 μM (13, 43, 58, 60))—an aminosteroid membrane-permeant PLC inhibitor; IC_{50} = 0.2–1 μM (61); U-73343 is a structural analogue compound which is a very weak inhibitor of PLC and is useful as a negative control (1–10 μM (13, 43, 61)). *Drawbacks*: inhibits IP_3 production and may cause Ca^{2+} release from intracellular stores (61); both U73122 and U73343 directly inhibit G protein coupled inwardly rectifying K^+ channels (60); this action has been attributed to alkylation of G proteins, as U-73122 is a *N*-substituted maleimide compound similar to NEM (57).

(ii) NCDC (Tocris, 10–100 μM (60, 62), 2-nitro-4-carboxyphenyl-*N*,*N*-diphenylcarbamate)—membrane-permeant PLC inhibitor.

(iii) Neomycin (Sigma-Aldrich, 50 μM (12))—an aminoglycoside antibiotic which is poorly permeant and inhibits IP_3 formation (1 mM in pipette solution (12)). Drawback: also inhibits Ca^{2+} channels and ryanodine-sensitive Ca^{2+} release channels (31).

(iv) Doxorubicin (Sigma-Aldrich, 100 μM, also known as adriamycin)—an anthracycline antibiotic used in cancer chemotherapy which inhibits IP_3 formation.

(v) PMSF (Sigma-Aldrich, 100 μM, phenylmethylsulfonyl fluoride)—also inhibits serine proteases, including acetylcholinesterase (63), inhibits metabolism of anandamide and binds to the cannabinoid 1 receptor (64).

(vi) ET-18-OCH_3 (Calbiochem, 10–40 μM (57, 65), 1-*O*-octadecyl-2-*O*-methyl-*rac*-glycero-3-phosphorylcholine, also called edelfosine)—IC_{50} = 15 μM. *Drawback*: also inhibits PKC.

(vii) D609 (Tocris, 10–50 μM (34, 66, 67))—selective competitive phosphatidylcholine-specific PLC inhibitor (K_i = 6.4 μM) and also an antiviral and antitumor agent. It suppresses lipopolysaccharide- and interferon γ-induced NO production (IC_{50} = 20 mg/ml) and blocks oxidative glutamate toxicity in nerve cells (68).

(C) Phospholipase A_2 (PLA_2).

Most likely if a PTX-sensitive G protein is involved and arachidonic acid is produced. Cytosolic PLA_2 requires Ca^{2+} for its activity and preferentially hydrolyzes the sn-2 acyl bond of phospholipids, releasing arachidonic acid and lysophospholipids (69). Arachidonic acid may have direct effects (70) or may be modified by cyclooxygenases into the eicosanoids, which include prostaglandins and leukotrienes (see Fig. 2). βγ subunits may also directly activate PLA_2 (70). PLA_2 can also activate platelet activating factor (PAF), whose neuronal functions are still ill-defined (69).

Positive tests

(a) Activators of PLA_2 mimic (and occlude) response.

(i) Melittin (Sigma-Aldrich, 1 μM (25, 26))—a polypeptide from bee venom, which reversibly activates PLA_2 (26). *Drawbacks*: may activate other types of PLA_2 and has a variety of other effects (see entries for melittin above and below).

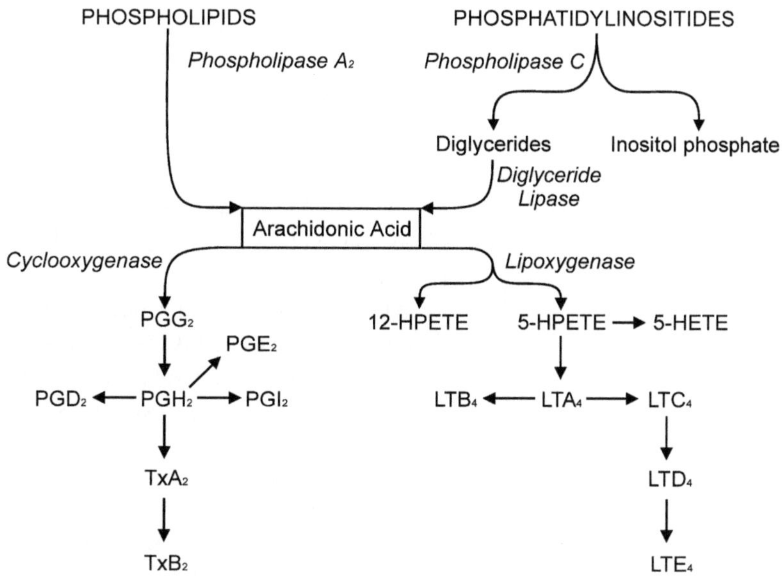

Fig. 2. Synthesis of eicosanoids from arachidonic acid: prostaglandins (PG), thromboxanes (Tx), leukotrienes (LT), hydroxyeicosatetraenoic acid (HETE).

Negative tests

(a) Inhibitors of PLA_2 reduce the response.

(i) Arachidonyl trifluoromethyl ketone (Tocris, $AACOCF_3$, 5–20 μM (71))—potent and selective cytosolic PLA_2 inhibitor which is readily membrane permeant, with IC_{50} of 1.5 μM for inhibition of cytosolic PLA_2 (71). *Drawback*: also inhibits cyclooxygenase activity (71).

(ii) Methyl arachidonyl fluorophosphonate (Biomol, 1–250 μM (72, 73), MAFP)—an irreversible and potent inhibitor of cytosolic PLA_2 (IC_{50} 0.5 μM) but also inhibits other members of the PLA_2 superfamily. *Drawbacks*: also inhibits metabolism of anandamide and acts as irreversible antagonist of the cannabinoid 1 receptor (74).

(iii) 4-Bromophenacyl bromide (Sigma-Aldrich 10–50 μM (26, 75))– it is slowly membrane permeant; preincubation of brain slices for >60 min at ~20°C is needed to reach steady state inhibition (M. Bellingham, unpublished observations). *Drawback*: also inhibits PLC.

(iv) Chloroquine, quinacrine and quinine (widely available, >50 μM (75))—IC_{50} for inhibition of cytosolic PLA_2 are 125, 200 and 250 μM (71). All are slowly membrane permeant; preincubation of brain slices for >60 min at ~20°C is needed to reach steady state inhibition. *Drawback*: also inhibit PLC.

(v) Manoalide (Biomol, 10 μM (11))—potent, irreversible inhibitor of PLA_2 (IC_{50} = 20–200 nM). Pretreat tissue for 10 min. at 37°C. *Drawback*: also inhibits PLC (IC_{50} = 1.5 μM) and Ca^{2+} channels.

(D) Guanylyl cyclase (GC) and cGMP-specific phosphodiesterases (cGMP-PDE).

Most likely if $G\alpha_t$ is involved and cGMP levels are altered. The soluble form of GC is ubiquitously expressed in many cell types and may be activated by a number of retrograde transmembrane messenger molecules, including nitric oxide, carbon monoxide (a weaker activator than NO) and arachidonic acid; cGMP is produced by GC activity and is broken down by cGMP-PDEs (several of which also break down cAMP) (76). cGMP can directly activate some ion channels, activate protein kinase G (PKG) or PKA or regulate PDE activity (77), changing cGMP or cAMP. Thus, this signalling pathway has considerable potential for crosstalk with the AC-cAMP-PKA signalling pathway (77).

Positive tests

(a) Activators of GC mimic (and occlude) response.

(i) NO/CO donors and generators—due to their short half-life and rapid oxidation, NO and CO are very difficult to deliver directly as pure gases. Many donor compounds generate NO or CO directly (78); activators of nitric oxide synthase generate NO, while activators of haem oxygenase generate CO. The pharmacological manipulation of these gaseous messengers is beyond the scope of this chapter; the reader is directed to recent reviews of the literature (79–81). *Drawbacks*: NO/CO may also directly regulate proteins through thiol groups or nitrosylation; they are freely diffusible and thus can act at sites different to those intended, producing indirect effects (78). NO/CO acts through high affinity binding to a haem moiety in GC, but full activation of GC may require higher NO/CO levels with available GTP and can be inhibited by physiological levels of ATP (82). NO can directly activate COX, due to the haem-containing structure in the latter molecule (81).

(ii) A350619 hydrochloride (widely available)—NO-independent activator of soluble GC which requires reduction of the haem moiety; structurally unrelated to YC-1 with a similar EC_{50} of ~30 μM (82).

(iii) YC-1 (Cayman Chemicals, 10–200 μM (83), 1-benzyl-3-(5′-hydroxymethyl-2′-furyl) indazole)—activates GC independently of NO (EC_{50} ~ 20 μM (82)) but requires reduction of the haem moiety. *Drawbacks*: potentiates GC responses to NO and CO and also inhibits cGMP-PDEs (82).

(iv) BAY 41–2272 and BAY 41–8543 (Calbiochem and Cayman Chemicals, 1–10 μM (83))—structurally related and more potent (~2 orders of magnitude; EC_{50} ~ 0.3 and 1 μM, respectively) activators of GC than YC-1 (82). *Drawbacks*: may directly activate Ca^{2+}-dependent K^{+} ion channels and can inhibit PDE5 at higher concentrations than required for GC activation(82).

(v) BAY 58–2667 (Bayer, AdipoGen, Cinaciguat)—a NO- and haem moiety-independent potent and selective activator of GC with an EC_{50} of ~1 nM without inhibition of PDE at <10 μM (82).

Negative tests

(a) Inhibition of guanylate cyclase decreases response.

(i) LY-83,583 (widely available, 1 μM (84), 6-(phenylamino)-5,8-quinolenedione)—inhibits soluble GC at 10 μM (50% inhibition at 2 μM); it is not readily membrane-permeant and application (10 μM (85)) in the whole-cell pipette solution is useful, otherwise preincubation of tissue with higher concentrations (100 μM) or direct application (20 μM (86)) is required. *Drawbacks*: also inhibits NO synthase and can inhibit cyclic nucleotide-gated channels.

(ii) ODQ (widely available, 1–10 μM (83, 84, 87–89), 1H-[1, 2, 4]oxadiazolo[4,3-*a*]quinoxalin-1-one)—selective inhibitor of NO-sensitive guanylyl cyclase which acts by oxidising GC, producing a NO/CO-insensitive form of the enzyme (82).

(iii) NS2028 (widely available, 1 μM (90, 91), 4H-8-bromo-1,2,4-oxadiazolo(3,4-d)benz(b)(1,4)oxazin-1-one)—structurally similar to ODQ, but ~fivefold more potent, inhibiting soluble GC in brain slices with IC_{50} of 30 nM (91).

(iv) Methylene blue (widely available, 100 μM in pipette solution (92))—a dye which acts as a NO scavenger, thus indirectly inhibiting GC (93); it is not readily membrane permeant. *Drawback*: antagonist of purinergic ATP receptors (37).

(E) ADP-ribosyl cyclase.
Most likely if increased intracellular Ca^{2+} and Ca^{2+}-calmodulin are generated and cGMP levels increase. Cyclic ADP-ribose (cADPR) has been proposed to be a ubiquitous Ca^{2+} signalling molecule in neurons, with direct and indirect effects on ryanodine-sensitive Ca^{2+} release channels (94).

Negative tests

(a) Inhibition of ADP-ribosyl cyclases decreases response.

(i) Benzamide and 3-aminobenzamide (widely available, 100 μM–1 mM (95, 96))—selectively inhibits polyADP-ribosyl cyclases.

(ii) 1,5-Isoquinolediol (widely available, 10–100 μM (96), *aka* 1,5-dihydroxyisoquinoline)—selectively inhibits polyADP-ribosyl cyclases.

(iii) Phylloquinone (widely available, 100 μM (95))—selectively inhibits monoADP ribosyl cyclases.

(iv) Nicotinamide (widely available, 10 mM (95))—inhibits mono- and polyADP-ribosyl cyclases.

2.3. What Is the Second Messenger?

(A) cAMP.
Most likely if the G proteins $G\alpha_s$ or $G\alpha_i$ and AC are involved. cAMP has three intracellular targets: PKA, cyclic nucleotide-gated channels (97) and the exchange protein activated by cAMP (Epac), of which the best studied is PKA-mediated signalling (98). It must also be noted that many ion channels contain amino acid consensus sequences susceptible to cAMP/PKA-mediated phosphorylation (7). Thus, showing that responses can be replicated by cAMP or AC activation or enhanced by PDE inhibitors does not establish this as the route mediating the response. Inhibition of response by prevention of AC activation, or direct demonstration of PKA-mediated phosphorylation, is a necessary test.

Positive tests

(a) cAMP analogues mimic the response.

(i) cAMP (widely available, 50–100 μM in pipette solution (51), 1 mM in external solution (99))—natural product of AC activity; best applied with inhibitors of cAMP-PDE to maximise cAMP effects; as it is non-membrane permeant, application in the pipette solution is required.

(ii) 8-Bromo-cAMP (widely available, 100 μM–1 mM (66, 75, 93, 97))—membrane permeant, with greater resistance to cAMP-PDE activity than cAMP.

(iii) Dibutyryl-cAMP (widely available, 1 mM (75))—membrane-permeant cAMP analogue.

(iv) 8-(4-chlorophenylthio)-cAMP (widely available, 1–3 μM (52, 97))—more lipophilic and membrane permeant than dibutyryl-cAMP, slower metabolism by PDE compared to cAMP but not resistant to PDE. *Drawbacks*: also a potent inhibitor of cGMP-PDE (IC_{50} = 900 nM), thus elevating cGMP levels, and can inhibit cAMP-PDE (IC_{50} = 25 μM).

(v) Sp-cAMPS (widely available, 1 mM (97))—membrane-permeant activator of PKA and resistant to PDE.

(vi) N^6,$O^{2'}$-dioctanoyl-cAMP (Santa Cruz, 50 μM)—cell permeable, with ~100-fold greater activity than cAMP. Highly resistant to cAMP-PDE and can increase IP_3 in some preparations (100).

vii) Rp-cAMPS (widely available, 100 μM–1 mM (51, 97)) and Rp-8-Cl-cAMPS (70 μM)—mimics action of cAMP on the catalytic subunit of PKA; both are somewhat membrane permeant, and Rp-8-cAMPS is more lipophilic (50).

(b) Phosphodiesterase (PDE) inhibitors prolong or enhance response (by prevention of cAMP breakdown); cAMP-specific PDEs are PDEs 4, 7 and 8, while PDEs 1–3 and 10–11 can act on both cGMP and cAMP (101).

(i) IBMX (widely available, 100 μM (36, 99, 102), 3-isobutyl-1-methylxanthine)—inhibits both cAMP- and cGMP-PDEs (IC_{50} = 2–50 μM (103, 104)). *Drawback*: also an adenosine receptor antagonist at similar concentrations (104).

(ii) Theophylline (widely available, 1,3-dimethylxanthine)—IC_{50} for PDEs1–5 ranges between 150 and 600 μM (104). *Drawback*: also an adenosine receptor antagonist at 10 μM (104).

(iii) Rolipram (widely available, 10–100 μM (101, 105), 4-(3-cyclopentyloxy-4-methoxy-phenyl)-2-pyrrolidone)—the prototypical selective cAMP-PDE4 inhibitor (IC_{50} = 1–5 μM (101, 104)) with over 100-fold selectivity for PDE4 over other PDE; examples of Rolipram-insensitive cAMP-PDE effects exist.

(iv) RO-20-1724 (widely available, 30 μM (105) (4-3-butoxy-4-methoxybenxyl-2-imidazolidine))—a selective inhibitor of cAMP-PDE4 (IC_{50} = 2 μM (101)).

(c) Non-hydrolysable cAMP analogues prolong response.

(i) ATPγS (widely available, 1 mM in pipette solution (51))—a non-hydrolysable activator of PKA which needs to be applied intracellularly. *Drawback*: also a potent agonist of ATP purinergic receptors.

(B) Inositol triphosphate (IP_3).
Most likely involved if PLC activity is required and intracellular Ca^{2+} increases. IP_3 generated by PLC binds to the IP_3 receptor (IP_3R) located on smooth endoplasmic reticulum and releases intracellular Ca^{2+} stores through the IP_3R channel; note that the IP_3R has phosphorylation sites for cAMP-dependent phosphorylation, leading to potential crosstalk with the AC-CAMP signalling pathway (106).

Positive tests

(a) Intracellular increases in IP_3 or analogues mimic response.

(i) IP_3 (K^+ salt, Sigma, 50 μM–1 mM in pipette solution (19, 35)).

(ii) Heparin (widely available, 100 μM–1 mM in pipette solution (19))—not membrane permeable and is a low-affinity, competitive inhibitor at the IP_3R binding site. Negative control is *N*-desulphonated heparin (107). *Drawbacks*: in intact cells, it may also act to inhibit IP_3 generation (106) and may stimulate ryanodine receptor-sensitive Ca^{2+} release (108). Heparin can bind to and inhibit protein phosphatase PP1 and appears to stimulate PP2A (109).

(iii) Adenophostins A and B (Calbiochem, 1 μM in pipette solution (110))—potent activators of the IP_3R which are not degraded by enzymes which break down IP_3 (30, 107).

(b) Blockade of IP_3 phosphatase prolongs response by decreasing metabolism of IP_3.

(i) 2,3-Diphosphoglyceric acid (widely available, 1 mM (111))—inhibits both IP_3 phosphatase and IP_3 kinase (112).

(C) Diacylglycerol (DAG).
Most likely if PLC activity and PKC activation are required. The production of DAG following PLC stimulation represents a branch point (see Fig. 1), after which DAG can either directly activate PKC, or can be further metabolised by DAG lipase to AA (56).

Positive tests

(a) DAG or analogues mimic the response.

(i) OAG (widely available, 5–100 μM (67, 113), 1-oleoyl-2-acetylglycerol)—OAG may have both PKC-dependent and PKC-independent actions, with the former occurring at concentrations above 10 μM and the latter occurring at lower concentrations (67).

(ii) SAG (widely available, 50 μM (113), 1-stearoyl-2-arachidonyl-*sn*-glycerol)—DAG derivative, endogenous activator of PKC produced by PIP hydrolysis.

(iii) 1-Stearoyl-2-linoleoyl-*sn*-glycerol—similar to (ii).

(b) Inhibition of DAG lipase increases response (seemingly paradoxical, but occurs because loss of DAG as a substrate for AA generation by DAG lipase acts as a "sink" decreasing DAG effects on PKC).

(i) RHC80267 (widely available, 10–70 μM (67, 114))—IC_{50} ~4 μM for inhibition of DAG lipase (67); preincubation of brain slices may be required (50–70 μM for 30 min (114)). It also inhibits a number of other serine hydrolases, most notably acetylcholinesterase and enzymes involved in endocannabinoid metabolism (115).

(ii) Tetrahydrolipstatin (Cayman Chemicals and Sigma-Aldrich, 0.5–2 μM in pipette solution or 10 μM for preincubated brain slices (114), Orlistat). Inhibits a number of serine hydrolases (115), most potently DAG lipase α (IC_{50} 1 μM in intact cells (116)). There is little overlap in the serine hydrolases inhibited by tetrahydrolipstatin and RHC80267 so that similar inhibition by both compounds provides good evidence of responses modulated by DAG (115).

Negative tests

(a) Inhibition of PKC activation decreases response (see below for PKC inhibitors).

(D) Arachidonic acid (AA).

Most likely if PLC is required but PKC inhibition does not block the downstream response or if PLA_2 activity and AA generation is required. Arachidonic acid normally exists in esterified form in membrane phospholipids; free arachidonic acid is a transient, critical substrate for the biosynthesis of many eicosanoid second messengers (see below) (56). In addition, the metabolism of endocannabinoids has many common points with AA metabolism (117), so it is wise to rule out endocannabinoid contributions to responses.

Positive tests

(a) AA or analogues mimic the response.

(i) Arachidonic acid (Cayman Chemicals, Helena Labs, 50–200 μM (67, 75, 118, 119))—*Drawback*: AA can be readily metabolised via the pathways shown in

Fig. 2; thus, a negative result could be due to rapid AA metabolism by cyclooxygenase and lipoxygenase activities, whose inhibition (see below for inhibitors) may thus reveal direct effects of AA analogues. AA is also an effective activator of PKC (120).

(ii) 5,8,11,14-eicosatetraynoic acid (widely available, 10–50 μM (121) ETYA)—an AA analogue which is not metabolised via lipoxygenase or cyclooxygenase pathway and which also inhibits these pathways, possibly by acting as a suicide substrate (122). This compound is useful to distinguish whether AA can act directly on channels without further metabolism (121); if the response to ETYA is equivalent to AA, then AA does not cause responses via further breakdown to metabolites. *Drawbacks*: also inhibits lipoxygenase activity in this concentration range or lower, has inhibitory effects on PLA_2 and cytochrome P-450 and is a potent inhibitor of Ca^{2+} entry into cells.

Negative tests

(a) Inhibition of DAG lipase decreases response (see above entries for compounds inhibiting DAG lipase).

(E) cGMP.

Most likely if GC is activated or NO is produced and PKG is involved.

Positive tests

(a) Intracellular increases in cGMP or analogues mimic response.

(i) 8-Bromo-cGMP (widely available, 100 μM to 2 mM (90, 95, 97, 123–125))—membrane permeant (~2.5-fold cf. GMP) cGMP analogue with slightly higher potency (EC_{50} 10 nM to 1 μM) (126, 127).

(ii) Dibutyryl-cGMP (widely available, 100–250 μM (95))—membrane-permeant cGMP analogue.

(iii) 8-pCPT-cGMP and 8-Br-PET-cGMP (BioLog)—more potent (EC_{50} 4–900 nM) and lipophilic activators of PKG than (i) and (ii) (126). *Drawback*: both compounds activate PKA with EC_{50} range of 0.5–50 μM (126).

(b) cGMP-PDE inhibitors prolong or enhance response (by prevention of cGMP breakdown); cGMP-specific PDEs are PDEs 5, 6 and 9, while PDEs 1–3 and 10–11 can act on both cGMP and cAMP (101).

(i) Dipyridamole (widely available, 20–30 μM (102, 125))—selective inhibitor of several cGMP-PDEs

(IC_{50} = 900 nM). *Drawback*: also inhibits nucleoside (e.g. adenosine) transport into or out of cell.

(ii) MBAM (widely available, 0.5 μM (128), 4-([3′,4′-(methylenedioxy)benzyl)amino)-6-methoxyquinazoline)—potent inhibitor of cGMP-PDE (IC_{50} = 230 nM).

(iii) Milrinone (widely available, 10 μM (129))—selective inhibitor of PDE3 (IC_{50} = 150 nM (101)).

(iv) MY-5445 (widely available, 5 μM (87) [1-(3-chlorophenylamino)-4-phenylphthalazine])—selective inhibitor of cGMP-PDE5 (IC_{50} = 600 nM).

(v) Quazinone (widely available, 30 μM (105), Ro-13-6438)—selective inhibitor of cGMP-PDE (IC_{50} = 600 nM).

(vi) Trequinsin HCl (widely available, 10 nM (105))—selective and extremely potent inhibitor of PDE3 (IC_{50} = 0.3 nM (101)).

(vii) Zinprinast(widelyavailable,2–10μM(93,102,128))—selective inhibitor of cGMP-PDE5 (IC_{50} = 130 nM (101)).

(viii) Sildenafil, vardenafil and tadalafil (widely available, 100 nM–1 μM in pipette solution (130))—selective inhibitors of the cGMP-PDE5 isoform, with IC_{50} of 10, 1 and 10 nM, respectively (77, 101), although all have inhibitory effects on PDE6 as well (130).

(ix) BAY 73–6691 (Bayer, 1-(2-chlorophenyl)-6-[(2R)-3-,3,3-trifluoro-2-methylpropyl)-1,5-dihydro-4H-pyrazolo[3,4-d)pyrimidine-4-one)—selective inhibitor of cGMP-PDE9, with IC_{50} of 100 nM (101).

(F) Ca^{2+}-calmodulin (CAM).

Most likely if Ca^{2+} increases and PLC/PKC activation occurs. CAM can activate a number of CAM-dependent kinases (phosphorylase kinase, myosin light chain kinase and CAM kinase I to IV) (131) but also can modulate both cAMP and cGMP responses by potent activation of phosphodiesterases (132).

Negative tests

(a) Inhibitors of CAM decrease or block response. *Drawbacks*: calmodulin directly inhibits the IP_3R; many inhibitors of calmodulin thus can elicit IP_3R-mediated Ca^{2+} release (24, 107). Recently, many calmodulin antagonists have been also been shown to directly block transient receptor potential channels by binding to the extracellular pore region (133).

(i) W-7 and W-13 (widely available, 100 nM (134)–100 μM)—binds to Ca^{2+}-calmodulin and inhibits

CAM activation of phosphodiesterase (IC_{50} = 28 μM) and of myosin light chain kinase (IC_{50} = 51 μM). W13, W12 and W5 have similar but less potent effects, inhibiting CAM activation of phosphodiesterase (IC_{50} = 70–260 μM) and of myosin light chain kinase (IC_{50} = 60–300 μM). *Drawbacks*: also inhibits PKC with IC_{50} of 50–150 μM (135), may directly inhibit voltage-gated Ca^{2+} (IC_{50} 29 μM) and K^{+} (IC_{50} range 8–34 μM) currents (136) and directly inhibits cAMP-gated cationic current in olfactory neurons (IC_{50} 5 μM (137)).

(ii) Tricyclic antipsychotics (fluoperazine, chlorpromazine, fluphenazine) (widely available, trifluoperazine 30 μM (138))—several tricyclic antipsychotics, including trifluoperazine and chlorpromazine, inhibit CAM. Fluphenazine *N*-mustard (SKF-7172-A2) causes irreversible antagonism of CAM activation of phosphodiesterase (10 μM); its action is partially Ca-dependent. *Drawbacks*: trifluoperazine also inhibits PKC (135), is a dopaminergic D_2 antagonist (138) and directly inhibits P_2X_7 receptors (139). IC_{50} for direct inhibition of cAMP-gated cationic current in olfactory neurons is 13 μM (137). Chlorpromazine and fluphenazine inhibit the SERCA pump with IC_{50} 15–25 μM (24).

(iii) Calmidazolium chloride (widely available, 20 μM (140), R24571)—a potent inhibitor of calmodulin activation of phosphodiesterase (IC_{50} = 10 nM) and strongly inhibits calmodulin-dependent calcium ATPase at 50–200 μM. *Drawbacks*: also inhibits P_2X ATP receptors (139), may directly inhibit voltage-gated Ca^{2+} channels (140), irreversibly inhibits cAMP-gated cationic current in olfactory neurons (10 μM, (137)) and inhibits SERCA pump with IC_{50} 0.5 μM (24).

(iv) Camstatin peptide (Tocris, 10 μM in pipette solution (141))—an analogue of Pep-19 and an endogenous IF-motif peptide highly expressed in neurons which accelerates Ca^{2+} association and dissociation from CAM (141).

2.4. What Is the Secondary Effector Enzyme?

(A) Protein Kinase A (PKA).
Most likely if AC activation and cAMP generation occurs. PKA exists as a tetramer of two regulatory and two catalytic subunits in its inactive state (98). Each regulatory subunit has two cAMP binding sites; cAMP binding then releases the catalytic subunits which in turn bind to ATP and can then

phosphorylate serine/threonine residues with an appropriate consensus sequence (98).

Positive tests

(a) Activators of PKA mimic (and occlude) response.

(i) cAMP (widely available). Non-membrane permeant, susceptible to PDE breakdown and useful for direct application to membranes; EC_{50} is 2–6 nM (126). *Drawback*: also activates PKG at >10 μM (126).

(ii) 8-Br-cAMP (widely available, 50 μM in pipette solution (142), 100–500 μM in bath (143–145)). Slightly membrane permeant (twofold cf. AMP) with similar EC_{50} at PKA (126). *Drawback*: also activates PKG at >5 μM (126).

(iii) Sp-cAMP (widely available, 100 μM (146))—membrane-permeant cAMP analogue resistant to PDE breakdown.

Negative tests

(a) Inhibitors of PKA decrease or block response.

(i) Protein kinase A inhibitory peptides (Sigma-Aldrich, 20–100 μM in pipette solution (12, 51, 52))—an endogenous 20-amino-acid-peptide sequence which binds to the free catalytic subunit of PKA, preventing target phosphorylation; endogenous PKI is thought to be highly specific for PKA (98). Synthetic forms such as PKI-(6–22)-amide, PKI-(Myr-14-22)-amide and Wiptide (Bachem, PKI-(14–22)-amide) have also been developed, of which the last is most potent and specific for PKA cf. PKG (147).

(ii) Rp-cAMPS and Rp-8-Br-cAMPS (widely available, 100 μM (143))—cAMP analogues which act as competitive inhibitors of the cAMP binding site of the regulatory subunit. *Drawback*: high endogenous cAMP levels may diminish effects of Rp-cAMPS and related analogues(98).

(iii) H-7, H-8, H-88 and H-89 (widely available, H-89 at 1–2 μM (143, 144))—all are isoquinoline derivatives which competitively bind to the ATP site of the PKA catalytic subunit (98). H-8 is relatively non-specific, as it is roughly equipotent in inhibiting PKG and PKA activity. H-7 is also non-specific, as it also inhibits PKG and PKC. HA-1004 also inhibits PKG but has low specificity for PKC and is thus a good negative control for H-7. H-89 is widely used as a selective PKA inhibitor, as it is thought to be more selective for PKA cf. H-8, with a binding constant of

125 nM and almost complete suppression of PKA activity at 10 μM (148). *Drawbacks*: H-89 has a number of effects independent of PKA inhibition; it can act as a potent β-adrenergic receptor antagonist, with a receptor binding affinity <0.5 μM (149) and can decrease voltage-gated K^+ currents (IC_{50} ~2 μM) (98). H-89 also inhibits a wide range of other protein kinases, often at concentrations below those which inhibit PKA, and some of these kinases can indirectly regulate PKA (148). H-89 inhibition of responses should thus be verified by use of other PKA inhibitors.

(iv) KT5720 (Tocris, 100 nM–1 μM (35, 134, 143, 145))—blocks the activity of by competitive inhibition of the ATP site of the PKA catalytic subunit and relatively selective for PKA (IC_{50} of 60 nM–3 μM (98)) over PKC or cGMP-dependent kinase (>2 μM) (50). *Drawbacks*: KT5720 also inhibits a number of other protein kinases by factors of up to >100 more potent than its PKA inhibition and is a potent allosteric modulator of the muscarinic M1 receptors at concentrations as low as 10 nM (98, 148). KT5720 inhibition of responses should thus be verified by use of other PKA inhibitors.

(v) PKA catalytic subunit (Sigma and Promega, 50–250 U/ml in pipette solution (51, 97, 146))—directly phosphorylates proteins.

(B) Protein kinase C (PKC).

Most likely if PLC activation and DAG generation occurs. Multiple isoforms of PKC exist in mammals; all are activated by DAG, most are also Ca^{2+} sensitive and specific isoforms may also be separately activated by PLA_2 or phospholipase D (135). Inhibitors and activators can interact with PKC at several sites, including sites that bind phorbol esters, phospholipids and ATP (the catalytic subunit) (135). Cytosolic PKC is normally inactive, due to interactions between its pseudosubstrate region and the substrate-binding catalytic region; PKC translocation to the membrane in response to DAG causes release of the pseudosubstrate and binding of Ca^{2+} (120).

Positive tests

(a) Activators of PKC mimic response.

(i) Phorbol esters (widely available, 5–10 μM phorbol-12,13-diacetate (134, 150–152), 100 nM–10 μM phorbol dibutyrate (11, 151, 152); see (153) for comprehensive review of phorbol ester effects on neuronal responses)—are tetracyclic diterpenoids which mimic DAG. Phorbol esters are readily membrane permeant; phorbol ester binding causes PKC to translocate to plasma membranes in a way similar to DAG activation (120). 4α-phorbol-12,13-didecanoate is a weak PKC activator that can be used as negative control (10 μM) for effects other than PKC activation (11, 150, 151). *Drawbacks*: can directly activate voltage-gated Ca^{2+} channels and increase transmitter release (13, 154, 155) and are also potent tumour promoters and carcinogens (120).

(ii) Naphthalenesulphonamides (Sigma, SC-10 at 100–200 μM (152), also SC-9)—activate PKC in a Ca^{2+}-dependent manner; structurally similar to W-7 but without effects on CAM (120).

(iii) DAG derivatives and unsaturated fatty acids (AA and other long chain polyunsaturated fatty acids)—see DAG and AA above.

(iv) (−)-Indolactam V (widely available, 3 μM (152))—is an alkaloid lactam, which binds to the DAG binding site. An inactive analogue, (+)-indolactam V, serves as a good control. *Drawback*: causes tumour promotion and initiation (120).

(v) Bryostatin and analogues (widely available, 100 nM (156))—a membrane-permeant macrolide lactone, which competes for the DAG binding site with very high affinity (~1 nM) causing a brief activation of PKC followed by prolonged downregulation (120). *Drawback*: higher concentrations of bryostatin (>10 nM) can cause less net activation than lower concentrations (120).

Negative tests

(a) Inhibitors of PKC decrease or block response.

(i) H7 (widely available, 100–400 μM (157, 158))—inhibits PKC with IC_{50} of 60 μM but relatively unselective as it also inhibits PKA (IC_{50} of ~170 μM (159)).

(ii) Polymyxin B (widely available, 20 μM in pipette solution (160))—an aminoglycoside antibiotic which acts at the phospholipid binding site, with an IC_{50} of 1–30 μM (135).

(iii) Staurosporin (widely available, 0.1–2 μM (12, 53, 161))—is an indolocarbazole natural product of *Streptomyces* sp.; it is a potent inhibitor (IC_{50} 3 nM) of PKC, but is relatively unselective, also inhibiting PKA (IC_{50} 8 nM) (162). Chemical derivatives such as midostaurin (PKC412), ruboxistaurin (LY333531), enzataurin (LY317615) and sotrastaurin (AEB071,

IC_{50} 2 nM) have similar nM potency and some selectivity for certain PKC isoforms.

(iv) K-252a (widely available, 50 μM (158))—cell-permeable and potent PKC inhibitor (IC_{50} 25 nM) (163) but relatively unselective, also inhibiting PKA, PKC, PKG and CAM KII at similar potency (162).

(v) Tamoxifen citrate (widely available, 10 μM (164))—acts at ATP binding site and is relatively selective for PKC (IC_{50} 10–100 μM) (135).

(vi) Calphostin C (widely available, 0.5–1 μM (161, 164))—potent inhibitor, acting at the phorbol ester binding site and selective for PKC over PKA, but needs fluorescent light for PKC inhibition (135). No effect on PKA at a concentration of 50 μM.

(vii) Bisindolylmaleimides (widely available, 1–5 μM (35, 44, 113, 161), Bis-1 also called GF 109203X or Gö 6850)—a family of membrane-permeant selective inhibitors of PKC (IC_{50} = 10 nM (135) cf. PKA (2 μM) (135)) acting at the ATP binding site (165). Bis-1 and Bis-3 strongly inhibit PKC at 1 μM (148). *Drawbacks*: also strongly inhibit MAP kinases and directly inhibit voltage-gated Na^+ ion channels(148).

(viii) Chelerythrine Cl (widely available, 3–10 μM (13, 67, 102))—inhibitor at catalytic domain on PKC with an IC_{50} of 660 nM (135). *Drawback*: reported to have no inhibitory effect on PKC up to 10 μM (148).

(ix) NPC15437 (Tocris, 30 μM (166))—weaker inhibition of PKC (IC50 ~20 μM) but no effect on PKA or Ca^{2+}-CAM kinase II up to 300 μM (159).

(x) Ro-32-0432 (Calbiochem, 3 μM (13))—a membrane-permeant PKC inhibitor.

(C) Cyclooxygenase (COX).

Most likely if PLC activation and DAG generation occurs but PKC activation is not required. To date, 3 COX isoforms (COX1–3) have been discovered, but COX3, which is selectively inhibited by acetaminophen, is only expressed in brain tissue in a few species, which do not include human, mouse or rat (167, 168).

Negative tests

(a) Inhibitors of COX decrease or block response.

(i) Indomethacin (widely available, 5–10 μM (75, 118, 169))—is a more potent inhibitor of COX1 (IC_{50} is 0.5–1 μM (118) to 0.006 μM (170)) but also inhibits COX2 (IC_{50} is 0.01 μM (170)). *Drawback*: can also inhibit lipoxygenase (IC_{50} > 10 μM (118)) and PLA_2 activity at higher concentrations (IC_{50} = 145 μM).

(ii) Aspirin (widely available, 100 μM (119))—a relatively selective COX1 inhibitor (IC_{50} ~2 μM cf. 280 μM for COX2 (171)), which acetylates the cyclooxygenase enzyme, irreversibly inhibiting it (122).

(iii) Eicosatetraynoic acid (ETYA; see above under AA).

(iv) NS398 (widely available, 5–30 μM (73, 119) *N*-(2-cyclohexyloxy-4-nitrophenyl) methanesulphonamide)—selective inhibitor of COX2 (IC_{50} = 100 nM cf. 17 μM for COX1 (171)).

(v) Celecoxib (LC Laboratories, 10 μM (172) Celebrex, SC58236) ~375-fold more potent at COX2 *vs* COX1 inhibition (IC_{50} 40 nM cf. 15 μM (171)). *Drawback*: may inhibit voltage-gated K^+ and Na^+ ion channels in a COX-independent manner (IC_{50} range 1–10 μM (172).

(vi) FR122047 (widely available, 10 μM)—a selective inhibitor of COX1 (IC_{50} 0.028 cf. 65 μM for COX2 (173)).

(vii) Valeroylsalicylate (Cayman Chemicals, 100–300 μM (119, 174))—a selective irreversible inhibitor of COX1, which requires preincubation of >40 min at 37°C (175).

(D) Lipoxygenases (LOX).

Most likely if PLC activation and DAG generation occurs but PKC activation is not required. Five LOXs (5*S*-LOX, 12*R*-LOX, 12*S*-LOX and two distinct 15*R*-LOXs) oxygenate AA at different positions along the carbon chain, generating the corresponding 5*S*, 12*S*-, 12*R*- or 15*R*-hydroperoxides, respectively (168, 176).

Negative tests

(a) Inhibitors of LOXs decrease or block response.

(i) Nordihydroguaiaretic acid (widely available, 10 μM (11, 75, 169) NDGA)—a nonselective LOX inhibitor (IC_{50} 0.2–3 μM (118, 177)). *Drawback*: may directly inhibit transient receptor potential channels(169).

(ii) Esculetin (R&D Chemicals, Santa Cruz, 10 μM (178), 6,7-dihydroxycoumarin)—a relatively nonselective LOX inhibitor, inhibits 5-LOX (IC_{50} = 1–4 μM(177) (178)). *Drawbacks*: also inhibits COX (IC_{50} = 52 μM) and decreases free radicals (178).

(iii) Baicalein (widely available, 2–4 μM (75, 169), 5,6,7-trihydroxyflavone)—a selective inhibitor of 12-LOX (IC_{50} = 120 nM) but also inhibits 15-LOX. *Drawbacks*: also inhibits cellular Ca^{2+} uptake and mobilisation and may have inhibitory effects on PKC and protein tyrosine kinase.

(iv) CDC (Enzo Life Sciences, cinnamyl-3,4-dihydroxy-alpha-cyanocinnamate)—often used as a selective 12/15LOX inhibitor (IC_{50} range 0.5–2 μM (177, 179) but also a potent 5*S*-LOX inhibitor (IC_{50} range 10–40 nM (179)).

(v) Zileuton (Sequoia Research Chemicals, 50 μM (169) [*N*-(1-benzo[b)thien-2-ylethyl)-*N*-hydroxyurea))—a selective 5*S*-LOX inhibitor (IC_{50} ~2 μM (179)).

(vi) PD146176 (widely available, 2–10 μM (169, 180), 6,11-dihydro(1)benzothiopyrano[4,3-b]indole)—a selective inhibitor of 12/15LOX which requires pre-incubation of brain slices for 2 hours.

(vii) Eicosapentaenoic acid (widely available, 4–40 μM (181))—inhibits 5-LOX. *Drawback*: directly inhibits voltage-gated Na^{+} and Ca^{2+} channels at 4–40 μM (181).

(viii) ETYA—see above. IC_{50} for 15-LOX is 1.7 μM; 12-LOX is also inhibited.

(ix) L-655,238 (Santa Cruz, 25 μM (182), REV 5901, *para*-isomer)—potent, selective inhibitor of 5-LOX-activating protein (IC_{50} = 100 nM). *Drawback*: can inhibit Ca^{2+}-dependent K^{+} BK channels at 10 μM (183).

(x) MK-886 (Cayman Chemicals, 20–25 μM (169, 182))—potent, selective inhibitor of 5-LOX-activating protein, without effect on other arachidonic acid-metabolising enzymes. *Drawback*: may directly inhibit transient receptor potential channels (169).

(E) cGMP kinase (PKG).

Most likely if GC is activated and cGMP is produced. PKG exists in three isoforms (PKGIα, PKGIβ and PKGII); most activators and inhibitors display relatively low selectivity for individual isoforms (126). A significant drawback is that all cGMP analogues can modulate PKA at concentrations ranging from a few-fold to a few hundredfold higher than those used to modulate PKG; careful selection of the cGMP analogue concentration, and checking for crosstalk by using PKA modulators should be done (126).

Positive tests

(a) Activators of PKG mimic (and occlude) response.

(i) cGMP (widely available)—non-membrane permeant, susceptible to PDE breakdown, useful for direct application to membranes and also activates PKA (126).

(ii) 8-pCPT-cGMP (widely available, 50 μM (88, 89), 8-(4-chlorophenylthio)-cGMP 3′,5′-cyclic monophosphate)—potent membrane-permeant PKG activator (EC_{50} range 4–900 nM (126)) which is 20 times more lipophilic than iii) and more resistant to hydrolysis (88, 127). *Drawback*: inhibits several PDEs at >5 μM (126).

(iii) 8-Br-cGMP (widely available, 50–500 μM (88, 89, 145, 184))—membrane-permeant PKG activator (IC_{50} range 0.1–1 μM (126)), more susceptible to hydrolysis than (i). *Drawbacks*: inhibits several PDEs at >5 μM and can activate PKA at similar concentrations to PKG activation(126).

(iv) PET-cGMP(10 μM, b-phenyl-1,N^2-ethenoguanosine-3′,5′-cyclic monophosphate).

(v) 8-Br-PET-cGMP (BioLog, 50–100 μM (184))—potent (EC_{50} range 9–20 nM (126)), highly membrane permeant (>50-fold cf. 8-Br-cGMP (126)), resistant to PDE breakdown and fairly selective for PKG cf. PKA (EC_{50} for PKA activation is >420 nM (126)).

(vi) Sp-8-pCPT-PET-cGMPS (BioLog, 8-(4Chlorophenylthio)-β-phenyl-1,*N*2-ethenoguanosine-3′, 5′-cyclic monophosphorothioate, Sp-isomer)—membrane-permeant PKG activator which is resistant to PDE hydrolysis.

(vii) Sp-cGMPS (BioLog, 1 μM in pipette solution (92))—non-membrane-permeant, irreversible activator of PKG. *Drawback*: relatively nonselective for PKG over PKA.

Negative tests

(a) Inhibitors of PKG decrease or block response.

(i) H-7, H-8, H-9 (widely available, 1–2 μM (95))—*drawbacks*: also inhibit PKA and PKC (IC_{50} of 0.48, 1.5 and 15 μM for PKG, PKA and PKC (95)). HA-1004 inhibits both PKG and PKA but has low specificity for PKC and is a good negative control for H-7.

(ii) Rp-8-Br-cGMPS (BioLog Life Sciences Institute, 0.1–10 μM (88, 97, 128))—membrane-permeant inhibitor (127). Also Rp-cGMPS (10 μM) is less membrane permeant and less potent.

(iii) Rp-8-pCPT-cGMPS and Rp-8-Br-PET-cGMPS (BioLog, 10 μM (145))—more potent and membrane-permeant PKG inhibitors; the latter is ~tenfold more potent than the former (126). *Drawbacks*: can inhibit some PDEs at similar concentrations (126).

(iv) KT5823 (widely available, 100 nM–5 μM in pipette solution (85, 102))—Highly specific for PKG (K_i 230 nM) over other protein kinases, competes at ATP binding site and is reversible; it is not readily membrane permeant (85) but can preincubate cultures or brain slices for 30 min at 2 μM to achieve PKG inhibition (145), structurally different to (i)–(iii).

(v) Protein kinase G inhibitory peptides (Merck, Tocris, 100 U/ml in pipette solution (185), DT2 and DT3 from BioLog, Merck)—PKG inhibitory peptide (RKRARKE; H-Arg-Lys-Arg-Ala-Arg-Lys-Glu-OH) is non-membrane permeable and is relatively nonselective for PKG cf. PKA (K_i 86 μM for PKG cf. 550 μM for PKA) (186). DT2 and DT3 are the same octapeptide fused to the membrane translocation signals from the HIV-1 Tat protein (DT2) or the *Drosophila* Antennapedia homeodomain (DT3), with IC_{50} values of 12 and 25 nM, respectively, for PKG and IC_{50} values >100 μM for inhibition of PKA (127, 186).

(F) Ca^{2+}-calmodulin kinases (CAMK).

Most likely if Ca^{2+} increase and calmodulin are required. The family of CAMKs include CAMKI, CAMKII and CAMKIV; all are multimeric proteins containing 6 catalytic units which are normally inhibited by autoregulatory domains (131). Ca^{2+}/CAM binding maximally activates CAMKII, while the other CAMKs require additional activation by phosphorylation for maximal activity (131). CAMKII can undergo autophosphorylation which results in persistent enzymatic activity in the absence of Ca^{2+}/CAM (131).

Positive tests

(a) Activators of CAM kinase mimic response.

(i) CAM kinase II substrate—synthetic peptide.

Negative tests

(a) Inhibitors of Ca^{2+}-calmodulin kinase decrease or block response.

(i) W7—see CAM.

(ii) Trifluoperazine (TFP)—see CAM.

(iii) HA-1077—see above for PKA and PKG.

(iv) KN-62 (widely available, 1–5 μM (50, 134, 158), 1-[*N*,*O*,-bis-(5-isoquinolinesulfonyl)-*N*-methyl-L-tyrosyl]-4-phenylpiperazine)—potent, selective inhibitor of CAM kinase II; KN-62 binds directly to the calmodulin binding site of the CAM kinase II enzyme with an IC_{50} of 500 nM (148), without effects on PKC or PKA at up to 100 μM (187). KN-04 is a structurally similar molecule which has no effect on CAM kinase II activity at 100 μM and is a good negative control. *Drawback*: both KN-62 and KN-04 inhibit human P_2X_7 ATP receptors with an IC_{50} of 40 nM (139).

(v) Pimozide (widely available, 30 μM (138))—inhibits CAM kinase II with IC_{50} of 15 μM. *Drawback*: also inhibits the dopamine D_2 receptor.

(vi) CAM kinase II inhibitor peptides (Merck and others)—[Ala286]Ca^{2+}-CAM kinase II inhibitor is a synthetic peptide corresponding to residues 281–301 of the CAM kinase II α subunit, which inhibits the catalytic fragment with an IC_{50} of 2 μM (188); it can also inhibit PKC (189). Autocamtide-related inhibitory peptide is a non-phosphorylatable analogue of Autocamtide-2, which is a highly specific and potent CAM kinase II inhibitor (IC_{50} = 40 nM); it can be applied in the pipette solution (5–50 μM (188, 190)) and is also available in various membrane-permeant forms. Calmodulin kinase IINtide is a peptide sequence corresponding to the inhibitory domain of CAM kinase II, which is very potent (IC_{50} = 50 nM) and without effects on other CAM kinases, PKA or PKC (131); it is available in a myristoylated form to improve membrane permeability.

(vii) KN-93 (widely available, 5 μM (191), *N*-[2-[[[3-(4-Chlorophenyl)-2-propenyl]methylamino]methyl] phenyl]-*N*-(2-hydroxyethyl)-4-methoxybenzenesulphonamide—cell-permeable, potent, selective inhibitor of CAM kinase II; it binds directly to the calmodulin site of the kinase enzyme. IC_{50} is 0.37 μM. *Drawback*: directly inhibits IP_3R-mediated release of Ca^{2+} (107).

(G) IP_3 receptor.

Most likely if PLC activation, Ca^{2+} increase and IP_3 generation is required. The IP_3R comprises three homotetrameric isoforms, which differ in their sensitivity to Ca^{2+}; IP_3R1 has a bell-shaped dependence, with maximal activation at 300–400 nM, while IP_3R2 and IP_3R3 are stimulated by Ca^{2+} but not as readily inhibited by higher Ca^{2+} increases (30).

Regulation of IP_3Rs is principally through phosphorylation/dephosphorylation of the receptor molecule (30). PKA, PKC and PKG can phosphorylate IP_3Rs, enhancing activity and thus potentially providing a source of significant crosstalk between upstream signalling pathways and IP_3Rs (30). Dephosphorylation is catalysed by PP2B/calcineurin, which also downregulates IP_3R expression (30). The type 1 IP_3R has a binding site for calmodulin, which inhibits receptor activity (30).

Positive tests

(a) IP_3 or analogues mimic response.

(i) IP_3 (widely available, 20–100 μM (30, 192); see above)—increasing IP_3 in neurons, by direct injection, via permeabilised membranes or by photo release causes a rise in intracellular Ca^{2+} due to the depletion of IP_3R sensitive stores (30); maximal depletion is observed at 20–100 μM, which is higher than the affinity of purified IP_3R for IP_3 (EC_{50} 0.2–2 μM)(30).

(ii) Adenophostins A and B (see above).

(b) Increasing free intracellular Ca^{2+} mimics response.

(i) Thapsigargin (widely available, 1–10 μM (193))—a sesquiterpene lactone which is a cell-permeable, potent and selective irreversible inhibitor of the sarcoplasmic/endoplasmic reticulum Ca^{2+}-ATPase (SERCA) pump, resulting in release of Ca^{2+} from ER stores; IC_{50} is ~20 nM (24). *Drawbacks*: at concentrations above 1–10 μM, it can also inhibit other ATPase pumps, and it blocks voltage-gated Ca^{2+} channels at 0.2–2 μM (30).

(ii) Cyclopiazonic acid (widely available, 20–30 μM (193, 194))—a cell-permeable reversible inhibitor of SERCA pump (IC_{50} 90–2,500 nM) (24).

(iii) Calcium ionophores (widely available, ionomycin 5 μM with 2 mM Ca^{2+} present (195))—such as A23187 or ionomycin; the latter selectively increases release from internal stores without increasing plasma permeability to Ca^{2+} at <1 μM (ED_{50} 50 nM); 1–10 μM will also cause permeability of the plasma membrane, ER and mitochondria to Ca^{2+} ions(196).

Negative tests

(a) Inhibition of IP_3 binding to the receptor decreases response.

(i) Heparin (widely available, 100 μM–1 mM or 2 mg/ml in pipette solution (12, 193))—see above.

(ii) Caffeine (widely available, 0.1–10 mM (107, 193))—potent inhibitor of IP_3Rs (107). *Drawbacks*: also activates ryanodine receptors, acts as an antagonist at adenosine receptors and inhibits PDEs (107).

(iii) Ethanol, butanol and other long-chain aliphatic alcohols (widely available, 2%v/v ethanol, 0.1%v/v butanol (197))—reduces amount of IP_3-releasable Ca^{2+} and IP_3 binding sites without altering IP_3 binding (107).

(iv) Xestospongin C (AG Scientific, 1 μM in pipette solution (35))—cell-permeant blocker of IP_3R-mediated Ca^{2+} release (IC_{50} 360 nM (108)), which does not bind to the IP_3R site, at 5–20 μM (108). *Drawbacks*: equipotent blocker of SERCA pump (107) and can inhibit nitric oxide synthase (107) and ryanodine receptor-mediated Ca^{2+} release at >10 μM (108).

(v) 2-APB (widely available, 100 μM in pipette solution or extracellular solution (193), 2-aminoethoxydiphenyl borate)—membrane-permeant inhibitor of IP_3R-mediated Ca^{2+} release (IC_{50} 40–70 μM) (107, 198). *Drawback*: also inhibits store-operated Ca^{2+} release and some transient receptor potential channels (107).

(vi) IP_3R monoclonal antibody 18A10 (T. Michikawa,160 μg/ml in pipette solution (199))—an antibody against type 1 IP_3R which blocks IP_3-dependent long-term synaptic plasticity.

(b) Strong intracellular chelation of free Ca^{2+} decreases response.

(i) EGTA and BAPTA (widely available, BAPTA 20 mM in pipette solution (9, 11) or 25 μM BAPTA-AM in bathing solution (195), EGTA 1 mM in pipette solution (12))—BAPTA is a Ca^{2+} chelator with faster binding kinetics than EGTA; the AM ester of either is cell permeable. *Drawback*: BAPTA can directly inhibit some ion channels (195).

(H) Ryanodine receptors.

Most likely if Ca^{2+} increase is required. Ryanodine receptors are actually Ca^{2+}-activated Ca^{2+} release channels (CICR), whose regulation overlaps significantly with regulation of IP_3Rs (30). There are three distinct CICRs, all of which have been found in neurons with type 2 being the predominant type; type 1 is highly expressed in cerebellar Purkinje neurons, and type 3 is found in the hippocampus, striatum and diencephalon (30). Classically, the CICR type 1 is the "skeletal muscle" CICR, type 2 is the "cardiac" CICR and type 3 is the "brain" CICR (30). All CICRs are activated by Ca^{2+} from the cytosolic side with a rank order of Ca^{2+} sensitivity of type 1 > type 2 > type 3 (30). CICRs can be phosphorylated by most protein kinases, with PKA phosphorylation increasing their activity, and Ca^{2+}-CAM kinase II phosphorylation decreasing their activity, and they are sensitive to CAM, which enhances CICR activity at low (<0.1 μM) cytosolic Ca^{2+} concentrations and inhibits it at higher (>1 μM) concentrations (31).

Positive tests

(a) Ryanodine or analogues mimic response.

(i) Ryanodine (widely available, 1–100 μM (30, 193, 194, 200–202); see above)—ryanodine binds to Ca^{2+}-activated Ca^{2+} release channels (CICRs) and elicits multiple concentration-dependent effects (5–40 nM increases CICR opening, 1–5 μM increases CICR open time in a subconducting state and 50–>100 μM irreversible blocks CICRs) (30). Ryanodine effects on CICRs exhibit use dependence, i.e. its action is facilitated by other activators of CICRs, such as caffeine or Ca^{2+} increase (30).

(ii) Caffeine (see IP_3R above).

(iii) Volatile anaesthetics (widely available, 10–50 μM (194))—halothane and enflurane enhance CICR release in low μM (subtherapeutic) concentrations (30). *Drawbacks*: also modulate synaptic transmission (30), enhance $GABA_A$ channel opening (203) and activate two-pore K^+ channels (204).

(iv) Suramin (widely available, 10 μM–1 mM)—see above under G proteins; suramin can activate CICRs at μM concentration, stabilising the CICR in an open state (31).

Negative tests

(a) Inhibition of CICRs decreases response.

(i) Ryanodine (at > 100 μM; see above)—high concentrations of ryanodine irreversibly block CICRs.

(ii) Dantrolene (widely available, 25–100 μM (12, 30, 194, 201); see above)—a potent muscle relaxant which reduces the open time of CICRs; effective concentrations for this range between 10 and 90 μM (30).

(iii) Ruthenium red (see above under G proteins)—often used as a CICR receptor blocker, as it prolongs the closed state of the channel; effective concentrations in muscle are 1–20 μM (30). *Drawback*: many off-target effects (see under G proteins).

(iv) TMB-8 (widely available, 10–100 μM (202, 205), 3,4,5-trimethoxybenzoic acid 8-diethylamino-octyl ester) a membrane-permeant inhibitor of intracellular Ca^{2+} increases. *Drawbacks*: while TMB-8 suppresses metabotropic glutamate receptor-elicited increases in intracellular Ca^{2+} and neuronal death(202), its specificity for inhibition of CICRs has been questioned (206); it also inhibits nicotinic acetylcholine receptors (but see (205)) and the ATP-sensitive K^{+} channel and modulates muscarinic receptors(206).

2.5. What Is the Secondary Regulator?

(A) Direct gating of ionic channels.
Most likely if protein kinase activity is not required and Ca^{2+} is not involved.

Positive tests

(a) Response to agonist in cell-free patch or in reconstituted system (see above under G proteins).

(b) Cell-attached patch—response to agonist in pipette and no response to externally applied agonist. *Drawbacks*: if second messenger is membrane-delimited (e.g. PLA_2-AA or PLC-DAG), then it will probably not diffuse to the membrane under the patch.

(B) Protein phosphorylation.
Proteins are phosphorylated at specific serine, threonine or tyrosine sites by protein kinases (see above for testing specific protein kinases).

(C) Protein phosphatases (PP).
Most likely if PKA/PKC/PKC/Ca^{2+}-CAM kinase II are activated. PPs dephosphorylate phosphorylated protein kinases. There are many PPs; expression in brain tissue is predominantly of PP1, PP2A, PP2B (also called calcineurin) and PP2C, although low levels of PP3-7 are also found (109). As shown below, selective inhibition of individual PPs is difficult to achieve, and results with a single inhibitor should be confirmed with structurally different inhibitors with similar selectivity profiles; a useful scheme for such overlapping positive controls can be found in (109).

Negative tests

(a) Inhibition of protein phosphatases abolishes response.

(i) Okadaic acid and derivatives (widely available, 500 nM–1 μM in recording patch pipette solution (97, 158, 207) or 100 nM with slice preincubation for 2 h (208)). Okadaic acid is a polyether carboxylic acid derived from marine sponges; other derivatives are also from marine sp (dinophysistoxin-1, acantifolcicin) (109). Okadaic acids is only moderately membrane permeant; reports of membrane permeation may be due to ionophore formation. An inactive analogue of okadaic acid, 1-nor-okadone (500 nM–1 μM (209)), is a useful control compound which should not abolish responses (109). It is most potent at PP2A (~2 nM) > PP1 (IC_{50} = ~270 nM) ≫ PP2B (3 μM) with no effect on PP2C up to 10 μM (109). Several water-soluble forms are available. *Drawback*: Okadaic acid also inhibits PP3-6 at <2 nM (109, 210).

(ii) Calyculin A (widely available, 100–300 nM with preincubation of slices for 2 h (158, 211))—an octamethylpolyhydroxylyate fatty acid isolated from marine sponges, which is ~100-fold more cell permeable than okadaic acid (109); it is equipotent at PP1 and PP2A (IC_{50} = 1–14 nM) and competes with okadaic acid for the same inhibitory site (109). *Drawback*: also potently inhibits PP3 (210).

(iii) Microcystins (widely available, 2–10 μM in recording patch pipette solution (97, 158))—a family of over 40 cyclic heptapeptides which are biological toxins produced by various algae; as five peptides are constant, microcystins are denoted by the two variable peptides (e.g. microcystin-LR, microcystin-YR and microcystin-RR) (109). They are not membrane permeable, but are the most potent PP inhibitors known, competing with okadaic acid for the same inhibitory site; microcystin-LR inhibits PP1 (IC_{50} 0.1 nM) = PP2A > PP2B (200 nM) > PP2C (>4 μM) (109). *Drawback*: also potently inhibit PP3 (210), PP4 and PP5 (IC_{50} < 2 nM) (109).

(iv) Nodularin (widely available)—a cyclic pentapeptide isolated from toxic cyanobacteria, which is not cell permeable (109). Similar potency to the microcystins, but somewhat selective (~70-fold) for PP2A (IC_{50} 0.3 nM) > PP1 (1.6 nM) ≫ PP2B (9 μM) >>> PP2C (no effect) (109). *Drawback*: inhibits PP3 with similar potency to PP2A (210).

(v) Tautomycin (widely available, 1 μM in pipette solution (207))—a membrane-permeant polyketide isolated from *Streptomyces* bacteria which has antibiotic activity (109). It is structurally similar to okadaic acid, with PP1 (IC_{50} 0.4–0.7 nM) ~= PP2A (0.7–34 nM) ≫ PP2B (100 μM) >>> PP2C (no effect) (109). *Drawback*: potently inhibits PP3 (210).

(vi) Cantharidin and analogues—(widely available, 100 nM with slice preincubation for 2 h (208) or bath perfusion with 50–100 μM (142, 212))—

cell-permeant terpenoid which is a vesicant extracted from blister beetles and Spanish flies (crude extracts are known as Spanish fly with reputed aphrodisiac properties) with selectivity order of PP2A (IC_{50} 40 nM) > PP1 (IC_{50} 500 nM)□PP2B (IC_{50} 30 μM) ≫ PP2C (IC_{50} > 1 mM) (109). Cantharidic acid is a non-cell-permeant terpenoid with similar selectivity; other structural analogues are palasonin and the less potent (IC_{50} 1 and 5 μM for PP2A and PP1, respectively) endothall (109).

(vii) Type II synthetic pyrethroids (Calbiochem, cypermethrin 40 nM with preincubation for 30 min (213), deltamethrin 1 μM with preincubation for 3 min (214))—potent inhibitors of PP2B (IC_{50} = 40 pM–1 nM (cypermethrin), 100 pM (deltamethrin), 2–4 nM (fenvalerate) (109, 215)). The type I synthetic pyrethroid, resmethrin, is a weakly active negative control. *Drawbacks*: may have varied effects on ion channels (all), increase neurotransmitter release and intrasynaptosomal Ca^{2+} levels (deltamethrin) or open Na^{+} channels (fenvalerate) (216). An inability to inhibit PP2A has been reported (217).

(viii) Cyclosporin A (widely available, 1–10 μM (142, 209, 218) (214), cicolsporin)—a membrane-permeant cyclic undecapeptide which inhibits PP2B (IC_{50} 100 nM) by an indirect mechanism involving binding to the endogenous cyclophilin which then interacts with PP2B to inhibit its catalytic activity (109). *Drawback*: inhibits the SERCA pump with IC_{50} ~60 μM (24).

(ix) FK-506 (LC Laboratories but also widely available, 1–5 μM (218–220), tacrolimus or fujimycin)—a macrocyclic lactone immunosuppressant which is a specific inhibitor of PP2B by an indirect mechanism involving binding to the endogenous FK-506-binding protein which then interacts with PP2B to inhibit its catalytic activity (109). Drawback: can have direct inhibitory effects on ion channels which are PP2B independent (213).

(x) Fostriecin (BioAustralis, but also widely available, 250 nM in pipette solution (221))—a water-soluble polyene lactone with a phosphate ester isolated from *Streptomyces* bacteria, which inhibits PP1 and PP2A (IC_{50} 0.1–4 μM and 3–40 nM, respectively, and appears to be actively transported across membrane (109). Drawback: also inhibits PP4 (221).

(xi) DARPP-32 (0.5 mg/ml in pipette solution (207))—an endogenous inhibitory peptide which is a potent and specific inhibitor of PP1 found predominantly in brain when phosphorylated (IC_{50} 1 μM to 1–2 nM upon phosphorylation) (109).

(xii) Inhibitory peptides (Calbiochem, 100 μM in pipette solution (213))—endogenous peptides which act as selective pseudosubstrates for PP1, PP2A or PP2B (213): inhibitor 1 for PP1 has IC_{50} of 1 nM cf. 21 μM for PP2A (109). Some require phosphorylation to effectively inhibit (e.g. inhibitor 1 for PP1) (109).

(D) Eicosanoids.

Most likely if PLC is activated, DAG is generated and neither PKC nor IP_3 is required. All are synthesised from the common precursor, arachidonic acid by unrelated enzymatic pathways. Prostaglandins (PGs) and thromboxanes (TXs) are produced by cyclooxygenase (171), while leukotrienes, lipoxins and HETEs or HPETEs are produced by lipoxygenase (176). Investigating the effects of these bioactive lipids in neurons has largely limited to application of analogues (75) and is beyond the scope of this method review.

3. Summary

Pharmacological probing of G protein-coupled signalling pathways remains an effective first-line strategy for understanding receptor coupling to effector mechanisms in neurons, even though transgenic gene manipulation in vivo and in vitro and the increasing use of RNA interference allows targeted ablation or suppression of specific gene products. Without a well-grounded knowledge of the signalling pathway molecules and both on- and off- target effects of the pharmacological tools utilised, the researcher may make costly, time-consuming and invalid interpretation of pharmacological experiments. I hope that this outline of the actions, application and pitfalls of pharmacological ligands commonly used to dissect the major second messenger signalling cascades activated by G proteins in neurons will provide a useful knowledge base to avoid these mistakes, allowing the fruitful exploration and definition of these pathways at all levels, despite the limitations of many currently used pharmacological tools.

Acknowledgments

Funded by grants to MCB from the NHMRC, ARC, Motor Neuron Disease Research Institute of Australia, the Australian Brain Foundation and the Clive and Vera Ramaciotti Foundation.

References

1. Hille B (1992) G protein-coupled mechanisms and nervous signaling. Neuron 9:187–195
2. Klinker JF, Hagelüken A, Grünbaum L, Heilmann I, Nürnberg B, Harhammer R, Offermanns S, Schwaner I, Ervens J, Wenzel-Seifert K, Müller T, Seifert R (1994) Mastoparan may activate GTP hydrolysis by G_i-proteins in HL-60 membranes indirectly through interaction with nucleoside diphosphate kinase. Biochem J 304:377–383
3. Ross EM (1989) Signal sorting and amplification through G protein-coupled receptors. Neuron 3:141–152
4. Brown AM, Birnbaumer L (1988) Direct G protein gating of ion channels. Am J Physiol 254:H401–H410
5. Sternweis PC, Pang IH (1990) The G protein-channel connection. Trends Neurosci 13:122–126
6. Pfaffinger PJ, Martin JM, Hunter DD, Nathanson NM, Hille B (1985) GTP-binding proteins couple cardiac muscarinic receptors to a K channel. Nature 317:536–538
7. Dai S, Hall DD, Hell JW (2009) Supramolecular assemblies and localized regulation of voltage-gated ion channels. Physiol Rev 89:411–452
8. Zhou J, Shapiro MS, Hille B (1997) Speed of Ca^{2+} channel modulation by neurotransmitters in rat sympathetic neurons. J Neurophysiol 77:2040–2048
9. Bernheim L, Beech DJ, Hille B (1991) A diffusible second messenger mediates one of the pathways coupling receptors to calcium channels in rat sympathetic neurons. Neuron 6:859–867
10. Lamb TD, Pugh ENJ (1992) G-protein cascades: gain and kinetics. Trends Neurosci 15:291–298
11. Pan ZZ, Grudt TJ, Williams JT (1994) α_1-adrenoceptors in rat dorsal raphe neurons: regulation of two potassium conductances. J Physiol 478:437–447
12. Abdul-Ghani MA, Valiante TA, Carlen PL, Pennefather PS (1996) Metabotropic glutamate receptors coupled to IP_3 production mediate inhibition of I_{AHP} in rat dentate granule neurons. J Neurophysiol 76:2691–2700
13. Aman TK, Shen RY, Haj-Dahmane S (2007) D2-like dopamine receptors depolarize dorsal raphe serotonin neurons through the activation of nonselective cationic conductance. J Pharmacol Exp Ther 320:376–385. doi:10.1124/jpet.106.111690
14. Allen TGJ, Brown DA (1993) M_2 muscarinic receptor-mediated inhibition of the Ca^{2+} current in rat magnocellular cholinergic basal forebrain neurones. J Physiol 466:173–189
15. Nürnberg B, Tögel W, Krause G, Storm R, Breitweg-Lehmann E, Schunack W (1999) Non-peptide G-protein activators as promising tools in cell biology and potential drug leads. Eur J Med Chem 34:5–30
16. Barritt GJ, Gregory RB (1997) An evaluation of strategies available for the identification of GTP-binding proteins required in intracellular signalling pathways. Cell Signal 9:207–218
17. Dubyak GR, El-Moatassim C (1993) Signal transduction via P_2 -purinergic receptors for extracellular ATP and other nucleotides. Am J Physiol 265:C577–C606
18. Miller LD, Petrozzino JJ, Connor JA (1995) G protein-coupled receptors mediate a fast excitatory postsynaptic current in CA3 pyramidal neurons in hippocampal slices. J Neurosci 15:8320–8330
19. Okada Y, Fujiyama R, Miyamoto T, Sato T (2001) Saccharin activates cation conductance via inositol 1,4,5-trisphosphate production in a subset of isolated rod taste cells in the frog. Eur J Neurosci 13:308–314
20. Mizuno N, Itoh H (2009) Functions and regulatory mechanisms of G_q-signaling pathways. Neurosignals 17:42–54. doi:10.1159/000186689
21. Pfaffinger P (1988) Muscarine and t-LHRH suppress M-current by activating an IAP-insensitive G-protein. J Neurosci 8:3343–3353
22. Blackmore PF, Bocckino SB, Waynick LE, Exton JH (1985) Role of a guanine nucleotide-binding regulatory protein in the hydrolysis of hepatocyte phosphatidylinositol 4,5-bisphosphate by calcium-mobilizing hormones and the control of cell calcium. J Biol Chem 260:14477–14483
23. Liu QY, Chang YH, Schaffner AE, Smith SV, Barker JL (2002) Allopregnanolone activates $GABA_A$ receptor/Cl^- channels in a multiphasic manner in embryonic rat hippocampal neurons. J Neurophysiol 88:1147–1158. doi:10.1152/jn.00942.2001
24. Michelangeli F, East JM (2011) A diversity of SERCA Ca^{2+} pump inhibitors. Biochem Soc Trans 39:789–797. doi:10.1042/bst0390789
25. Liu T, Fujita T, Kumamoto E (2011) Acetylcholine and norepinephrine mediate GABAergic but not glycinergic transmission enhancement by melittin in adult rat substantia gelatinosa neurons. J Neurophysiol 106:233–246. doi:10.1152/jn.00838.2010
26. Yue HY, Fujita T, Kumamoto E (2005) Phospholipase A_2 activation by melittin

enhances spontaneous glutamatergic excitatory transmission in rat substantia gelatinosa neurons. Neuroscience 135:485–495. doi:10.1016/j.neuroscience.2005.05.040

27. Freissmuth M, Waldhoer M, Bofill-Cardona E, Nanoff C (1999) G protein antagonists. Trends Pharmacol Sci 20:237–245
28. Trudeau LE, Doyle RT, Emery DG, Haydon PG (1996) Calcium-independent activation of the secretory apparatus by ruthenium red in hippocampal neurons: a new tool to assess modulation of presynaptic function. J Neurosci 16:46–54
29. Wu LJ, Sweet TB, Clapham DE (2010) International union of basic and clinical pharmacology. LXXVI. Current progress in the mammalian TRP ion channel family. Pharmacol Rev 62:381–404. doi:10.1124/pr.110.002725
30. Verkhratsky A (2005) Physiology and pathophysiology of the calcium store in the endoplasmic reticulum of neurons. Physiol Rev 85:201–279. doi:10.1152/physrev.00004.2004
31. Zucchi R, Ronca-Testoni S (1997) The sarcoplasmic reticulum Ca^{2+} channel/ryanodine receptor: modulation by endogenous effectors, drugs and disease states. Pharmacol Rev 49:1–51
32. Schubert B, VanDongen AM, Kirsch GE, Brown AM (1989) β-adrenergic inhibition of cardiac sodium channels by dual G-protein pathways. Science 245:516–519
33. Shapiro MS, Wollmuth LP, Hille B (1994) Modulation of Ca^{2+} channels by PTX-sensitive G-proteins is blocked by *N*-ethylmaleimide in rat sympathetic neurons. J Neurosci 14:7109–7116
34. Yang B, Samson WK, Ferguson AV (2003) Excitatory effects of orexin-A on nucleus tractus solitarius neurons are mediated by phospholipase C and protein kinase C. J Neurosci 23:6215–6222
35. Li Z, Ji G, Neugebauer V (2011) Mitochondrial reactive oxygen species are activated by mGluR5 through IP_3 and activate ERK and PKA to increase excitability of amygdala neurons and pain behavior. J Neurosci 31:1114–1127. doi:10.1523/jneurosci.5387-10.2011
36. Ukhanov K, Brunert D, Corey EA, Ache BW (2011) Phosphoinositide 3-kinase-dependent antagonism in mammalian olfactory receptor neurons. J Neurosci 31:273–280. doi:10.1523/jneurosci.3698-10.2011
37. Ireland MF, Noakes PG, Bellingham MC (2004) $P2X_7$-like receptor subunits enhance excitatory synaptic transmission at central synapses by presynaptic mechanisms. Neuroscience 128:269–280
38. Jeong SW, Ikeda SR (1999) G protein α subunit $G\alpha_z$ couples neurotransmitter receptors to ion channels in sympathetic neurons. Neuron 21:1201–1212
39. Sodickson DL, Bean BP (1996) $GABA_B$ receptor-activated inwardly rectifying potassium current in dissociated hippocampal CA3 neurons. J Neurosci 16:6374–6385
40. Morishita W, Kirov SA, Pitler TA, Martin LA, Lenz RA, Alger BE (1997) *N*-ethylmaleimide blocks depolarization-induced suppression of inhibition and enhances GABA release in the rat hippocampal slice *in vitro*. J Neurosci 17:941–950
41. Asano T, Ogasawara N (1986) Uncoupling of γ-aminobutyric acid B receptors from GTP-binding proteins by *N*-ethylmaleimide: effect of *N*-ethylmaleimide on purified GTP-binding proteins. Mol Pharmacol 29:244–249
42. Egorova A, Hoshi N, Knijnik R, Shahidullah M, Hashii M, Noda M, Higashida H (1997) Sulfhydryl modification inhibits K^+ (M) current with kinetics close to acetylcholine in rodent NG108-15 cells. Neurosci Res 27:35–44
43. Besser L, Chorin E, Sekler I, Silverman WF, Atkin S, Russell JT, Hershfinkel M (2009) Synaptically released zinc triggers metabotropic signaling via a zinc-sensing receptor in the hippocampus. J Neurosci 29(9):2890–2901. doi:10.1523/jneurosci.5093-08.2009
44. Veale EL, Kennard LE, Sutton GL, MacKenzie G, Sandu C, Mathie A (2007) $G\alpha_q$-mediated regulation of TASK3 two-pore domain potassium channels: the role of protein kinase C. Mol Pharmacol 71:1666–1675. doi:10.1124/mol.106.033241
45. Lehmann DM, Seneviratne AMPB, Smrcka AV (2008) Small molecule disruption of G protein βγ subunit signaling inhibits neutrophil chemotaxis and inflammation. Mol Pharmacol 73:410–418. doi:10.1124/mol.107.041780
46. Bonacci TM, Mathews JL, Yuan C, Lehmann DM, Malik S, Wu D, Font JL, Bidlack JM, Smrcka AV (2006) Differential targeting of Gβγ-subunit signaling with small molecules. Science 312:443–446. doi:10.1126/science.1120378
47. Willoughby D, Cooper DM (2007) Organization and Ca^{2+} regulation of adenylyl cyclases in cAMP microdomains. Physiol Rev 87:965–1010. doi:10.1152/physrev.00049.2006
48. Chavez-Noriega LE, Stevens CF (1994) Increased transmitter release at excitatory synapses produced by direct activation of

adenylate cyclase in rat hippocampal slices. J Neurosci 14:310–317

49. Chavez-Noriega LE, Stevens CF (1992) Modulation of synaptic efficacy in field CA1 of the rat hippocampus by forskolin. Brain Res 574:85–92
50. Huang YY, Li XC, Kandel ER (1994) cAMP contributes to mossy fiber LTP by initiating both a covalently mediated early phase and macromolecular synthesis- dependent late phase. Cell 79:69–79
51. Vogalis F, Harvey JR, Furness JB (2003) PKA-mediated inhibition of a novel K^+ channel underlies the slow after-hyperpolarization in enteric AH neurons. J Physiol 548:801–814. doi:10.1113/jphysiol.2002.037325
52. Ingram SL, Williams JT (1994) Opioid inhibition of I_h via adenylyl cyclase. Neuron 13:179–186
53. Larkman PM, Kelly JS (1997) Modulation of I_H by 5-HT in neonatal rat motoneurones *in vitro*: mediation through a phosphorylation independent action of cAMP. Neuropharmacology 36:721–733
54. Onda T, Hashimoto Y, Nagai M, Kuramochi H, Saito S, Yamazaki H, Toya Y, Sakai I, Homcy CJ, Nishikawa K, Ishikawa Y (2001) Type-specific regulation of adenylyl cyclase. Selective pharmacological stimulation and inhibition of adenylyl cyclase isoforms. J Biol Chem 276:47785–47793. doi:10.1074/jbc.M107233200
55. Rangel-Barajas C, Silva I, Lopéz-Santiago LM, Aceves J, Erlij D, Florán B (2011) l-DOPA-induced dyskinesia in hemiparkinsonian rats is associated with up-regulation of adenylyl cyclase type V/VI and increased GABA release in the substantia nigra reticulata. Neurobiol Dis 41:51–61. doi:10.1016/j.nbd.2010.08.018
56. Rebecchi MJ, Pentyala SN (2000) Structure, function, and control of phosphoinositide-specific phospholipase C. Physiol Rev 80: 1291–1335
57. Horowitz LF, Hirdes W, Suh BC, Hilgemann DW, Mackie K, Hille B (2005) Phospholipase C in living cells: activation, inhibition, Ca^{2+} requirement, and regulation of M current. J Gen Physiol 126:243–262. doi:10.1085/jgp.200509309
58. Zhang P, Yang C, Delay RJ (2010) Odors activate dual pathways, a TRPC2 and a AA-dependent pathway, in mouse vomeronasal neurons. Am J Physiol 298:C1253–C1264. doi:10.1152/ajpcell.00271.2009
59. Krjukova J, Holmqvist T, Danis AS, Akerman KE, Kukkonen JP (2004) Phospholipase C activator *m*-3M3FBS affects Ca^{2+} homeostasis independently of phospholipase C activation. Br J Pharmacol 143(1):3–7. doi:10.1038/sj.bjp. 0705911
60. Sickmann T, Klose A, Huth T, Alzheimer C (2008) Unexpected suppression of neuronal G protein-activated, inwardly rectifying K^+ current by common phospholipase C inhibitor. Neurosci Lett 436:102–106. doi:10.1016/j.neulet.2008.02.067
61. Jin W, Lo TM, Loh HH, Thayer SA (1994) U73122 inhibits phospholipase C-dependent calcium mobilization in neuronal cells. Brain Res 642:237–243
62. Naramatsu M, Yamashita T, Kokubun S (1997) The signalling pathway which causes contraction via P2-purinoceptors in rat urinary bladder smooth muscle. Br J Pharmacol 122:558–562. doi:10.1038/sj.bjp. 0700157
63. Umemura A, Mabe H, Nagai H (1992) A phospholipase C inhibitor ameliorates postischemic neuronal damage in rats. Stroke 23:1163–1166
64. Deutsch DG, Lin S, Hill WA, Morse KL, Salehani D, Arreaza G, Omeir RL, Makriyannis A (1997) Fatty acid sulfonyl fluorides inhibit anandamide metabolism and bind to the cannabinoid receptor. Biochem Biophys Res Commun 231:217–221. doi:10.1006/bbrc.1997.6072
65. Llansola M, Monfort P, Felipo V (2000) Inhibitors of phospholipase C prevent glutamate neurotoxicity in primary cultures of cerebellar neurons. J Pharmacol Exp Ther 292:870–876
66. Kim J, Nakajima K, Oomura Y, Wayner MJ, Sasaki K (2009) Orexin-A and ghrelin depolarize the same pedunculopontine tegmental neurons in rats: an *in vitro* study. Peptides 30:1328–1335
67. Albert AP, Piper AS, Large WA (2005) Role of phospholipase D and diacylglycerol in activating constitutive TRPC-like cation channels in rabbit ear artery myocytes. J Physiol 566:769–780. doi:10.1113/jphysiol.2005.090852
68. Li Y, Maher P, Schubert D (1998) Phosphatidylcholine-specific phospholipase C regulates glutamate-induced nerve cell death. Proc Natl Acad Sci USA 95:7748–7753
69. Burke JE, Dennis EA (2009) Phospholipase A_2 structure/function, mechanism, and signaling. J Lipid Res 50(Suppl):S237–S242. doi:10.1194/jlr.R800033-JLR200
70. Ordway RW, Singer JJ, Walsh JVJ (1991) Direct regulation of ion channels by fatty acids. Trends Neurosci 14:96–100
71. Farooqui AA, Ong WY, Horrocks LA (2006) Inhibitors of brain phospholipase A_2 activity: their neuropharmacological effects and

therapeutic importance for the treatment of neurologic disorders. Pharmacol Rev 58:591–620. doi:10.1124/pr.58.3.7

72. Mendes CT, Gattaz WF, Schaeffer EL, Forlenza OV (2005) Modulation of phospholipase A_2 activity in primary cultures of rat cortical neurons. J Neural Transm 112:1297–1308. doi:10.1007/s00702-004-0271-3
73. Windelborn JA, Lipton P (2008) Lysosomal release of cathepsins causes ischemic damage in the rat hippocampal slice and depends on NMDA-mediated calcium influx, arachidonic acid metabolism, and free radical production. J Neurochem 106:56–69. doi:10.1111/j.1471-4159.2008.05349.x
74. Fernando SR, Pertwee RG (1997) Evidence that methyl arachidonyl fluorophosphonate is an irreversible cannabinoid receptor antagonist. Br J Pharmacol 121:1716–1720. doi:10.1038/sj.bjp. 0701303
75. Schweitzer P, Madamba S, Champagnat J, Siggins GR (1993) Somatostatin inhibition of hippocampal CA1 pyramidal neurons: Mediation by arachidonic acid and its metabolites. J Neurosci 13:2033–2049
76. Lucas KA, Pitari GM, Kazerounian S, Ruiz-Stewart I, Park J, Schulz S, Chepenik KP, Waldman SA (2000) Guanylyl cyclases and signaling by cyclic GMP. Pharmacol Rev 52:375–414
77. Francis SH, Busch JL, Corbin JD (2010) cGMP-dependent protein kinases and cGMP phosphodiesterases in nitric oxide and cGMP action. Pharmacol Rev 62:525–563. doi:10.1124/pr.110.002907
78. Megson IL, Miller MR (2009) NO and sGC-stimulating NO donors. In: Schmidt HHHW, Hofmann F, Stasch J-P (eds) cGMP: generators, effectors and therapeutic implications, vol 191, Handbook of experimental pharmacology. Springer, Berlin, pp 247–276. doi:10.1007/978-3-540-68964-5_12
79. Ryter SW, Alam J, Choi AMK (2006) Heme oxygenase-1/carbon monoxide: from basic science to therapeutic applications. Physiol Rev 86:583–650
80. Syapin PJ (2008) Regulation of haeme oxygenase-1 for treatment of neuroinflammation and brain disorders. Br J Pharmacol 155:623–640. doi:10.1038/bjp. 2008.342
81. Calabrese V, Mancuso C, Calvani M, Rizzarelli E, Butterfield DA, Giuffrida Stella AM (2007) Nitric oxide in the central nervous system: neuroprotection versus neurotoxicity. Nat Rev Neurosci 8:766–775
82. Evgenov OV, Pacher P, Schmidt PM, Hasko G, Schmidt HH, Stasch JP (2006) NO-independent stimulators and activators of soluble guanylate cyclase: discovery and therapeutic potential. Nat Rev Drug Discov 5:755–768. doi:10.1038/nrd2038
83. Liu YC, Wu SN (2005) BAY 41–2272, a potent activator of soluble guanylyl cyclase, stimulates calcium elevation and calcium-activated potassium current in pituitary GH cells. Clin Exp Pharmacol Physiol 32:1078–1087. doi:10.1111/j.1440-1681.2005.04315.x
84. Lev-Ram V, Jiang T, Wood J, Lawrence DS, Tsien RY (1997) Synergies and coincidence requirements between NO, cGMP, and Ca^{2+} in the induction of cerebellar long-term depression. Neuron 18:1025–1038
85. Wahler GM, Dollinger SJ (1995) Nitric oxide donor SIN-1 inhibits mammalian cardiac calcium current through cGMP-dependent protein kinase. Am J Physiol 268:C45–C54
86. Zufall F, Leinders-Zufall T (1997) Identification of a long-lasting form of odor adaptation that depends on the carbon monoxide/cGMP second-messenger system. J Neurosci 17:2703–2717
87. Brahmachari S, Fung YK, Pahan K (2006) Induction of glial fibrillary acidic protein expression in astrocytes by nitric oxide. J Neurosci 26:4930–4939. doi:10.1523/jneurosci.5480-05.2006
88. Petrov AM, Giniatullin AR, Sitdikova GF, Zefirov AL (2008) The role of cGMP-dependent signaling pathway in synaptic vesicle cycle at the frog motor nerve terminals. J Neurosci 28:13216–13222. doi:10.1523/jneurosci.2947-08.2008
89. Smith SL, Otis TS (2003) Persistent changes in spontaneous firing of Purkinje neurons triggered by the nitric oxide signaling cascade. J Neurosci 23:367–372
90. Pose I, Silveira V, Morales FR (2011) Inhibition of excitatory synaptic transmission in the trigeminal motor nucleus by the nitric oxide-cyclicGMP signaling pathway. Brain Res 1393:1–16. doi:10.1016/j.brainres.2011.03.002
91. Olesen SP, Drejer J, Axelsson O, Moldt P, Bang L, Nielsen-Kudsk JE, Busse R, Mulsch A (1998) Characterization of NS 2028 as a specific inhibitor of soluble guanylyl cyclase. Br J Pharmacol 123:299–309. doi:10.1038/sj.bjp. 0701603
92. Archer SL, Huang JMC, Hampl V, Nelson DP, Shultz PJ, Weir EK (1994) Nitric oxide and cGMP cause vasorelaxation by activation of a charybdotoxin-sensitive K channel by cGMP-dependent protein kinase. Proc Natl Acad Sci USA 91:7583–7587
93. Machado JE, Segura F, Brioso MA, Borges R (2000) Nitric oxide modulates a late step of exocytosis. J Biol Chem 275:20274–20279

94. Higashida H, Salmina AB, Olovyannikova RY, Hashii M, Yokoyama S, Koizumi K, Jin D, Liu HX, Lopatina O, Amina S, Islam MS, Huang JJ, Noda M (2007) Cyclic ADP-ribose as a universal calcium signal molecule in the nervous system. Neurochem Int 51:192–199. doi:10.1016/j.neuint.2007.06.023
95. Schuman EM, Meffert MK, Schulman H, Madison DV (1994) An ADP-ribosyltransferase as a potential target for nitric oxide action in hippocampal long-term potentiation. Proc Natl Acad Sci USA 91:11958–11962
96. Oka M, Hirouchi M, Itoh Y, Ukai Y (2000) Involvement of peroxynitrite and hydroxyradical generated from nitric oxide in hypoxia/reoxygenation injury in rat cerebrocortical slices. Neuropharmacology 39:1319–1330. doi:10.1016/s0028-3908(99)00197-5
97. Ingram SL, Williams JT (1996) Modulation of the hyperpolarization-activated current (I_h) by cyclic nucleotides in guinea pig primary afferent neurons. J Physiol 492:97–107
98. Murray AJ (2008) Pharmacological PKA inhibition: all may not be what it seems. Sci Signal 1(22):4. doi:10.1126/scisignal.122re4
99. Segal M (1981) The action of norepinephrine in the rat hippocampus: intracellular studies in the slice preparation. Brain Res 206:107–128. doi:10.1016/0006-8993(81)90104-9
100. Horn VJ, Baum BJ, Ambudkar IS (1988) β-adrenergic receptor stimulation induces inositol trisphosphate production and Ca^{2+} mobilization in rat parotid acinar cells. J Biol Chem 263:12454–12460
101. Bender AT, Beavo JA (2006) Cyclic nucleotide phosphodiesterases: molecular regulation to clinical use. Pharmacol Rev 58:488–520. doi:10.1124/pr.58.3.5
102. Hartell NA (1996) Inhibition of cGMP breakdown promotes the induction of cerebellar long-term depression. J Neurosci 16: 2881–2890
103. Francis SH, Sekhar KR, Ke H, Corbin JD (2011) Inhibition of cyclic nucleotide phosphodiesterases by methylxanthines and related compounds. In: Fredholm BB (ed) Methylxanthines, vol 200, Handbook of experimental pharmacology. Springer, Berlin, pp 93–133. doi:10.1007/978-3-642-13443-2_4
104. Ukena D, Schudt C, Sybrecht GW (1993) Adenosine receptor-blocking xanthines as inhibitors of phosphodiesterase isozymes. Biochem Pharmacol 45:847–851
105. Osinski MT, Schrör K (2000) Inhibition of platelet-derived growth factor-induced mitogenesis by phosphodiesterase 3 inhibitors: role of protein kinase a in vascular smooth muscle cell mitogenesis. Biochem Pharmacol 60:381–387. doi:10.1016/s0006-2952(00)00328-2
106. Berridge MJ (1993) Inositol trisphosphate and calcium signalling. Nature 361:315–325
107. Bultynck GB, Sienaert IS, Parys JP, Callewaert GC, De Smedt HDS, Boens NB, Dehaen WD, Missiaen LM (2003) Pharmacology of inositol trisphosphate receptors. Pflugers Arch 445:629–642. doi:10.1007/s00424-002-0971-1
108. Gafni J, Munsch JA, Lam TH, Catlin MC, Costa LG, Molinski TF, Pessah IN (1997) Xestospongins: potent membrane permeable blockers of the inositol 1,4,5-trisphosphate receptor. Neuron 19:723–733
109. Herzig S, Neumann J (2000) Effects of serine/threonine protein phosphatases on ion channels in excitable membranes. Physiol Rev 80:173–210
110. Maher BJ, Mackinnon RL 2nd, Bai J, Chapman ER, Kelly PT (2005) Activation of postsynaptic Ca^{2+} stores modulates glutamate receptor cycling in hippocampal neurons. J Neurophysiol 93:178–188. doi:10.1152/jn.00651.2004
111. Behnisch T, Reymann KG (1994) 2,3-diphosphoglyceric acid blocks long-term potentiation of excitatory postsynaptic currents in hippocampal CA1 neurons of the rat. Neurosci Lett 165:23–26
112. Markram H, Segal M (1992) The inositol 1,4,5-trisphosphate pathway mediates cholinergic potentiation of rat hippocampal neuronal responses to NMDA. J Physiol 447:513–533
113. Tu P, Kunert-Keil C, Lucke S, Brinkmeier H, Bouron A (2009) Diacylglycerol analogues activate second messenger-operated calcium channels exhibiting TRPC-like properties in cortical neurons. J Neurochem 108:126–138. doi:10.1111/j.1471-4159.2008.05752.x
114. Min R, Testa-Silva G, Heistek TS, Canto CB, Lodder JC, Bisogno T, Di Marzo V, Brussaard AB, Burnashev N, Mansvelder HD (2010) Diacylglycerol lipase is not involved in depolarization-induced suppression of inhibition at unitary inhibitory connections in mouse hippocampus. J Neurosci 30:2710–2715. doi:10.1523/jneurosci.bc-3622-09.2010
115. Hoover HS, Blankman JL, Niessen S, Cravatt BF (2008) Selectivity of inhibitors of endocannabinoid biosynthesis evaluated by activity-based protein profiling. Bioorg Med Chem Lett 18:5838–5841. doi:10.1016/j.bmcl.2008.06.091
116. Bisogno T, Cascio MG, Saha B, Mahadevan A, Urbani P, Minassi A, Appendino G, Saturnino C, Martin B, Razdan R, Di Marzo V

(2006) Development of the first potent and specific inhibitors of endocannabinoid biosynthesis. Biochim Biophys Acta 1761:205–212. doi:10.1016/j.bbalip. 2005.12.009

117. Di Marzo V (2009) The endocannabinoid system: its general strategy of action, tools for its pharmacological manipulation and potential therapeutic exploitation. Pharmacol Res 60:77–84. doi:10.1016/j.phrs.2009.02.010

118. Piomelli D, Volterra A, Dale N, Siegelbaum SA, Kandel ER, Schwartz JH, Belardetti F (1987) Lipoxygenase metabolites of arachidonic acid as second messengers for presynaptic inhibition of *Aplysia* sensory cells. Nature 328:38–43

119. Hewett SJ, Uliasz TF, Vidwans AS, Hewett JA (2000) Cyclooxygenase-2 contributes to *N*-methyl-D-aspartate-mediated neuronal cell death in primary cortical cell culture. J Pharmacol Exp Ther 293:417–425

120. Nelson TJ, Alkon DL (2009) Neuroprotective versus tumorigenic protein kinase C activators. Trends Biochem sci 34:136–145. doi:10.1016/j.tibs.2008.11.006

121. Villarroel A (1993) Suppression of neuronal potassium A-current by arachidonic acid. FEBS Lett 335:184–188

122. Needleman P, Turk J, Jakschik BA, Morrison AR, Lefkowith JB (1986) Arachidonic acid metabolism. Annu Rev Biochem 55:69–102

123. Pape HC, Mager R (1992) Nitric oxide controls oscillatory activity in thalamocortical neurons. Neuron 9:441–448

124. Glaum SR, Miller RJ (1993) Zinc protoporphyrin-IX blocks the effects of metabotropic glutamate receptor activation in the rat nucleus tractus solitarii. Mol Pharmacol 43:965–969

125. Nakamizo T, Kawamata J, Yoshida K, Kawai Y, Kanki R, Sawada H, Kihara T, Yamashita H, Shibasaki H, Akaike A, Shimohama S (2003) Phosphodiesterase inhibitors are neuroprotective to cultured spinal motor neurons. J Neurosci Res 71:485–495. doi:10.1002/jnr.10483

126. Poppe H, Rybalkin SD, Rehmann H, Hinds TR, Tang X-B, Christensen AE, Schwede F, Genieser H-G, Bos JL, Doskeland SO, Beavo JA, Butt E (2008) Cyclic nucleotide analogs as probes of signaling pathways. Nat Methods 5:277–278, doi:http://www.nature.com/nmeth/journal/v5/n4/suppinfo/nmeth0408-277_S1.html

127. Zhuo M, Hu Y, Schultz C, Kandel ER, Hawkins RD (1994) Role of guanylyl cyclase and cGMP-dependent protein kinase in long-term potentiation. Nature 368:635–639

128. Monfort P, Munoz MD, Kosenko E, Felipo V (2002) Long-term potentiation in hippocampus involves sequential activation of soluble guanylate cyclase, cGMP-dependent protein kinase, and cGMP-degrading phosphodiesterase. J Neurosci 22:10116–10122

129. Méry PF, Pavoine C, Belhassen L, Pecker F, Fischmeister R (1993) Nitric oxide regulates cardiac Ca^{2+} current. Involvement of cGMP-inhibited and cGMP-stimulated phosphodiesterases through guanylyl cyclase activation. J Biol Chem 268:26286–26295

130. Zhang Z, Klyachko V, Jackson MB (2007) Blockade of phosphodiesterase Type 5 enhances rat neurohypophysial excitability and electrically evoked oxytocin release. J Physiol 584:137–147. doi:10.1113/jphysiol.2007.139303

131. Hudmon A, Schulman H (2002) Structure-function of the multifunctional Ca^{2+}/calmodulin-dependent protein kinase II. Biochem J 364:593–611

132. Swulius M, Waxham M (2008) Ca^{2+}/calmodulin-dependent protein kinases. Cell Mol Life Sci 65:2637–2657. doi:10.1007/s00018-008-8086-2

133. Oláh Z, Jósvay K, Pecze L, Letoha T, Babai N, Budai D, Ötvös F, Szalma S, Vizler C (2007) Anti-calmodulins and tricyclic adjuvants in pain therapy block the TRPV1 channel. PLoS One 2:e545. doi:10.1371/journal.pone.0000545

134. Wang ZM, Katsurabayashi S, Rhee JS, Brodwick M, Akaike N (2001) Substance P abolishes the facilitatory effect of ATP on spontaneous glycine release in neurons of the trigeminal nucleus pars caudalis. J Neurosci 21:2983–2991

135. Gordge PC, Ryves WJ (1994) Inhibitors of protein kinase C. Cell Signal 6:871–882

136. Caulfield MP, Robbins J, Sim JA, Brown DA, Mac Neil S, Blackburn GM (1991) The naphthalenesulphonamide calmodulin antagonist W7 and its 5-iodo-1-C8 analogue inhibit potassium and calcium currents in NG108-15 neuroblastoma x glioma cells in a manner possibly unrelated to their antagonism of calmodulin. Neurosci Lett 125:57–61. doi:10.1016/0304-3940(91)90130-I

137. Kleene SJ (1994) Inhibition of olfactory cyclic nucleotide-activated current by calmodulin antagonists. Br J Pharmacol 111:469–472

138. Dulon D, Zajic G, Schacht J (1990) Increasing intracellular free calcium induces circumferential contractions in isolated cochlear outer hair cells. J Neurosci 10:1388–1397

139. Humphreys BD, Virginio C, Surprenant A, Rice J, Dubyak GR (1998) Isoquinolines as antagonists of the $P2X_7$ nucleotide receptor: high selectivity for the human versus rat receptor homologues. Mol Pharmacol 54:22–32

140. Wang Y, Manis PB (2008) Short-term synaptic depression and recovery at the mature mammalian endbulb of Held synapse in mice. J Neurophysiol 100:1255–1264. doi:10.1152/jn.90715.2008
141. Simons SB, Escobedo Y, Yasuda R, Dudek SM (2009) Regional differences in hippocampal calcium handling provide a cellular mechanism for limiting plasticity. Proc Natl Acad Sci USA 106:14080–14084. doi:10.1073/pnas.0904775106
142. Gervasi N, Hepp R, Tricoire L, Zhang J, Lambolez B, Paupardin-Tritsch D, Vincent P (2007) Dynamics of protein kinase A signaling at the membrane, in the cytosol, and in the nucleus of neurons in mouse brain slices. J Neurosci 27:2744–2750. doi:10.1523/jneurosci.5352-06.2007
143. Yang SK, Wang K, Parkington H, Chen C (2008) Involvement of tetrodotoxin-resistant Na^+ current and protein kinase C in the action of growth hormone (GH)-releasing hormone on primary cultured somatotropes from GH-green fluorescent protein transgenic mice. Endocrinology 149:4726–4735. doi:10.1210/en.2008-0405
144. Tanaka S, Chandler SH (2006) Serotonergic modulation of persistent sodium currents and membrane excitability via cyclic AMP-protein kinase A cascade in mesencephalic V neurons. J Neurosci Res 83:1362–1372
145. Serulle Y, Zhang S, Ninan I, Puzzo D, McCarthy M, Khatri L, Arancio O, Ziff EB (2007) A GluR1-cGKII interaction regulates AMPA receptor trafficking. Neuron 56:670–688. doi:10.1016/j.neuron.2007.09.016
146. Saywell SA, Feldman JL (2004) Dynamic interactions of excitatory and inhibitory inputs in hypoglossal motoneurones: respiratory phasing and modulation by PKA. J Physiol 554:879–889
147. Dalton GD, Dewey WL (2006) Protein kinase inhibitor peptide (PKI): a family of endogenous neuropeptides that modulate neuronal cAMP-dependent protein kinase function. Neuropeptides 40:23–34. doi:10.1016/j.npep. 2005.10.002
148. Davies SP, Reddy H, Caivano M, Cohen P (2000) Specificity and mechanism of action of some commonly used protein kinase inhibitors. Biochem J 351:95–105
149. Penn RB, Parent JL, Pronin AN, Panettieri RA, Benovic JL (1999) Pharmacological inhibition of protein kinases in intact cells: antagonism of β-adrenergic receptor ligand binding by H-89 reveals limitation of usefulness. J Pharmacol Exp Ther 288:428–437
150. Agopyan N, Agopyan I (1991) Effects of protein kinase C activators and inhibitors on membrane properties, synaptic responses, and cholinergic actions in CA1 subfield of rat hippocampus *in situ* and *in vitro*. Synapse 7:193–206
151. Colbert CM, Johnston D (1998) Protein kinase C activation decreases activity-dependent attenuation of dendritic Na^+ current in hippocampal CA1 pyramidal neurons. J Neurophysiol 79:491–495
152. Papp A, Hoyer J (1996) PKC-dependent reduction of the acetylcholine-evoked inward Na current in *Aplysia* D-neurons: effect of injected PKC and PKC activators. Brain Res 708:123–127. doi:10.1016/0006-8993(95)01299-0
153. Shearman MS, Sekiguchi K, Nishizuka Y (1989) Modulation of ion channel activity: a key function of the protein kinase C enzyme family. Pharmacol Rev 41:211–237
154. Stocca G, Lovinger DM (2003) Phorbol ester uncouples adenosine inhibition of presynaptic Ca^{2+} transients at hippocampal synapses. Hippocampus 13:355–360
155. Parfitt KD, Madison DV (1993) Phorbol esters enhance synaptic transmission by a presynaptic, calcium-dependent mechanism in rat hippocampus. J Physiol 471:245–268
156. Pascale A, Amadio M, Scapagnini G, Lanni C, Racchi M, Provenzani A, Govoni S, Alkon DL, Quattrone A (2005) Neuronal ELAV proteins enhance mRNA stability by a PKCα-dependent pathway. Proc Natl Acad Sci USA 102:12065–12070. doi:10.1073/pnas.0504702102
157. Ghirardi M, Braha O, Hochner B, Montarolo PG, Kandel ER, Dale N (1992) Roles of PKA and PKC in facilitation of evoked and spontaneous transmitter release at depressed and nondepressed synapses in *Aplysia* sensory neurons. Neuron 9:479–489
158. Wyllie DJ, Nicoll RA (1994) A role for protein kinases and phosphatases in the Ca^{2+}-induced enhancement of hippocampal AMPA receptor-mediated synaptic responses. Neuron 13:635–643
159. Sullivan JP, Connor JR, Shearer BG, Burch RM (1991) 2,6-diamino-N-([1-(1-oxotridecyl)-2-piperidinyl] methyl) hexanamide (NPC 15437): a novel inhibitor of protein kinase C interacting at the regulatory domain. Mol Pharmacol 41:38–44
160. Dolphin AC, McGuirk SM, Scott RH (1989) An investigation into the mechanisms of inhibition of calcium channel currents in cultured sensory neurones of the rat by guanine nucleotide analogues and (−)-baclofen. Br J Pharmacol 97:263–273

161. Shapiro MS, Zhou J, Hille B (1996) Selective disruption by protein kinases of G-protein-mediated Ca^{2+} channel modulation. J Neurophysiol 76:311–320
162. Rüegg UT, Burgess GM (1989) Staurosporine, K-252 and UCN-01: potent but nonspecific inhibitors of protein kinases. Trends Pharmacol Sci 10:218–220
163. Kase H, Iwahashi K, Nakanishi S, Matsuda Y, Yamada K, Takahashi M, Murakata C, Sato A, Kaneko M (1987) K-252 compounds, novel and potent inhibitors of protein kinase C and cyclic nucleotide-dependent protein kinases. Biochem Biophys Res Commun 142:436–440
164. Bandrowski AE, Ashe JH, Crawford CA (2001) Tetanic stimulation and metabotropic glutamate receptor agonists modify synaptic responses and protein kinase activity in rat auditory cortex. Brain Res 894:218–232. doi:10.1016/s0006-8993(01)02052-2
165. Toullec D, Pianet P, Coste H, Bellevergue P, Grand-Perret T, Ajakane M, Baudet V, Boissin P, Loriolle F, Duhamel L, Charon D, Kirilovsky J (1991) The bisindolylmaleimide GF 109203X is a potent and selective inhibitor of protein kinase C. J Biol Chem 266: 15771–15781
166. Kloc R, Luchowska E, Wielosz M, Owe-Larsson B, Urbanska EM (2008) Memantine increases brain production of kynurenic acid via protein kinase A-dependent mechanism. Neurosci Lett 435(2):169–173. doi:10.1016/j.neulet.2008.02.027
167. Sang N, Chen C (2006) Lipid signaling and synaptic plasticity. Neuroscientist 12:425–434. doi:10.1177/1073858406290794
168. Phillis JW, Horrocks LA, Farooqui AA (2006) Cyclooxygenases, lipoxygenases, and epoxygenases in CNS: their role and involvement in neurological disorders. Brain Res Rev 52:201–243. doi:10.1016/j.brainresrev.2006.02.002
169. Chen HC, Xie J, Zhang Z, Su LT, Yue L, Runnels LW (2010) Blockade of TRPM7 channel activity and cell death by inhibitors of 5-lipoxygenase. PLoS One 5(6):e11161. doi:10.1371/journal.pone.0011161
170. Blobaum AL, Marnett LJ (2007) Structural and functional basis of cyclooxygenase inhibition. J Med Chem 50:1425–1441. doi:10.1021/jm0613166
171. Vane JR, Bakhle YS, Botting RM (1998) Cyclooxygenases 1 and 2. Annu Rev Pharmacol Toxicol 38:97–120
172. Frolov RV, Slaughter MM, Singh S (2008) Effects of celecoxib on ionic currents and spontaneous firing in rat retinal neurons. Neuroscience 154:1525–1532. doi:10.1016/j.neuroscience.2008.05.004
173. Ochi T, Goto T (2002) Differential effect of FR122047, a selective cyclo-oxygenase-1 inhibitor, in rat chronic models of arthritis. Br J Pharmacol 135:782–788. doi:10.1038/sj.bjp. 0704511
174. Candelario-Jalil E, de Oliveira AC, Graf S, Bhatia HS, Hull M, Muñoz E, Fiebich BL (2007) Resveratrol potently reduces prostaglandin E_2 production and free radical formation in lipopolysaccharide-activated primary rat microglia. J Neuroinflammation 4:25. doi:10.1186/1742-2094-4-25
175. Bhattacharyya DK, Lecomte M, Dunn J, Morgans DJ, Smith WL (1995) Selective inhibition of prostaglandin endoperoxide synthase-1 (cyclooxygenase-1) by valerylsalicylic acid. Arch Biochem Biophys 317:19–24. doi:10.1006/abbi.1995.1130
176. Piomelli D, Greengard P (1990) Lipoxygenase metabolites of arachidonic acid in neuronal transmembrane signalling. Trends Pharmacol Sci 11:367–373
177. Du L, Zhang Z, Luo X, Chen K, Shen X, Jiang H (2006) Binding investigation of human 5-lipoxygenase with its inhibitors by SPR technology correlating with molecular docking simulation. J Biochem 139:715–723. doi:10.1093/jb/mvj084
178. SnC B, Higginbottom A, Mead RJ, Barber S, Shaw PJ (2009) An *in vitro* screening cascade to identify neuroprotective antioxidants in ALS. Free Radic Biol Med 46(8):1127–1138. doi:10.1016/j.freeradbiomed.2009.01.019
179. Pergola C, Jazzar B, Rossi A, Buehring U, Luderer S, Dehm F, Northoff H, Sautebin L, Werz O (2011) Cinnamyl-3,4-dihydroxy-alpha-cyanocinnamate is a potent inhibitor of 5-lipoxygenase. J Pharmacol Exp Ther 338:205–213. doi:10.1124/jpet.111.180794
180. DeCostanzo AJ, Voloshyna I, Rosen ZB, Feinmark SJ, Siegelbaum SA (2010) 12-Lipoxygenase regulates hippocampal long-term potentiation by modulating L-type Ca^{2+} channels. J Neurosci 30:1822–1831. doi:10.1523/jneurosci.2168-09.2010
181. Vreugdenhil M, Bruehl C, Voskuyl RA, Kang JX, Leaf A, Wadman WJ (1996) Polyunsaturated fatty acids modulate sodium and calcium currents in CA1 neurons. Proc Natl Acad Sci USA 93:12559–12563
182. Uz T, Manev R, Manev H (2001) 5-Lipoxygenase is required for proliferation of immature cerebellar granule neurons *in vitro*. Eur J Pharmacol 418:15–22. doi:10.1016/s0014-2999(01)00924-4
183. Li P-C, Liang J-T, Huang H-T, Lin P-H, Wu S-N (2002) Enhanced activity of Ca^{2+}-activated K^+ channels by 1-[2-hydroxy-3-

propyl-4-[(1H-tetrazol-5-yl)butoxyl]phenyl] ethanone (LY-171883) in neuroendocrine and neuroblastoma cell lines. J Cell Physiol 192:188–199. doi:10.1002/jcp. 10127

184. Wei J-Y, Jin X, Cohen ED, Daw NW, Barnstable CJ (2002) cGMP-induced presynaptic depression and postsynaptic facilitation at glutamatergic synapses in visual cortex. Brain Res 927:42–54. doi:10.1016/s0006-8993(01)03323-6

185. Saywell SA, Babiec WE, Neverova NV, Feldman JL (2010) Protein kinase G-dependent mechanisms modulate hypoglossal motoneuronal excitability and long-term facilitation. J Physiol 588:4431–4439

186. Dostmann WR, Taylor MS, Nickl CK, Brayden JE, Frank R, Tegge WJ (2000) Highly specific, membrane-permeant peptide blockers of cGMP-dependent protein kinase Ialpha inhibit NO-induced cerebral dilation. Proc Natl Acad Sci USA 97:14772–14777. doi:10.1073/pnas.97.26.14772

187. Tokumitsu H, Chijiwa T, Hagiwara M, Mizutani A, Terasawa M, Hidaka H (1990) KN-62, 1-[N, O-bis(5-isoquinolinesulfonyl)-N-methyl-L-tyrosyl]-4-phenylpiperazi ne, a specific inhibitor of Ca2+/calmodulin-dependent protein kinase II. J Biol Chem 265:4315–4320

188. Jiang X, Lautermilch NJ, Watari H, Westenbroek RE, Scheuer T, Catterall WA (2008) Modulation of $Ca_V2.1$ channels by Ca^{2+}/calmodulin-dependent protein kinase II bound to the C-terminal domain. Proc Natl Acad Sci USA 105:341–346. doi:10.1073/pnas.0710213105

189. Hvalby O, Hemmings HC Jr, Paulsen O, Czernik AJ, Nairn AC, Godfraind JM, Jensen V, Raastad M, Storm JF, Andersen P, Greengard P (1994) Specificity of protein kinase inhibitor peptides and induction of long-term potentiation. Proc Natl Acad Sci USA 91:4761–4765

190. Mockett BG, Guevremont D, Wutte M, Hulme SR, Williams JM, Abraham WC (2011) Calcium/calmodulin-dependent protein kinase II mediates group I metabotropic glutamate receptor-dependent protein synthesis and long-term depression in rat hippocampus. J Neurosci 31(20):7380–7391. doi:10.1523/jneurosci.6656-10.2011

191. Strauss O, Steinhausen K, Mergler S, Stumpff F, Wiederholt M (1999) Involvement of protein tyrosine kinase in the $InsP_3$-induced activation of Ca^{2+}-dependent Cl^- currents in cultured cells of the rat retinal pigment epithelium. J Membr Biol 169(3):141–153

192. Higashida H, Brown DA (1986) Two polyphosphatidylinositide metabolites control two K^+ currents in a neuronal cell. Nature 323:333–335

193. Rankovic V, Ehling P, Coulon P, Landgraf P, Kreutz MR, Munsch T, Budde T (2010) Intracellular Ca^{2+} release-dependent inactivation of Ca^{2+} currents in thalamocortical relay neurons. Eur J Neurosci 31:439–449. doi:10.1111/j.1460-9568.2010.07081.x

194. Gomez RS, Guatimosim C, Barbosa J Jr, Massensini AR, Gomez MV, Prado MA (2001) Halothane-induced intracellular calcium release in cholinergic cells. Brain Res 921:106–114

195. Boyd DF, Millar JA, Watkins CS, Mathie A (2000) The role of Ca^{2+} stores in the muscarinic inhibition of the K^+ current $I_{K(SO)}$ in neonatal rat cerebellar granule cells. J Physiol 529:321–331. doi:10.1111/j.1469-7793.2000.00321.x

196. Bird GS, DeHaven WI, Smyth JT, Putney JW Jr (2008) Methods for studying store-operated calcium entry. Methods 46:204–212. doi:10.1016/j.ymeth.2008.09.009

197. Mezna M, Patchick T, Tovey S, Michelangeli F (1996) Inhibition of the cerebellar inositol 1,4,5-trisphosphate-sensitive Ca^{2+} channel by ethanol and other aliphatic alcohols. Biochem J 314:175–179

198. Maruyama T, Kanaji T, Nakade S, Kanno T, Mikoshiba K (1997) 2APB, 2-aminoethoxydiphenyl borate, a membrane-penetrable modulator of Ins(l,4,5)P_3-induced Ca^{2+} release. J Biochem 122:498–505

199. Nishiyama M, Hong K, Mikoshiba K, Poo MM, Kato K (2000) Calcium stores regulate the polarity and input specificity of synaptic modification. Nature 408:584–588

200. Bowden SE, Selyanko AA, Robbins J (1999) The role of ryanodine receptors in the cyclic ADP ribose modulation of the M-like current in rodent m1 muscarinic receptor-transformed NG108-15 cells. J Physiol 519:23–34

201. Sah P, McLachlan EM (1991) Ca^{2+}-activated K^+ currents underlying the afterhyperpolarization in guinea pig vagal neurons: a role for Ca^{2+}-activated Ca^{2+} release. Neuron 7:257–264

202. Kato BM, Rubel EW (1999) Glutamate regulates IP_3-type and CICR stores in the avian cochlear nucleus. J Neurophysiol 81:1587–1596

203. Lingamaneni R, Hemmings HC Jr (2003) Differential interaction of anaesthetics and antiepileptic drugs with neuronal Na^+ channels, Ca^{2+} channels, and $GABA_A$ receptors. Br J Anaesth 90:199–211

204. Talley EM, Bayliss DA (2002) Modulation of TASK-1 (Kcnk3) and TASK-3 (Kcnk9) potassium channels: volatile anesthetics and neurotransmitters share a molecular site of action. J Biol Chem 277:17733–17742

205. Shimizu H, Fukaya M, Yamasaki M, Watanabe M, Manabe T, Kamiya H (2008) Use-dependent amplification of presynaptic Ca^{2+} signaling by axonal ryanodine receptors at the hippocampal mossy fiber synapse. Proc Natl Acad Sci USA 105:11998–12003. doi:10.1073/pnas.0802175105
206. Doutheil J, Paschen W (1999) Effect of 3,4,5-trimethoxybenzoic acid 8-diethylamino-octyl ester (TMB-8) on neuronal calcium homeostasis, protein synthesis, and energy metabolism. Pharmacol Toxicol 85:85–91
207. Schiffmann SN, Desdouits F, Menu R, Greengard P, Vincent JD, Vanderhaeghen JJ, Girault JA (1998) Modulation of the voltage-gated sodium current in rat striatal neurons by DARPP-32, an inhibitor of protein phosphatase. Eur J Neurosci 10:1312–1320
208. Koss DJ, Hindley KP, Riedel G, Platt B (2007) Modulation of hippocampal calcium signalling and plasticity by serine/threonine protein phosphatases. J Neurochem 102:1009–1023. doi:10.1111/j.1471-4159.2007.04579.x
209. Randriamampita C, Tsien RY (1995) Degradation of a calcium influx factor (CIF) can be blocked by phosphatase inhibitors or chelation of Ca^{2+}. J Biol Chem 270:29–32
210. Honkanen RE, Codispoti BA, Tse K, Boynton AL, Honkanan RE (1994) Characterization of natural toxins with inhibitory activity against serine/threonine protein phosphatases. Toxicon 32:339–350
211. Yu SP (1995) Roles of arachidonic acid, lipoxygenases and phosphatases in calcium-dependent modulation of M-current in bullfrog sympathetic neurons. J Physiol 487:797–811
212. Butt SJ, Pitman RM (2005) Indirect phosphorylation-dependent modulation of postsynaptic nicotinic acetylcholine responses by 5-hydroxytryptamine. Eur J Neurosci 21(5):1181–1188. doi:10.1111/j.1460-9568.2005.03947.x
213. Ahn HS, Kim SE, Choi BH, Choi J-S, Kim M-J, Rhie D-J, Yoon SH, Jo Y-H, Kim M-S, Sung K-W, Kwon O-J, Hahn SJ (2007) Calcineurin-independent inhibition of $K_V1.3$ by FK-506 (tacrolimus): a novel pharmacological property. Am J Physiol 292:C1714–C1722. doi:10.1152/ajpcell.00258.2006
214. Misonou H, Mohapatra DP, Park EW, Leung V, Zhen D, Misonou K, Anderson AE, Trimmer JS (2004) Regulation of ion channel localization and phosphorylation by neuronal activity. Nat Neurosci 7:711–718. doi:10.1038/nn1260
215. Enan E, Matsumura F (1992) Specific inhibition of calcineurin by type II synthetic pyrethroid insecticides. Biochem Pharmacol 43:1777–1784
216. Wang SY, Wang GK (2003) Voltage-gated sodium channels as primary targets of diverse lipid-soluble neurotoxins. Cell Signal 15:151–159
217. Fakata KL, Swanson SA, Vorce RL, Stemmer PM (1998) Pyrethroid insecticides as phosphatase inhibitors. Biochem Pharmacol 55:2017–2022
218. Kim JY, Saffen D (2005) Activation of M1 muscarinic acetylcholine receptors stimulates the formation of a multiprotein complex centered on TRPC6 channels. J Biol Chem 280(36):32035–32047. doi:10.1074/jbc.M500429200
219. Hongpaisan J, Winters CA, Andrews SB (2004) Strong calcium entry activates mitochondrial superoxide generation, upregulating kinase signaling in hippocampal neurons. J Neurosci 24:10878–10887
220. Kourrich S, Glasgow SD, Caruana DA, Chapman CA (2008) Postsynaptic signals mediating induction of long-term synaptic depression in the entorhinal cortex. Neural Plast 2008:840374
221. Petrik D, Wang B, Brenner R (2011) Modulation by the BK accessory β4 subunit of phosphorylation-dependent changes in excitability of dentate gyrus granule neurons. Eur J Neurosci 34:695–704. doi:10.1111/j.1460-9568.2011.07799.x

Chapter 6

In Vivo Manipulation of Intracellular Signalling Pathways

V.J. Tallapragada

Abstract

Manipulation of intracellular signalling pathways and ligand protein interactions can lead to new therapeutic strategies in many cardiovascular diseases. One of the most commonly used methods, performed on different disease models of rat, is the microinjection of various agonist and antagonists of ligand and proteins into specific regions of the brain that are implicated in a variety of cardiovascular diseases. This chapter will describe a detailed step-by-step method for manipulation of intracellular signalling pathways in specific regions of the rat brainstem. In this method, the rostral ventrolateral medulla, an area of the brainstem which is essential for sympathetic activity and blood pressure, is studied with regard to the cyclic adenosine 3′-5′-monophosphate pathway. With this method, a more detailed interpretation of the physiological state of the rat in vivo can be achieved.

Key words: Sympathetic nerve activity, Blood pressure, cAMP

1. Introduction

Understanding the diverse intracellular pathways within a neuron will eventually provide a detailed schematic of how these neurons act under normal conditions. Relatively few laboratories in neuroscience have addressed questions like: What are the actions of G-protein-coupled receptors on protein kinases, phospholipids and calcium ions/channels? Or what are the actions of gases such as nitric oxide as intracellular transducers in specific regions of the brain? What effects do over-expression of these protein kinases, or gases, have on homeostasis or how does a specific region of the brain react to such intracellular inconsistencies? In this chapter I shall use the rostral ventrolateral medulla (RVLM) as the target for manipulating intracellular signalling pathways and the subsequent effects on different parameters such as sympathetic nerve activity (SNA), blood pressure (BP) and heart rate (HR).

The RVLM is crucial for the maintenance of sympathetic tone and reflex responses, which are lost following destruction of these

Paul M. Pilowsky et al. (eds.), *Stimulation and Inhibition of Neurons*, Neuromethods, vol. 78,
DOI 10.1007/978-1-62703-233-9_6, © Springer Science+Business Media, LLC 2013

neurons (1–3). Thus, a basal level of activity of these neurons maintains resting level of blood pressure and SNA. The RVLM contains presympathetic neurons that have monosynaptic inputs to the sympathetic preganglionic vasomotor neurons in the IML (4–7), are directly associated with blood pressure homeostasis and are exquisitely barosensitive (1, 2, 6).

Traditionally, the effect on SNA, BP and HR is studied by pharmacological manipulations such as the microinjection of peptides, drugs, hormones or their antagonists into the RVLM (8–11) and by measuring sympathetic output at different vascular beds (such as the splanchnic, lumbar or the renal sympathetic nerves). However, few studies have looked at the effects of intracellular pathways in the RVLM. Only a few laboratories, including ours, have tried to investigate cellular signalling pharmacologically by using various cell permeable agonists or antagonists against intracellular signal transduction proteins (12–15).

Pharmacological studies in vivo are not without their pitfalls and are often constrained by the specificity of cell permeable drugs that most studies use. For example, injection in the lateral, 3rd or 4th ventricles of dibutyryl-cAMP, an analogue of cyclic adenosine 3′-5′-monophosphate (cAMP) evoked an increase in blood pressure (16–18), whereas, a biphasic effect was evoked when it was injected into the NTS, and a small decrease in blood pressure was elicited from the locus coeruleus (19). Similarly, microinjection of H-89 (commonly used as a selective and potent inhibitor of protein kinase A (PKA)) into the RVLM of an adult rat evoked a modest increase in BP and heart rate (20). It should be noted here that by using dibutyl-cAMP, the authors did not perform control experiments to test the influence of butyrate as dibutyryl-cAMP breaks down into butyryl groups prior to the activation of cAMP. Similarly, H-89 not only inhibits PKA but also inhibits other kinases including PKG, PKC, MSK and ROCKII (21, 22).

Recent advances in pharmacology, in particular, in the development of specific cell permeable intracellular drugs, have given us more freedom in our interpretation of cell signalling pathways in vivo. For example, studies of microinjecting drugs such as PD098059 (antagonist of MAP kinase activator, MAP kinase kinase) and wortmannin (a selective irreversible PI3 kinase inhibitor) into the RVLM have shown that MAP kinase pathways are necessary in the tonic regulation of arterial pressure in hypertensive and normotensive animals, whereas PI3 kinase pathways are upregulated only in the hypertensive states (12). Such studies show that by using specific antagonists, one can clearly demonstrate a difference between specific intracellular pathway in normotensive and disease states in vivo.

A similar approach to the Seyedabadi et al. (12) study is described in this article on the effect of cAMP and its downstream effectors in modulating splanchnic sympathetic responses and arterial pressure via the RVLM. There are several pharmacological cell permeable drugs that differentiate between cAMP and its downstream signalling pathways. One agent used to study cAMP pathways is the Sp analogue of cAMP (Sp-cAMP). Sp-cAMP is a phosphodiesterase (PDE) resistant synthetic form of cAMP that has a single sulphur substitution in the exocyclic axial oxygen position. This property helps the Sp-cAMP to act as a potent non-specific cAMP agonist (23, 24) that effectively activates all downstream effectors of cAMP such as PKA, effector protein activated by cAMP (EPAC) and hyperpolarisation-activated cation channels (HCN). Therefore, the net effect of cAMP activation by Sp-cAMP is not just by PKA, EPAC or HCN channels but also the sum of all these pathways. Due to this property of Sp-cAMP, it becomes critical to consider, which cAMP effects are mediated by EPAC, PKA or HCN channels. For example, studies in mitogenic HEK cells show that PKA and EPAC exert opposing effects on the protein kinase B (PKB)/AKT signalling pathway. PKA suppresses PKB phosphorylation activity, whereas EPAC increases PKB phosphorylation (25). In contrast both PKA and EPAC exert synergistic inhibitory effects on the Na^+–H^+ exchanger-3 in the rodent renal proximal tubules (26).

Selective activation of EPAC is conferred by the substitution of the O–Me group for the –OH group present at the 2′ carbon of the ribose moiety of cAMP. It is this 2′-O–Me substitution that impairs the interaction of cAMP with PKA but allows this analogue (8-(4-chlorophenylthio)-2′-*O*-methyladenosine 3′,5′-cyclic monophosphate monosodium hydrate) to act as a specific agonist of EPAC without effecting PKA (27, 28). Thus, this study effectively showed that cAMP microinjection in the RVLM causes sympathoexcitation, hypertension and tachycardia, but microinjection of an EPAC agonist (8-pCPT) only increased splanchnic SNA.

This chapter will deal with all the equipment, reagents and surgical techniques necessary to perform studies dealing with intracellular signalling pathways in the RVLM. It is important to note that this method does not address signal transduction pathways at pre- or postsynaptic sites, but is an example of how signal transduction pathways can be studied using specific agonists and antagonists against particular protein kinases. Furthermore, the effects seen on HR are likely to be mediated by direct effects on sympathetic function as the rats were vagotomised.

2. Materials

1. Urethane (ethyl carbamate; 1.3–1.5 mg/kg, 10% solution in saline, i.p., e.g. Sigma-Aldrich): make fresh on experiment day. Urethane should be stored in an airtight container, in the dark.

Urethane is highly deliquescent and is therefore best purchased in small amounts and stored with desiccant. Urethane also causes chemical burns at concentrations exceeding 10%.

2. 0.9% sodium chloride (NaCl, saline).
3. Heparinised 0.9% saline (~40 units/ml) to be made fresh on the day of experiment. We achieve this by adding three drops of 5,000 U/ml heparin (e.g. DBL Heparin Sodium, Hospira) to 100 ml of 0.9% saline. This prevents clotting in the arterial cannula.
4. 26-gauge needles.
5. Syringes (1, 2, 10, 20 ml).
6. Borosilicate glass microcapillary (A-M Systems, Inc., Sequim, WA 98382).
7. L-glutamate (Sigma-Aldrich, Australia).
8. Atropine sulfate (100 μg/kg; e.g. Pfizer): administered along with urethane. This reduces bronchial secretions and assists ventilation.
9. Surgical tools—e.g. scalpel, curved haemostats, iris scissors, fine forceps (straight and curved), a microvascular clip, small scissors and large-toothed forceps.
10. Silk sutures (e.g. Pearsalls, UK): 5/0 for tracheotomy and 2/0 for ligating nerves and blood vessels.
11. Cyanoacrylate adhesive (e.g. Super Glue, Krazy Glue, etc.).
12. Polyethylene tubing (PE9658; o.d. 0.96 mm; i.d. 0.58 mm) about 10-cm lengths for cannulation of blood vessels.
13. 18-gauge drawing up needle for all cannulas—used to connect the cannula to the 3-way tap (vascular).
14. 3-way taps—for construction of vascular cannula.
15. 14-gauge cannula (e.g. Optiva) for tracheotomy, needle removed and sheath shortened to 1 cm.
16. Neuromuscular blocker—e.g. pancuronium bromide (0.8 mg/kg i.v. initially, then 0.8 mg/kg/h, i.v., e.g. Astra Pharmaceuticals).
17. Silver bipolar electrode—either purchased or self-made.
18. Liquid paraffin oil.
19. Sp-cAMPs—Sp-diastereomer of adenosine 3′,5′-cyclic monophosphorothioate (membrane-permeable activator of cAMP-dependent protein kinase I and II that mimics the effects of cAMP as a second messenger in numerous systems while being resistant to cyclic nucleotide PDE).
20. EPAC—Exchange protein directly activated by cAMP—analogue of natural cAMP, potent and specific membrane-permeant

activator of exchange factors directly activated by cAMP (EPAC or cAMP-GEF)—also known as 8-pCPT-2′-O-Me-cAMP or 8-pCPT.

3. Methods

All experiments described were conducted in accordance with the Australian Code of Practice for the Care and Use of Animals for Scientific Purposes (29). The protocols for the experimental procedures were approved by the Macquarie University Animal Ethics Committee (AEC).

3.1. Induction of Anaesthesia

1. Rats are anaesthetised with 10% (w/v) urethane injected intraperitoneally (i.p.) using a 26-G needle. Rats were checked for withdrawal reflexes with a strong hindpaw pinch and a corneal touch reflex to ensure adequate anaesthetic depth.
2. Temperature was maintained between 36.5 and 37.5°C using a rectal probe connected to a thermoregulated heating blanket.
3. Surgical procedures did not commence until withdrawal reflexes to nociceptive stimuli, e.g. paw and tail pinch were absent.
4. The depth of anaesthesia was monitored by observing reflex responses to nociceptive and tactile stimuli (periodic tail/paw pinch), and the corneal touch reflex. Additional anaesthetic (urethane 0.13 g/kg i.p.) was administered if responses to nociceptive stimuli caused increases or decreases in arterial blood pressure > 10 mmHg.

3.2. Cannulations

1. The femoral artery and right jugular vein are cannulated for administrations of drugs and fluids and also for recording the arterial blood pressure.
2. The femoral artery is located at the inner surface of the upper thighs, cleared of connective tissue, and a polyethylene tube (OD 0.96 mm; ID 0.58 mm) filled with heparinised (~10 IU heparin/ml) 0.9% saline is inserted into the lumen and secured in place with silk sutures.
3. The jugular vein (right external jugular vein) lies outside the sternocleidomastoid muscle in the neck and is cannulated in the same way; however, the polyethylene tube for the veins is filled with saline (0.9% (w/v) sodium chloride) only.

3.3. Tracheostomy

1. Tracheostomy is performed in all of the electrophysiological experiments to permit mechanical ventilation.
2. A 14-G catheter was inserted into the trachea just below the larynx and secured in place with a silk suture. The rats breathed

freely until vagotomised at which point the tracheal catheter was connected to the ventilator.

3. Rats are normally artificially ventilated with oxygen-enriched room air to avoid hypoxia, and inactivate peripheral chemoreceptors.
4. Rats are then paralysed with pancuronium bromide (0.4 mg given as 0.2-ml bolus i.v., then an infusion of 20% pancuronium in 0.5% glucose in 0.9% saline at the rate of 1.5 ml/h).
5. A CO_2 analyzer is connected to the expiratory line close to the rat for monitoring and measurement of end tidal CO_2.

3.4. Electrocardiogram

1. Heart rate is derived from the ventricular depolarisations (taken at the peak between the time of beginning and end of the QRS complex) of the electrocardiogram.
2. A silver electrode is inserted into both front paws, and a ground lead inserted into the exposed muscle on the back of the neck.
3. The signal was amplified and filtered between 10 and 1,000 Hz using CWE bioamplifier and acquired at 2 kHz using CED 1401 plus and Spike2 acquisition and analysis software (v 6.09) (CED, Cambridge, UK).

3.5. Nerve Dissections

1. In all electrophysiological experiments, a dam is created from the cut edges of the skin and muscle.
2. All incisions are cauterised to prevent further blood and heat loss.
3. The phrenic nerve and the left greater splanchnic nerve were dissected, and their neurograms were recorded.
4. The vagus nerve is transected.
5. Once prepared and ready for recording or stimulating each nerve was placed across a silver bipolar electrode and the dam filled with paraffin oil (30).
6. The recording electrodes are connected to a headstage connected to a bioamplifier. Data is recorded with a CED 1401, using Spike2 acquisition and analysis software (v 6.09) (CED, Cambridge, UK).

3.6. Splanchnic Sympathetic Nerve

1. The vascular beds are innervated by the splanchnic nerves and are considered to play an important role in the maintenance of blood pressure (31).
2. A midline incision is made on the dorsal surface of the rat, and the edges of the skin are cauterised.
3. A dam is created between the skin and muscle in the abdominal cavity.

4. The splanchnic nerve is located proximal to the celiac ganglion with the help of an operating microscope.
5. The nerve bundle should be carefully desheathed and cleared of any connective tissue and tied at the distal end with a silk suture very close to the celiac ganglion.
6. The nerve is then transected distal to the tie and the wound packed with saline-soaked cotton wool until the time of recording.

3.7. Ventilation

1. Rats are artificially ventilated with oxygen-enriched room air at a volume between 3.6 and 4.2 ml and at a rate of 68–75 cycles/s.
2. CO_2 was sampled with a CO_2 analyser and ventilation rate is adjusted to maintain expired CO_2 at 3.5–4.5%. This information is then digitised using a Cambridge Electronic Design ADC system and recorded on a computer using Spike2 acquisition and analysis software (v 6.09).
3. After initiation of ventilation and neuromuscular blockade, arterial blood gas analyses are performed to ensure $PaCO_2$, pH and PaO_2, stayed within the normal physiological range.

3.8. Stereotaxic Step-Up Prior to Microinjections

1. The rat is placed in ear bars in a stereotaxic frame and secured tightly so that the head does not move in the frame.
2. The arterial cannula is connected to a blood pressure transducer (e.g. Edwards Lifesciences) via the 3-way tap, which feeds into a bridge amplifier (e.g. Scientific Concepts).
3. At this point care should be taken to avoid any air bubbles in the cannula or the 3-way tap feeding the bridge amplifier. This will help in obtaining an accurate, undamped, blood pressure recording.
4. Once we make sure that the blood pressure trace is not damped, the tracheal cannula is then connected to the ventilator, and the rat is ventilated with room air supplemented with 100% O_2.
5. Neuromuscular blockade is then instituted (*-onium class drugs, e.g. pancuronium 0.4 mg given as a 0.2-ml bolus injection i.v.). Note that excessive doses can affect nicotinic receptors in the autonomic ganglia. This effect varies with different drugs.
6. Paralysis is maintained with a slow i.v. infusion of 10% pancuronium bromide in saline at a rate of 2 ml/h using a syringe pump (e.g. 11 Plus, Harvard Apparatus) connected to the venous cannula.
7. It is imperative at this stage to get a blood gas analysis from a small amount of blood drawn from the arterial cannula.

8. The ventilatory rate and depth are maintained so that the pH is between the range 7.38 and 7.42.
9. At this stage the retroperitoneal cavity is held open with silk ties to the surrounding skin edges.
10. The dissected greater splanchnic nerve is placed across a silver bipolar electrode for recording, which is connected to a head-stage (e.g. Super Z, CWE) and then a bioamplifier (e.g. BMA-400 AC/DC, CWE; sample rate 2 kHz, 10,000–20,000 gain, 0.1–3 Hz filtering).
11. The retroperitoneal cavity is filled with an electrically isolated material such as liquid paraffin oil, which should be at body temperature, i.e. 37°C.
12. At this point a hard setting Silgel (e.g. 604, 2 part A: part B, Wacker Chemie, Germany) can also be used to fill the cavity instead of paraffin oil.
13. In such in vivo electrophysiological preparations, the nerve activity is generally reduced by surgery, handling of the nerve and loss of body temperature. It is therefore necessary to allow the nerve recording to stabilise for no less than 60 min.
14. In our laboratory, data is acquired using a Cambridge Electronic Design ADC system (model 1401; Cambridge Electronic Design, Cambridge, UK) and Spike2 acquisition and analysis software.

3.9. Brainstem Microinjections

3.9.1. Occipital Craniotomy

1. A skin incision is made from the top of the skull to the vertebral level of T2, and the muscles overlying the occipital bone were removed to reveal the atlanto–occipital junction.
2. All the skin incisions are cauterised to prevent any further bleeding.
3. The occipital bone was removed with a pair of fine-tipped rongeurs, revealing the cerebellum and the brainstem.
4. The dura is carefully incised and reflected laterally, and the exposed brain tissue is covered with saline-soaked gauze until it was time for microinjection.

3.9.2. Micropipettes

1. Borosilicate glass microcapillary tubes were pulled using a laser puller machine (P2000, Sutter Instrument Company, USA) to make micropipettes.
2. Tips are broken to create an opening, and the pipettes were filled with the desired drug/vehicle using capillary action, by placing them tip-up in the solution for 5–10 min.
3. Micropipettes are attached to a piece of tubing prior to use. A 20-ml syringe was used to pressure-inject the solutions.

3.9.3. Locating Brainstem Sites via Drug Administration

1. The RVLM was typically found in an area 1.8–2.2 mm rostral and 1.6–2.0 mm lateral to calamus scriptorius and between 3.0 and 3.8 mm ventral to the brainstem surface. A site was considered to be within the RVLM if a 50-nl microinjection of 100 mM glutamate caused a rise in blood pressure > 30 mmHg as described in previous studies (11, 32) (see Gaede and Pilowsky, Chap. 3).
2. Care should be taken against repeated microinjection of glutamate as this might cause cytotoxicity leading to neuronal death.
3. Once the RVLM sites are found, the recording parameters such as the SNA, BP and HR should be allowed to return back to preinjection levels before further microinjections. This can take up to 15 min, after which control microinjections should be performed.
4. The control microinjections should comprise of the "vehicle", i.e. whatever the drug of interest is dissolved "in" (PBS, saline, DMSO, etc.). This is to ensure that the vehicle by itself does not have any effect, or in the case that it does (as might happen in case of DMSO), the effect of drug is greater than that of the vehicle.

Once the response from the vehicle control returns to baseline, the drug of interest can now be microinjected. At this point, the drug of interest is microinjected in the RVLM, and for the purpose of this chapter, it is the cAMP analogue Sp-cAMP. Microinjection of Sp-cAMP leads to profound sympathoexcitation, hypertension and tachycardia as shown in the representative trace (Fig. 1). The dose of Sp-cAMP (5 nmol in 50 nL) used here is similar to those used in other brain regions (33). Once the drug of interest is microinjected, various responses of the RVLM can be tested. Some of the examples are the somatosympathetic responses and the baroreflexes as shown in the representative trace at regular intervals after the drug microinjection. This allows the researcher to see if endogenous application of the intracellular protein analogue changes tonic and/or reflex activity within the RVLM.

It is now known that cAMP activates a defined number of downstream targets including the classic PKA pathway (34), the EPAC (35) and the HCN channels (36). At this point the entire cAMP pathway can be broken down into individual effects mediated by each downstream effector proteins. Using similar methodology as the Sp-cAMP microinjection, the representative trace in Fig. 2 shows the net effect of EPAC microinjection into the RVLM. Here, it can be clearly seen that the effect of EPAC activation causes sympathoexcitation but not changes in blood pressure. The assumption can be made at this point that EPAC is not necessary for maintaining blood pressure. Another way to test that

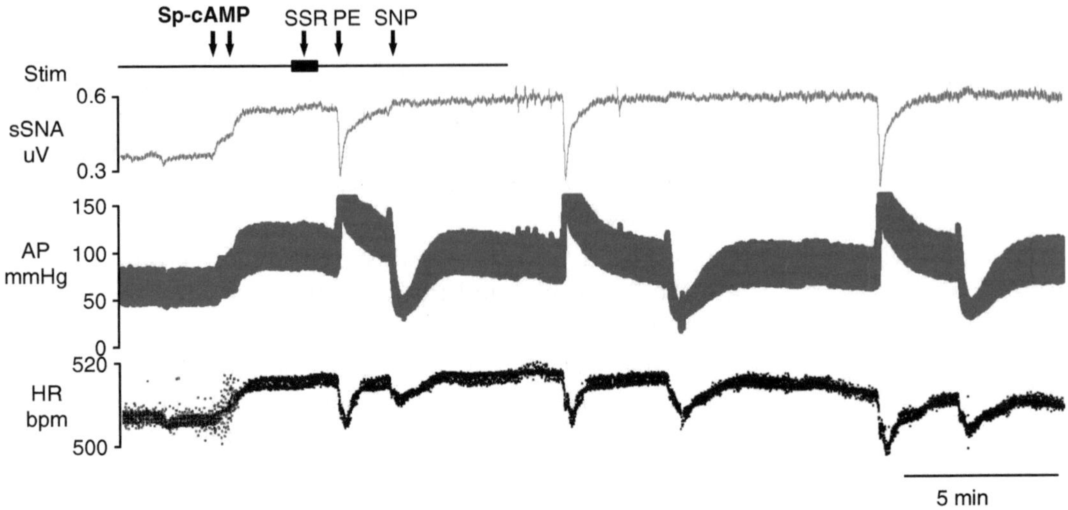

Fig. 1. Bilateral microinjection of Sp-cAMP (cAMP activator) in the RVLM. Representative trace of Sp-cAMPs microinjection in the RVLM showing sciatic nerve stimulus (*top panel* (Stim)), splanchnic sympathetic nerve activity (*second panel from top* (sSNA)), arterial blood pressure (*third panel from the top* (BP)), followed by heart rate (HR).

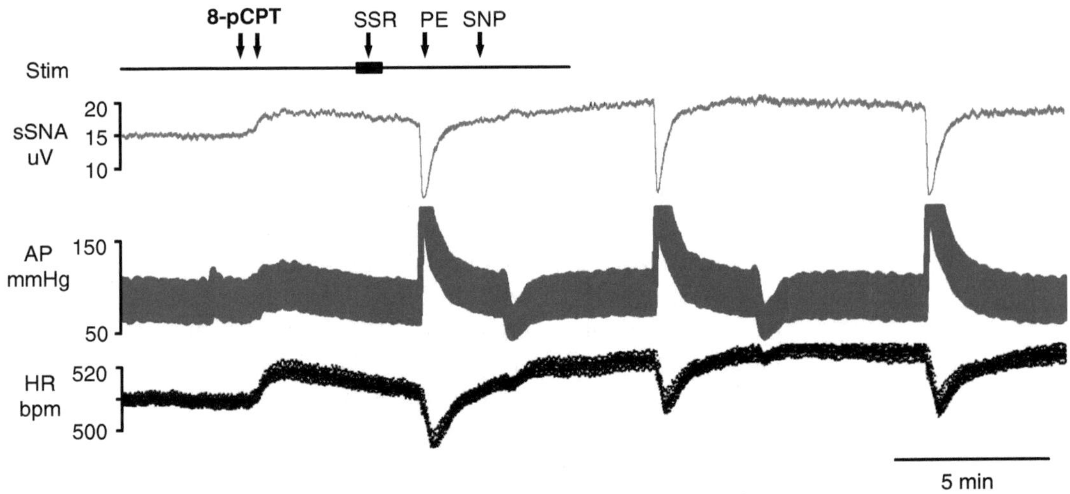

Fig. 2. Bilateral microinjection of 8-pCPT (EPAC activator) in the RVLM. Representative trace of 8-pCPT microinjection in the RVLM showing sciatic nerve stimulus (*top panel* (Stim)), splanchnic sympathetic nerve activity (*second panel from top* (sSNA)), arterial blood pressure (*third panel from the top* (BP)), followed by heart rate (HR). *MAP* mean arterial pressure, *bpm* beats per minute.

EPAC is not actively involved in maintaining basal blood pressure is by blocking its activity. This was achieved by microinjecting brefeldin A used in studies to block the effect of EPAC (37). Microinjecting BFA into the RVLM evoked no changes in any of the parameters measured (Fig. 3). Studies showing no responses when injected with antagonists often generate other questions, such as, "has the antagonist worked or not?" This can be addressed

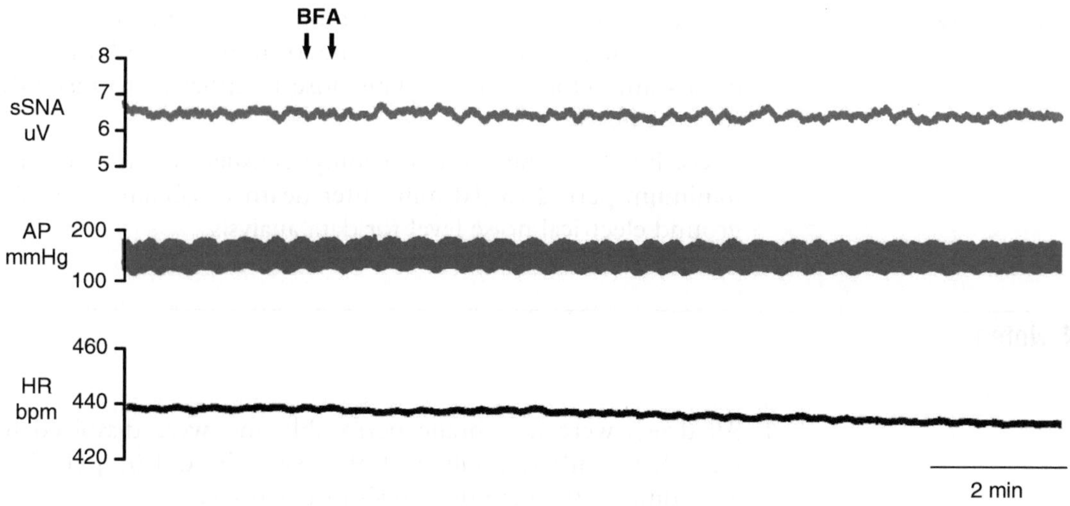

Fig. 3. Bilateral microinjection of BFA (inhibitor of EPAC) in the RVLM. Representative trace of BFA microinjection in the RVLM showing splanchnic sympathetic nerve activity (*first panel from top* (sSNA)), arterial blood pressure (*second panel from the top* (BP)), followed by heart rate (HR). *MAP* mean arterial pressure, *bpm* beats per minute.

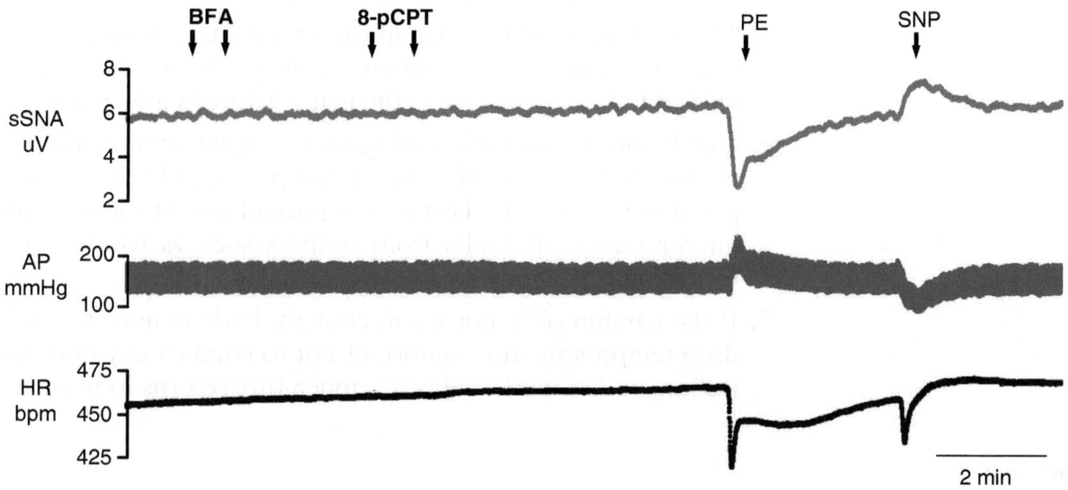

Fig. 4. Bilateral microinjection of BFA prior to 8-pCPT in the RVLM. Representative trace of BFA microinjection followed by 8-pCPT in the RVLM, splanchnic sympathetic nerve activity (*first panel from top* (sSNA)), arterial blood pressure (*second panel from the top* (BP)), followed by heart rate (HR). *MAP* mean arterial pressure, *bpm* beats per minute.

by microinjecting the antagonist prior to the agonist as shown in the representative trace (Fig. 4).

By performing a series of experiments as described above, one can break down a complete signal transduction pathway into individual components and study the role of individual intracellular signalling pathways. All the experiments are recorded for a minimum of 60 min to determine whether they are short or long acting in their effects on the cardiovascular system.

3.10. Euthanasia

1. At the conclusion of electrophysiological experiments, rats can be euthanised with a bolus injection of 0.5 ml of 3 M potassium chloride (KCl). This dose is sufficient to instantly stop the heart.
2. Recording from the nerve recordings is usually continued for a minimum period (5–10 min) after death to obtain the background electrical noise level for data analysis.

4. Notes

1. All drugs were membrane permeable and were dissolved in phosphate buffered saline (0.9% NaCl in dH_20, pH 7.4) according to the manufacturer's instructions.
2. Various cell permeable drugs are found to be in inactive forms in their native states and need to split from one of their groups to be active. Such drugs need to be avoided as they can lead to the manifestation of various cardiovascular changes, which might not be true for the targeted intracellular protein.
3. This method of in vivo manipulations of intracellular proteins requires higher concentrations of drugs, than what is prescribed for in vitro work, in the manufacturer's information.
4. cAMP and its synthetic analogues are highly membrane permeable and can modulate signalling pathways both pre- and postsynaptically (38). This is an important consideration in the interpretation of results from outputs such as whole nerve recordings.
5. If the paraffin oil is not warm enough, body temperature will drop temporarily. It is important not to conduct any microinjection studies until the body temperature returns to normal.

References

1. Dampney RA (1994) Functional organization of central pathways regulating the cardiovascular system. Physiol Rev 74:323–364
2. Guyenet PG (2006) The sympathetic control of blood pressure. Nat Rev Neurosci 7:335–346
3. Sved AF, Ito S, Sved JC (2003) Brainstem mechanisms of hypertension: role of the rostral ventrolateral medulla. Curr Hypertens Rep 5:262–268
4. Oshima N, Kumagai H, Onimaru H, Kawai A, Pilowsky PM, Iigaya K, Takimoto C, Hayashi K, Saruta T, Itoh H (2008) Monosynaptic excitatory connection from the rostral ventrolateral medulla to sympathetic preganglionic neurons revealed by simultaneous recordings. Hypertens Res 31:1445–1454
5. Oshima N, McMullan S, Goodchild AK, Pilowsky PM (2006) A monosynaptic connection between baroinhibited neurons in the RVLM and IML in Sprague-Dawley rats. Brain Res 1089:153–161
6. Lipski J, Kanjhan R, Kruszewska B, Smith M (1995) Barosensitive neurons in the rostral ventrolateral medulla of the rat in vivo: morphological properties and relationship to C1 adrenergic neurons. Neuroscience 69:601–618
7. Brown DL, Guyenet PG (1985) Electrophysiological study of cardiovascular

neurons in the rostral ventrolateral medulla in rats. Circ Res 56:359–369
8. Farnham MM, Li Q, Goodchild AK, Pilowsky PM (2008) PACAP is expressed in sympathoexcitatory bulbospinal C1 neurons of the brain stem and increases sympathetic nerve activity in vivo. Am J Physiol Regul Integr Comp Physiol 294:R1304–1311
9. Krieger MH, Moreira ED, Oliveira EM, Oliveira VL, Krieger EM, Krieger JE (2006) Dissociation of blood pressure and sympathetic activation of renin release in sinoaortic-denervated rats. Clin Exp Pharmacol Physiol 33:471–476
10. Miyawaki T, Goodchild AK, Pilowsky PM (2002) Activation of mu-opioid receptors in rat ventrolateral medulla selectively blocks baroreceptor reflexes while activation of delta opioid receptors blocks somato-sympathetic reflexes. Neuroscience 109:133–144
11. Gaede AH, Pilowsky PM (2010) Catestatin in rat RVLM is sympathoexcitatory, increases barosensitivity, and attenuates chemosensitivity and the somatosympathetic reflex. Am J Physiol Regul Integr Comp Physiol 299: R1538–R1545
12. Seyedabadi M, Goodchild AK, Pilowsky PM (2001) Differential role of kinases in brain stem of hypertensive and normotensive rats. Hypertension 38:1087–1092
13. Chan SH, Tai MH, Li CY, Chan JY (2006) Reduction in molecular synthesis or enzyme activity of superoxide dismutases and catalase contributes to oxidative stress and neurogenic hypertension in spontaneously hypertensive rats. Free Radic Biol Med 40:2028–2039
14. Miyawaki T, Suzuki S, Minson J, Arnolda L, Chalmers J, Llewellyn-Smith I, Pilowsky P (1997) Role of AMPA/kainate receptors in transmission of the sympathetic baroreflex in rat CVLM. Am J Physiol 272:R800–R812
15. Miyawaki T, Goodchild AK, Pilowsky PM (2003) Maintenance of sympathetic tone by a nickel chloride-sensitive mechanism in the rostral ventrolateral medulla of the adult rat. Neuroscience 116:455–464
16. White HS, Driver PS, Isom GE (1979) Studies on the central pressor activity of dibutyryl cyclic AMP. Eur J Pharmacol 57:107–113
17. Clipsham PJ, Hamilton TC, Hunt AA, Poyser RH (1980) Cyclic nucleotides and central cardiovascular control in the conscious cat. Eur J Pharmacol 65:193–200
18. Walland A (1975) cAMP as a second messenger in central blood pressure control. Naunyn Schmiedebergs Arch Pharmacol 290:419–423
19. Johnson JD, White HS, Isom GE (1983) Local application of cyclic AMP in the rat brain: characterization of the cardiovascular response. Eur J Pharmacol 91:343–351
20. Xu Y, Krukoff TL (2006) Adrenomedullin in the rostral ventrolateral medulla inhibits baroreflex control of heart rate: a role for protein kinase A. Br J Pharmacol 148: 70–77
21. Chijiwa T, Mishima A, Hagiwara M, Sano M, Hayashi K, Inoue T, Naito K, Toshioka T, Hidaka H (1990) Inhibition of forskolin-induced neurite outgrowth and protein phosphorylation by a newly synthesized selective inhibitor of cyclic AMP-dependent protein kinase, N-(2-(p-bromocinnamylamino)ethyl)-5-isoquinolinesulfonamide (H-89), of PC12D pheochromocytoma cells. J Biol Chem 265: 5267–5272
22. Johannes FJ, Prestle J, Dieterich S, Oberhagemann P, Link G, Pfizenmaier K (1995) Characterization of activators and inhibitors of protein kinase C mu. Eur J Biochem 227: 303–307
23. Van Haastert PJ, Van Driel R, Jastorff B, Baraniak J, Stec WJ, De Wit RJ (1984) Competitive cAMP antagonists for cAMP-receptor proteins. J Biol Chem 259:10020–10024
24. Yokozaki H, Tortora G, Pepe S, Maronde E, Genieser HG, Jastorff B, Cho-Chung YS (1992) Unhydrolyzable analogues of adenosine 3′:5′-monophosphate demonstrating growth inhibition and differentiation in human cancer cells. Cancer Res 52:2504–2508
25. Mei FC, Qiao J, Tsygankova OM, Meinkoth JL, Quilliam LA, Cheng X (2002) Differential signaling of cyclic AMP: opposing effects of exchange protein directly activated by cyclic AMP and cAMP-dependent protein kinase on protein kinase B activation. J Biol Chem 277:11497–11504
26. Honegger KJ, Capuano P, Winter C, Bacic D, Stange G, Wagner CA, Biber J, Murer H, Hernando N (2006) Regulation of sodium-proton exchanger isoform 3 (NHE3) by PKA and exchange protein directly activated by cAMP (EPAC). Proc Natl Acad Sci USA 103:803–808
27. Vliem MJ, Ponsioen B, Schwede F, Pannekoek WJ, Riedl J, Kooistra MR, Jalink K, Genieser HG, Bos JL, Rehmann H (2008) 8-pCPT-2′-O-Me-cAMP-AM: an improved Epac-selective cAMP analogue. Chembiochem 9: 2052–2054
28. Christensen AE, Selheim F, de Rooij J, Dremier S, Schwede F, Dao KK, Martinez A, Maenhaut C, Bos JL, Genieser HG, Doskeland SO (2003) cAMP analog mapping of Epac1 and cAMP kinase. Discriminating analogs demonstrate that Epac and cAMP kinase act synergistically

to promote PC-12 cell neurite extension. J Biol Chem 278:35394–35402

29. NHMRC AG (2004) Australian code of practise for the care and use of animals for scientific purposes (7th edn). This is a small booklet on the web put by the Australian Gov. The web URL is: http://www.nhrmc.gov.au/publications/eahome.htm
30. Morrison SF, Reis DJ (1989) Reticulospinal vasomotor neurons in the RVL mediate the somatosympathetic reflex. Am J Physiol 256: R1084–R1097
31. Gootman PM, Cohen MI (1970) Efferent splanchnic activity and systemic arterial pressure. Am J Physiol 219:897–903
32. Makeham JM, Goodchild AK, Pilowsky PM (2001) NK1 receptor and the ventral medulla of the rat: bulbospinal and catecholaminergic neurons. Neuroreport 12:3663–3667
33. Paine TA, Neve RL, Carlezon WA Jr (2009) Attention deficits and hyperactivity following inhibition of cAMP-dependent protein kinase within the medial prefrontal cortex of rats. Neuropsychopharmacology 34:2143–2155
34. Holz GG, Kang G, Harbeck M, Roe MW, Chepurny OG (2006) Cell physiology of cAMP sensor Epac. J Physiol 577:5–15
35. Bos JL (2003) Epac: a new cAMP target and new avenues in cAMP research. Nat Rev Mol Cell Biol 4:733–738
36. Kaupp UB, Seifert R (2002) Cyclic nucleotide-gated ion channels. Physiol Rev 82:769–824
37. Ster J, de Bock F, Bertaso F, Abitbol K, Daniel H, Bockaert J, Fagni L (2009) Epac mediates PACAP-dependent long-term depression in the hippocampus. J Physiol 587:101–113
38. Zhong N, Zucker RS (2005) cAMP acts on exchange protein activated by cAMP/cAMP-regulated guanine nucleotide exchange protein to regulate transmitter release at the crayfish neuromuscular junction. J Neurosci 25: 208–214

Part III

The Single Neurons

Chapter 7

Identification of Spinally Projecting Neurons in the Rostral Ventrolateral Medulla In Vivo

Simon McMullan

Abstract

Putative sympathetic premotor neurons in the rostral ventrolateral medulla are critically important in the regulation of sympathetic vasomotor tone and are responsible for mediating many cardiovascular reflexes. In the rat, these neurons lie within a small area of the brainstem immediately caudal to the facial nucleus and can be distinguished from neighbouring cells by their axonal projections to the thoracic spinal cord, where they are thought to form synapses with sympathetic preganglionic neurons.

This protocol describes the steps required for identification of sympathetic premotor neurons in acute experiments *in vivo*. It provides a detailed description of the methodology we use routinely to electrophysiologically map the topography of the facial nucleus and an account of the steps needed to conduct the antidromic collision test.

Key words: Concentric, Bipolar, Electrical stimulation, Antidromic, Field potential, Sympathetic premotor, RVLM

1. Introduction

The region of the medulla immediately caudal to the facial nucleus was identified as crucial for the maintenance of blood pressure over a century ago (1–3). However, the cellular substrate responsible for maintaining baseline sympathetic nerve activity has only emerged in the last 25 years. In the 1970s and early 1980s, a number of laboratories established that the ongoing activity of neurons within the rostral ventrolateral medulla (RVLM) is critical for normal autonomic function (4, 5) and proposed that neurons of the C1 cell group are responsible (6, 7).

In a defining series of experiments, Brown and Guyenet (8, 9) identified neurons in the RVLM that had functional properties consistent with vasomotor sympathetic premotor neurons. Using extracellular recording techniques in the anesthetised rat, they discovered

Paul M. Pilowsky et al. (eds.), *Stimulation and Inhibition of Neurons*, Neuromethods, vol. 78,
DOI 10.1007/978-1-62703-233-9_7, © Springer Science+Business Media, LLC 2013

a population of neurons that were spontaneously active, sensitive to stimuli known to alter sympathetic nerve activity, and projected to autonomic regions of the thoracic spinal cord.

Their methodology, which uses a combination of anatomical and functional criteria to identify putative sympathetic premotor neurons, has enabled researchers in many laboratories to examine the behaviour of these neurons in a variety of species and contexts, spawning a vast body of research. Their approach is extremely powerful, but can be difficult for other investigators to emulate.

In this chapter, I will describe in detail two key elements of the methodology described by Brown and Guyenet (8, 9) and used elsewhere, including our own laboratory. First, I will describe the protocol for recording field potentials in the facial nucleus in response to antidromic activation of the facial nerve. The caudal boundary of the facial nucleus is a key anatomical landmark, and electrophysiological mapping of the facial nucleus is a simple way to guide recordings to the area where most sympathetic premotor neurons lie. Concentrating recordings to within 0.5 mm of the caudal pole of the facial nucleus considerably enhances the success rate of experiments. The landmark provided by the facial nucleus can also prove useful for identifying other nearby cell groups such as the retrotrapezoid nucleus (10) and ventral respiratory groups (11, 12). Mapping of the facial nucleus is a crucial step because the chance of success using unguided stereotaxis is extremely low, even for highly experienced investigators. Using standard 'skull-flat' stereotaxis, RVLM neurons are located under ~8 mm of tissue, occupying an area of around 0.4 mm × 0.5 mm × 0.5 mm, and lie intermingled with other cells.

Secondly, I will detail the protocol we use for the identification of spinally projecting neurons recorded in the extracellular configuration using the antidromic collision test. Given the anatomical overlap of spinally projecting neurons with cells that project to rostral targets and the similarity in their functional properties (13), a formal protocol for the identification of bulbospinal neurons is required.

Although most studies of RVLM sympathetic premotor neurons concentrate on barosensitive, and therefore presumably vasomotor neurons, many RVLM sympathetic premotor neurons are non-barosensitive. These cells presumably subserve sympathetic outflows that are independent of blood pressure, such as changes in blood distribution in response to metabolic stimuli or the release of adrenaline (14–16). The reader is urged to choose the functional selection criteria most relevant to their research goals.

A brief consideration of the conceptual framework that underlies each technique, and its advantages and limitations, follows.

1.1. Field Potentials

Central concept: Low resistance electrodes, which are not sensitive enough to discriminate action potentials, can detect the synchronised

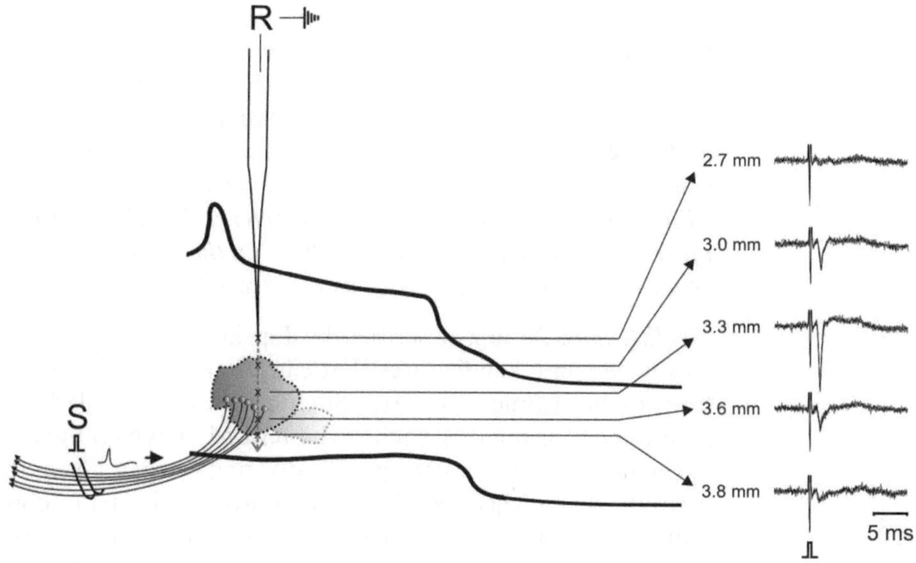

Fig. 1. Electrophysiological mapping of the facial nucleus (*grey area*) by recording field potentials evoked by antidromic stimulation (*S*) of the facial nerve. Recordings shown in the *right hand panels* are made via a low resistance electrode (*R*) positioned at different depths (*x*) in the brainstem. Field potentials are clearly visible as negative deflections occurring ~2 ms after the stimulus artefact at 3, 3.3 and 3.6 mm deep.

activity of pools of neurons. In the current example, motor neurons in the facial nucleus are activated by antidromic electrical activation of their axons, which lie in the facial nerve. In other applications, field potentials can also be evoked by activation of orthodromic excitatory pathways. The field potential is measured using a recording electrode positioned in the target nucleus. The magnitude of the potential corresponds with the proximity of the recording electrode to the centre of the nucleus. The anatomical boundaries of the nucleus are mapped by plotting the stereotaxic coordinates at which the field potential is detectable (see Fig. 1).

Applications: This approach is used to map the locations of clusters of brainstem neurons that receive excitatory input from cranial sensory nerves (17), to investigate the actions of sensory afferents that project to the spinal cord (18) and to guide electrophysiological recordings in and around the facial nucleus and nucleus ambiguus in the brainstem (9, 11).

Limitations: The sensitivity of the electrodes used to record field potentials is low, so best responses are seen in compact rather than diffuse nuclei.

1.2. Antidromic Collision Test

Central concept: In normal conditions, action potentials are generated in the cell body and are propagated along the axon followed by a wave of refractory voltage-gated sodium channels. The 'refractory period' of the membrane ultimately determines its maximum firing frequency.

Antidromic action potentials, generated by electrical stimulation of the axon, travel back towards the cell body in a retrograde direction. Should an antidromic action potential collide with an orthodromic potential, the refractory portion of axon that trails one action potential will block the propagation of the other, cancelling both. When a spontaneous orthodromic spike, recorded in the cell body, is used to trigger the stimulation of an antidromic action potential, such that the antidromic and orthodromic potentials 'collide' in the axon, the technique can be used to definitively prove the axonal trajectory of a cell.

Applications: As first described by Lipski (19), antidromic potentials have some key features that can be used to distinguish them from orthodromic responses. Such a tool is particularly useful in heterogeneous regions, such as the RVLM, in which the behaviour of the neuron is not a strong indicator of its target. Given that only neurons that project to the spinal cord can be truly considered sympathetic premotor neurons, Lipski's antidromic collision test (19) is essential for the unambiguous identification of these cells.

Three key electrophysiological features are exploited by the antidromic collision test. Firstly, the reproducibility of antidromic response latencies are high compared to orthodromic responses. Whereas the latencies of responses evoked by monosynaptic excitatory inputs may vary by ~1 ms, the standard deviation of antidromic responses is generally in the region of 0.1 ms (19), although this can become larger in some circumstances (20, 21).

Secondly, the fidelity of antidromic responses to high frequency stimulation is much greater than that observed following stimulation of presynaptic pathways. Whereas even the most faithful excitatory inputs fail to reliably evoke responses at stimulus frequencies exceeding 300 Hz (19), some neurons are capable of following trains of antidromic stimuli at up to 900 Hz (22).

Their final feature is also their most useful; as first described by Lipski (19), antidromic potentials annihilate and are annihilated by somatofugal orthodromic action potentials (see Fig. 2). This provides a tool that can unambiguously demonstrate the antidromic nature of an evoked response; if an antidromic response can be timed so that it meets with an orthodromic spike mid-axon, the resultant collision is irrefutable evidence of an axonal projection. The antidromic collision test is the gold standard for proof of axonal trajectory; Lipski's original description has, at the time of writing, been cited over 200 times.

Limitations: Although the presence of a constant latency spike that collides with spontaneous action potentials is considered definitive proof that the axon passes within the vicinity of the stimulating electrode, its *absence* is not considered evidence that the neuron does not project to the stimulating electrode: there are a number of confounding factors that could cause this effect. For example, if the stimulating electrode also activates afferent pathways that drive

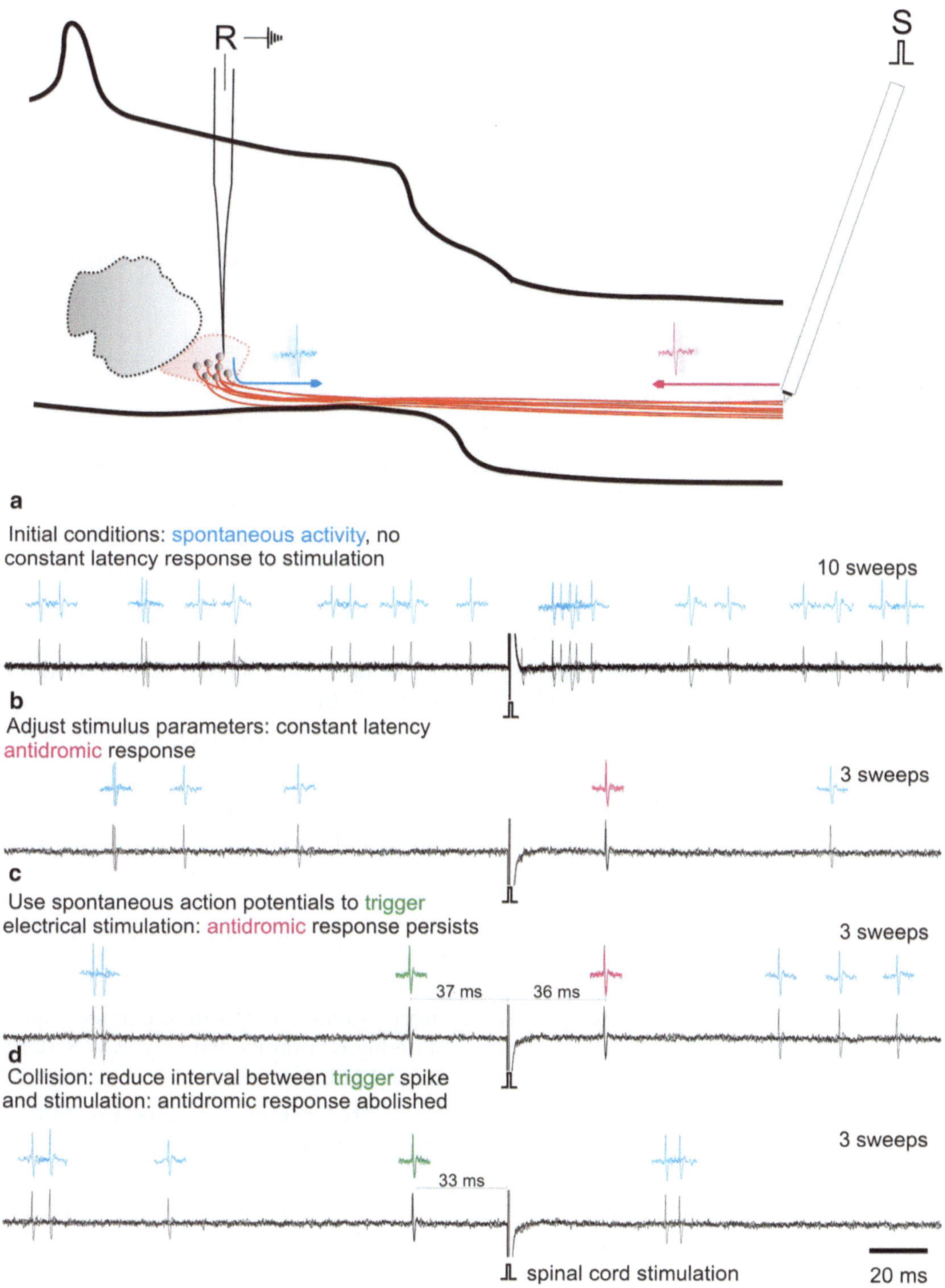

Fig. 2. The antidromic collision test. Extracellular somatic action potentials recorded in the rostral ventrolateral medulla (*R*) in response to stimulation of the IML column of the spinal cord (*S*) fulfil the criteria for consideration as antidromic. Raw data shown in continuous black trace, CED Spike2 'Wavemarks' shown above (**a**) Initial attempts to evoke an antidromic response are unsuccessful; spinal cord stimulation evokes short-latency orthodromic responses 10–30 ms after stimulation. (**b**) Adjusting the stimulus parameters (invert stimulus polarity, reduce intensity) causes reliable activation of antidromic responses (*red spikes*) at a latency of 36 ms. (**c**) Stimulator is triggered by spontaneous spikes (*green spikes*), initially set to occur 37 ms after the detection of a spike. Antidromic responses are evoked at the same latency as (**b**). (**d**) The interval between spike detection and triggering of the spinal stimulator is reduced; when stimulation occurs 33 ms after the spontaneous spike the response is abolished; the cell has fulfilled the criteria for consideration as spinally projecting.

short-latency orthodromic spikes in the cell, the orthodromic response will collide with and obscure any antidromic spikes. In our recordings of sympathetic premotor neurons in the RVLM, this is quite often the case (see Fig. 2a). Similarly, in cases where the spontaneous discharge rate of the cell is very high, the probability of a spontaneous spike colliding with an antidromic spike is also high. Strategies for avoiding these pitfalls are presented in the Notes (Sect. 4).

2. Materials

2.1. Laboratory Equipment

1. Operating microscope with light source, adjustable microscope angle and a long working distance (>20 cm) are extremely advantageous.
2. Steel-topped vibration-damping table for recordings (e.g. Newport).
3. Microelectrode puller (e.g. Sutter P2000).
4. Homeothermic heating blanket controlled by rectal thermometer (e.g. Harvard cat# 507220F).
5. Rodent ventilator (e.g. Ugo Basile) and CO_2 analyzer (CWE Capstar 100).
6. Cautery unit (e.g. RB Medical with JA351A tips).
7. Blood pressure transducer and bridge amplifier (e.g. Scientific Concepts QA1).
8. Syringe pump.
9. Stereotaxic frame (e.g. David Kopf—we use a custom built frame).
10. Isolated pulse stimulator(s) (a model with an integrated adjustable delay function is highly recommended, e.g. AM Systems 2100) with a maximum output of 1 mA.
11. Bioamplifiers for single unit recordings (e.g. Axoclamp 900A). We also use a mains noise eliminator (Humbug, Quest Scientific).
12. Analogue to digital converter and PC for data acquisition and stimulus control (e.g. CED power1401 Mark II running Spike2 version 7).
13. Micromanipulators
    (a) Single unit recording; three-dimensional stage with at least 0.05-mm resolution, 10-mm travel in the *x* and *y* dimensions (e.g. Physik Instrumente M-126.M0) and a motorised stepper with 1-μm resolution, 30-mm travel in the *z*-dimension (e.g. Physik Instrumente N381.3A).

(b) Stereotaxic CNS stimulation; three-dimensional manipulator with 0.1-mm resolution in all dimensions (e.g. Narishige MM-3).

14. Oscilloscope.
15. Audio amplifier (e.g. PC loudspeakers).

2.2. Tools and Consumables

1. Fur clippers.
2. Surgical instruments esp. fine forceps, iridectomy scissors and fine bone rongeurs.
3. Vascular (polyethylene 0.96 mm O.D., 0.5 mm I.D., PE 50) and endotracheal (14-gauge luer cannula) cannulae.
4. Stimulating electrodes: Bipolar concentric electrodes (e.g. Rhodes NE-100) for spinal cord stimulation or needle electrodes (home-made) for stimulation of the facial nerve.
5. Recording electrodes: Borosilicate glass microelectrodes (1 mm O.D., 0.5 mm I.D. with a ~1 μm tip diameter) containing 3 M NaCl for extracellular recordings (6–12 MΩ resistance); the tip can be broken down to give a resistance of <1 MΩ for field recordings.
6. ~ 5 ml Wacker Silgel 612.
7. Lab coat, gloves and mask.
8. Cotton buds, gauze and swabs.

2.3. Drug Solutions

1. Anaesthetic (e.g. urethane 1–1.3 g/kg i.p., dissolved to 10% in 0.9% NaCl)
2. 20 ml 4% glucose solution in a syringe
3. 3 M NaCl

3. Methods

All animal experiments must be approved by the appropriate local animal research ethics committee prior to the commencement of experimental work. As with all *in vivo* electrophysiology experiments, optimal physiological condition and mechanically stability of the preparation are paramount.

3.1. General Preparation

Appropriate anaesthesia should be ensured prior to surgery; we routinely use urethane (1–1.3 g/kg i.p.) in acute experiments. The anaesthetic agent should be selected according to the species of animal used and the requirements of the experiment.

Once the animal is anaesthetised, the fur from the top of the skull, back, sacral region and cheek should be clipped. Move the rat onto a homeothermic blanket to prevent heat loss. Cannulation of a vein and artery, permitting intravenous access and direct measurement of blood pressure respectively, is essential. In the rat, the jugular or femoral veins and carotid or femoral arteries are commonly used. The use of neuromuscular blockade (pancuronium bromide, 0.4 mg i.v. initially, supplemented with 0.2 mg hourly) is recommended to increase the stability of the preparation. In this case, cannulation of the trachea is required to permit artificial ventilation. In all experiments, but particularly in those in which neuromuscular blockade is used, careful and regular assessment of anaesthetic depth is imperative. Prior to neuromuscular blockade, limb withdrawal in response to noxious pinch (e.g. firm pinching of the paw or tail with forceps) indicates insufficient anaesthesia. Following administration of neuromuscular blockade, changes in autonomic parameters (blood pressure, heart rate, central respiratory drive) in response to the same stimulus should be monitored instead of motor reflexes. Changes in blood pressure >10 mmHg in response to a pinch indicate a requirement for a supplemental anaesthesia. We maintain hydration by constant infusion of 5% glucose solution (4 ml/kg/h) using a syringe pump attached to the intravenous line. Intravenous infusion of neuromuscular blockers is explicitly prohibited in some countries and is not recommended. Neuromuscular blockers should not be administered until required for mechanical stability (i.e. after surgical preparation) and should be administered after initiation of mechanical ventilation.

3.2. Preparation for Recording and Stimulation

3.2.1. Positioning of the Animal in the Stereotaxic Frame

Following cannulation of vessels and trachea, move the animal into the stereotaxic frame. We use the 'nose-down' position, in which the head fixed at about 45°. The precise angle is not crucial, but consistency between experiments greatly reduces the variability of stereotaxic surgery. Some investigators use a conventional skull-flat approach with a great deal of success, so use whatever you feel comfortable with.

Important: Correct positioning in the stereotaxic frame is a critical step; even small errors in alignment can result in large inaccuracies in electrode positioning. Correct positioning of the ear bars can be especially difficult—seek help from an experienced researcher if possible.

A small incision is made over the sacrum, and the sacral vertebral processes are clamped. The clamp is then attached to a magnetic stand and positioned so that the rear of the animal is in light contact with the heating pad and the spine is horizontal and stretched gently. This reduces movement artefact and pulls the vertebrae slightly apart, making it easier to expose the brainstem and spinal cord. Connect the blood pressure transducer to the arterial line and monitor blood pressure on-screen.

3.2.2. Exposure of the Brain

The top and rear of the skull is exposed by midline incision, and all muscle and connective tissue between the first cervical vertebra and the lower margin of the occipital bone is removed. The base of the occipital bone is removed using a fine rongeur, exposing the brainstem and cerebellum, and the atlanto-occipital membrane and dura are removed. The following landmarks must be clearly visible: obex, the large blood vessels that lie on the dorsal surface of the medulla (which lie at approximately 2 mm lateral to obex), and the caudal half of the cerebellum.

The mandibular branch of the facial nerve is exposed by performing a diagonal incision over the cheek. The nerve runs from near the ear down towards the mouth and lies under a sheath of connective tissue. Once exposed, cover the nerve with a small piece of saline-soaked tissue until ready to stimulate.

For stimulation of the nerve, insert a pair of needle electrodes immediately deep to the nerve or use a concentric stimulating electrode directly applied to the nerve bundle. Stimulate the nerve at low frequency (1 Hz, 0.2 ms pulse width, 0.5–1 mA)—the whiskers should twitch each time a pulse is delivered (if unparalysed). Once satisfied that the nerve is adequately stimulated, protect the nerve from dehydration by applying mineral oil (e.g. paraffin).

Important: Ensure that the rat is adequately anaesthetised, paralysed and ventilated before starting this step to avoid movement upon stimulation of the spinal cord.

The second thoracic vertebra is exposed by midline incision and the surrounding muscle carefully removed. The spinous process and the dorsal surface of the vertebra are removed with rongeurs to expose the cord. Any bleeding is treated with haemostat sponge—use of cautery in close proximity to exposed nervous tissue is not recommended. The dura are cut using a fine needle (e.g. 26 gauge) or iridectomy scissors and carefully removed.

The precise location of the stimulating electrode within the spinal cord is not critical; the objective is to activate all fibres of passage within the dorsolateral quadrant of the cord, in particular any bulbospinal axons passing through the dorsolateral funiculus to the interomediolateral (IML) column where they synapse with sympathetic preganglionic neurons. As shown by Morrison et al. (23), electrical stimulation of the spinal cord can activate axons up to one millimetre from the stimulating electrode (although note the stimulating configuration used in that study is slightly different from that used here). The vertical distance of the electrode penetration into the spinal cord is important but the lateral coordinates can be estimated from the medial margin of the dorsal root entry zone (DREZ).

Remove any excess fluid from the spinal cord using a small twist of tissue paper. Using a micromanipulator (e.g. Narishige MM3), position the concentric stimulating electrode so the electrode tip lies over the midline vessel. Move the electrode 0.75 mm lateral to the midline (the tip should now lie over the DREZ) and lower it so that it touches the surface of the cord. Make a small incision (using a 26-gauge needle mounted on a cotton tip) into the dorsal surface of the cord immediately under the electrode and lower the stimulating electrode into the incision. Position the electrode about 0.5 mm deep to the dorsal surface of the cord. Programme the isolated stimulator to deliver a train of 50 pulses at 20 ms intervals (pulse length 0.2 ms, pulse amplitude 500 μA, concentric tip negative) and monitor the effect on blood pressure. A transient (~3 s rising phase, 5–10 s falling phase) increase in blood pressure of 15–30 mmHg should occur upon stimulation of the IML. If there is no discernable effect on blood pressure, move the electrode deeper into the spinal cord in 0.2 mm steps; the maximal response should occur when the electrode is located 0.75–1 mm deep to the dorsal surface (see Notes 4.2).

Once pressor responses are reproducibly evoked by spinal cord stimulation, discontinue stimulation and protect the spinal cord from dehydration by covering with warmed mineral oil or similar. We use a quick-setting silicon polymer (Silgel), which also adds mechanical stability to the preparation.

3.3. Recording Field Potentials Evoked by Antidromic Activation of the Facial Nucleus

Position a low resistance (<1 MΩ if filled with 3 M NaCl) electrode in the micromanipulator (e.g. PI N381.3A) and position the electrode tip on obex. Ensure that the electrode is vertical. *Critical step: Place the reference electrode and a separate earth lead into nearby muscle.* Recheck the resistance of the electrode and note the *x*, *y* and *z* coordinates of the manipulator. If the resistance is too high, a small piece of tissue can be used to 'sweep' off a small piece of the electrode tip, until the desired impedance is reached. Move the electrode to the dorsal surface of the brain 2 mm lateral to midline and 2.5 mm rostral to obex—this position will lie over the caudal portion of the cerebellum and should lie over the centre of the facial nucleus. Configure the oscilloscope so that it triggers a sweep each time the facial nerve is stimulated; set the time base to 2 ms/division, and offset the sweep so that the stimulus artefact is clearly visible. Adjust the volume of the audio speakers so that the stimulus artefact is audible. The signal need not be filtered for recording field potentials, although we generally use the same band-pass (0.1–3 kHz) and gain (×50) settings used for extracellular recordings. Slowly (~100 μm/s) lower the electrode into the brain—unit firing should be audible as the pipette penetrates neural tissue. As the electrode approaches the facial nucleus (~ 3.5 mm deep), the sound of the evoked response will become deeper (changing from a 'tick' to a 'thud'), and a large downward deflection will emerge on the oscilloscope 2 ms after the stimulus artefact (see Fig. 1)—this is the field potential. Once a large field potential is consistently evoked, reduce the stimulus amplitude until the potential disappears—set the stimulus amplitude to twice the threshold value. Continue to lower the electrode through the brain—the magnitude of the field potential will increase as it approaches the centre of the facial nucleus and then start to diminish as it approaches the ventral surface. Make a note of the size of the maximum deflection and use 1/10th of this value as the threshold for determining when the potential is no longer present. Make a note of the range of vertical depths at which the field exceeds the threshold (normally between 3.2 and 4.2 mm at 2.5 mm rostral to obex if using the nose-down stereotaxic approach) and withdraw the electrode. Move the electrode to a position 0.3 mm caudal to the first track and repeat—the range of depths over which a field potential is visible will diminish as the caudal pole is approached. The caudal pole is defined as the most caudal position at which stimulation of the facial nerve evokes a field potential.

3.4. Antidromic Activation of Bulbospinal RVLM Neurons

3.4.1. Find a Cell

Once the caudal pole of the facial nucleus is mapped, withdraw the electrode and discard in a "sharps" bin. Replace with an electrode with a resistance appropriate for discriminating extracellular action potentials from single units (e.g. between 6 and 12 MΩ if filled with 3 M NaCl) and position 1.6–2.1 mm lateral to midline, 0–0.5 mm caudal to the caudal pole of the facial nucleus. Ensure that the amplifier gain and filters are appropriate for extracellular recordings (see settings above) and that signals are simultaneously played over a loudspeaker during data acquisition. Lower the electrode tip relatively quickly (100 μm/s) until it lies ~2.5 mm deep to the dorsal surface, then adjust the motor speed such that depth increases by 3 μm per step. Check the electrode resistance regularly—if the resistance drops suddenly, it indicates that the electrode tip has broken—withdraw and replace it if this happens. Blockage of the electrode tip, indicated by an increase in electrode resistance, can sometimes be cleared by passing a high current through the electrode.

Bulbospinal neurons lie between 3.6 and 4.1 mm deep to the dorsal surface, in the region ventral to the respiratory cells of the Bötzinger complex (13, 23–25). Likely bulbospinal neurons can be recognised by their ongoing irregular firing (0.5–40 Hz), which may be modulated by respiratory rhythm. Many bulbospinal RVLM neurons are barosensitive (13).

3.4.2. Antidromically Activate the Cell by Stimulating the Spinal Cord

Once a stable recording from a neuron of interest is established, configure the *x*-axis of the oscilloscope to 10–20 ms/division with the stimulus trigger offset to 50% and trigger the oscilloscope so that it refreshes each time the spinal stimulator is triggered. Each sweep should display about 100–200 ms of data with the stimulus artefact in the centre of the screen (e.g. Fig. 2). Stimulate the spinal cord at ~1 Hz (pulse width 0.2 ms, pulse amplitude 1 mA) and examine the oscilloscope sweep for antidromic action potentials. Antidromic action potentials can be distinguished from spontaneous or orthodromic action potentials by their constant latency

(Fig. 2b). Some adjustment of the stimulus intensity and electrode polarity (see Sect. 4.2.2.) may be required in order to reliably activate the axon.

3.4.3. Performing the Antidromic Collision Test

Once stimulation of the spinal cord reliably evokes an antidromic spike (Fig. 2b), use spontaneous action potentials to trigger the stimulator (see Sect. 4.1.5), introducing a delay between spike detection and stimulus output. An appropriate initial delay is one that is longer than the latency of the antidromic response. For example, if the latency of the antidromic spike is 36 ms (as in Fig. 2c), set the delay to ~40 ms. This should result in a response similar to that shown in Fig. 2c, in which a spontaneous 'trigger' spike occurs prior to the stimulus artefact, followed by an antidromic spike that occurs after the artefact. Gradually reduce the delay in 1-ms steps so that the interval between the spontaneous spike and the stimulus decreases incrementally. As the delay approaches the latency of the antidromic spike, a point may be reached at which the stimulus no longer evokes any antidromic response (Fig. 2d). This response is consistent with collision of the spontaneous and antidromic spikes within the axon, fulfilling the criteria of the antidromic collision test. Convention is to superimpose 3–5 sweeps of colliding (e.g. Fig. 2d) and non-colliding (e.g. Fig. 2c) responses to demonstrate the reproducibility of the response.

If responses to stimulation still occur despite stimulation occurring at a relatively short delay compared to the latency of the response (i.e. if the latency of the response is 10 ms but the stimulator is triggering 5 ms after the spontaneous spike), the unit has failed the antidromic collision test: this response is consistent with activation of excitatory afferent pathways rather than an antidromic response.

4. Notes

4.1. Before You Start: Configuration of Data Acquisition Hardware, Software and Stimulators

4.1.1. Stimulators

Ensure that your stimulator can be triggered by the electronic pulses that will be used to control stimulus timing (i.e. TTL pulse) and that it is easy to adjust the TTL delay while conducting the experiment. In our configuration, we use a Cambridge Electronic Design Power1401 Mark II data acquisition system (http://www.ced.co.uk) in conjunction with Spike2 software to generate 5 V trigger pulses. Many stimulators incorporate a delay function. If using a battery-powered stimulator, ensure that batteries are charged, including any internal batteries used to power the stimulator circuitry.

4.1.2. Earthing

Recordings of bioelectrical signals are susceptible to interference from mains electricity, which alternates at 50/60 Hz, depending

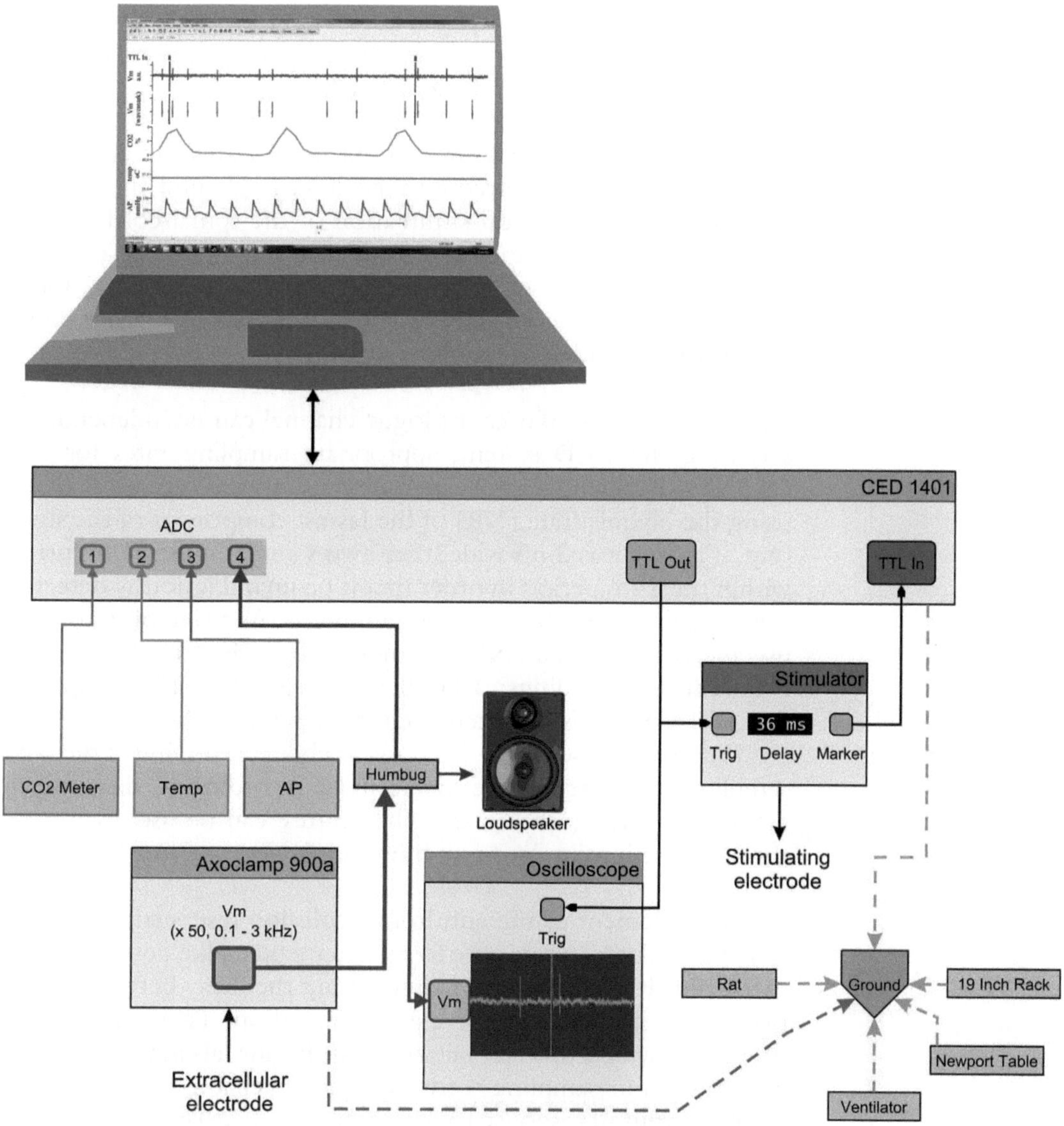

Fig. 3. Basic configuration used in our laboratory to conduct the experiments described in the current protocol.

on the country. The best way to minimise electrical interference is to connect potential sources of noise to the signal ground, as shown in Fig. 3. All electrical equipment connected to the experimental preparation (e.g. bioamplifiers, stimulators, analogue-digital converters), and the animal itself, should be connected to a common ground. Most electrophysiology equipment will feature a dedicated earth connection on the front or back panel. In our laboratory, we connect all earth leads to the earth pin of a three-point plug and connect this to the mains earth.

Caution: Make sure the earth pin is correctly identified; connection of earth leads to live mains electricity could lead to serious injury, destruction of equipment, fire and death.

4.1.3. Hardware Connections

The schematic diagram shown in Fig. 3 indicates the basic hardware configuration required for recording single unit/field potential responses to electrical stimulation of the spinal cord or facial nerve, respectively. We also record vital signs (arterial pressure, expired CO_2 and core temperature) in all experiments. Other channels, such as nerve activity, tracheal pressure or other digital inputs, can be added as required.

4.1.4. Sampling Configurations

The sampling rate of each analogue channel can be independently set using the CED system; appropriate sampling rates for each channel will depend on the experimental aims and can be calculated using the Nyquist rate (NR) of the fastest component of the signal (e.g. if an event is 2 ms wide, then two samples must be occurring within the 2 ms period in order that it be unambiguously detected: NR = 2 samples/0.002 s = 1,000 samples/s. As a rule of thumb, we use approximately 10× NR for smooth rendering of each channel: extracellular recordings: 10,000 samples/s; AP: 250 samples/s; CO_2: 50 samples/s, core temperature: 1 sample/s).

We also use a software 'Wavemark' channel running at the same sampling frequency as the extracellular recording to discriminate action potentials (see Fig. 2). This feature can be used to trigger the antidromic stimulation of bulbospinal neurons (see below).

4.1.5. Controlling Stimulator Timing

A critical element of the antidromic collision test is triggering the spinal stimulator in response to a spontaneous spike detected in the extracellular recording and manipulating the delay between detection of the spontaneous spike and stimulation. There are number of ways in which this can be arranged; in our laboratory, we have configured our sampling software, Spike2 version 7.10, to trigger a TTL output in response to detection of a spike on the Wavemark channel. The value of the delay is controlled by a thumbwheel integrated into the spinal stimulator.

To do this, we have modified the *Pulser* script, written my Malcolm Lidierth (26) and available at http://sourceforge.net/projects/sigtool/files/.

We have added the following text to the bottom of the *Pulser100.pls* sequence file:

```
EA: 'A DIGOUT (....0000)
DAC 0,0
DAC 1,0
DELAY s(0.0081)-1
CHAN DelayC,12
ADDI DelayC,1
RA0: CHAN Delay2,12 >Wait 1 events on 12
```

```
BLT Delay2,DelayC,RA0 > Wait 1 events on 12
LAA: DIGOUT (....0100)
DELAY s(0.0009)-1
DIGOUT (....0000)
DELAY s(0.01)-1
HALT; End of this sequence section
```

This inserts the button 'A' into the sequence control bar; when Pulser is running, pressing this button will trigger TTL outputs on DIGOUT 0 and 2 once a Wavemark (i.e. spike) is detected (the Wavemark channel is 12 in our configuration). We use output 0 to trigger the oscilloscope sweep and channel 2 to trigger the stimulator.

The system resets once triggered. To repeat the stimulus, the 'A' button must be pressed again. A delay could be incorporated into the code, but we find it easier to control using a thumbwheel integrated into our stimulator. Once the 'A' button has been pressed, *Pulser* must be restarted for normal function.

4.2. Troubleshooting

4.2.1. Facial Field Potentials

Stimulation issues: The facial nerve is relatively easy to stimulate; in our experience, problems tend to be related to incorrect stereotaxic technique (e.g. the head is not aligned properly, so the recording electrode is not actually in the facial nucleus at the appropriate stereotaxic coordinates). However, check that the stimulator is evoking twitching of the whiskers (if unparalysed) and that a stimulus artefact is visible on the oscilloscope. If stimulation evokes an artefact but no twitching, try reversing the electrode polarity, increasing the stimulus intensity (although a maximal response should be evoked by direct stimulation of the nerve at 1 mA) or repositioning the stimulating electrodes.

Recording issues: Always play bioelectrical signals over a loudspeaker to verify that recordings are set up properly—the contact of the electrode with the brain surface should result in a loud 'thud', and the clicking noises of neuronal activity should be audible as the electrode penetrates the brain. Check the electrode resistance by passing a 1 nA current through the electrode tip (consult the manual of the amplifier) and measuring the change in potential evoked. The resistance (in MΩ) is equal to the change in membrane potential (in mV, with no high-pass filtering) evoked by a 1 nA current. Check that the wire inside the electrode is in good contact with the electrolyte and that the reference and earth electrodes are connected. Turn off any unnecessary equipment to avoid mains interference. Ensure that the recording electrode is vertical.

Tissue issues: The facial nerve runs close to the ear canal and can be damaged if force is used to position the stereotaxic ear bar in the ear canal. In this case, the whiskers will still continue to twitch as usual because neuromuscular transmission is unaffected by a lesion proximal to the stimulating electrode.

4.2.2. The Antidromic Collision Test

Stimulation problems: Ensure that spinal cord stimulation evokes an increase in blood pressure (Sect. 3.2.4). If this is not the case, ensure that the electrode is connected to the stimulator and that the stimulator is producing a current when triggered; most stimulators indicate when current is being passed by flashing an LED and/or by producing a 'trigger' TTL output that can be captured by the data acquisition hardware. Ensure that the stimulus parameters are appropriate and that the tip of the electrode is attached to the negative (cathode) pole of the stimulator. Try changing the polarity of the current. If a bioelectric signal (e.g. extracellular recording) is being made, does spinal cord stimulation evoke a stimulus artefact?

Is the electrode definitely penetrating the tissue? Fine electrodes can become snagged on dura or vessels, or may bend rather than penetrate the dorsal surface. Withdraw the electrode and try making another incision in the cord surface. Poor penetration, and therefore stimulation at the cord surface, sometimes evokes depressor rather than pressor effects. If penetration is an issue, try overshooting the target and the pulling back to the appropriate depth.

Ensure the electrode tip is clean and undamaged prior to the experiment, especially if reusing concentric electrodes. Any residue from previous experiments may insulate the electrodes, preventing proper current delivery. If the electrode is insulated by glass (e.g. monopolar style), make sure the glass insulation comes down close to the electrode tip; otherwise, the current will be delivered at the point of contact with the tissue (i.e. the surface of the spinal cord), not at the electrode tip.

Ensure that the electrode does not change position during the experiment by maintaining neuromuscular blockade and avoiding contact with the manipulator and animal once positioned.

Recording issues: Stimulus artefact is too large: try placing an earth lead between the stimulating electrode and recording electrodes. Use a concentric rather than monopolar stimulation configuration.

No antidromic action potentials evoked: if everything is set up correctly (the spinal stimulator evokes an increase in blood pressure in response to tetanic stimulation, the facial field is accurately mapped, and a recording from a spontaneously active neuron in the correct region of the RVLM is established), it may just be that the cell is not bulbospinal or that the antidromic response is being concealed by orthodromic spikes that are spontaneous or result from activation of afferent pathways. The probability of the former occurring can be reduced by inhibiting the ongoing activity of the cell. In barosensitive neurons, this can be achieved by artificially raising blood pressure using an aortic snare or by injecting a bolus of vasoconstrictive drug (e.g. 5 μg phenylephrine i.v.: (8)). Good electrode position in the spinal cord reduces the threshold required to activate bulbospinal axons and therefore reduces the probability of activating other pathways in the latter situation.

References

1. Dittmar C (1870) Ein Neuer Beweis fur die Reizbarkeit der centripetalen Fasern des Ruckenmarks. J Sachs Ges Wiss Mathematischphysiche Klasse 22:18
2. Dittmar C (1873) Uber die Lage des sogenannten Gefasscentrums der Medulla oblongata. J Sachs Ges Wiss Mathematischphysiche Klasse 25:449–469
3. Owsjannikow P (1871) Die tonischen und reflektorischen centren der Gefassnerven. J Sachs Ges Wiss Mathematischphysiche Klasse 23:135–147
4. Guertzenstein PG, Silver A (1974) Fall in blood pressure produced from discrete regions of the ventral surface of the medulla by glycine and lesions. J Physiol 242:489–503
5. Ross CA, Ruggiero DA, Park DH, Joh TH, Sved AF, Fernandez-Pardal J, Saavedra JM, Reis DJ (1984) Tonic vasomotor control by the rostral ventrolateral medulla: effect of electrical or chemical stimulation of the area containing C1 adrenaline neurons on arterial pressure, heart rate, and plasma catecholamines and vasopressin. J Neurosci 4:474–494
6. Ross CA, Ruggiero DA, Joh TH, Park DH, Reis DJ (1984) Rostral ventrolateral medulla: selective projections to the thoracic autonomic cell column from the region containing C1 adrenaline neurons. J Comp Neurol 228: 168–185
7. Blessing WW, Goodchild AK, Dampney RA, Chalmers JP (1981) Cell groups in the lower brain stem of the rabbit projecting to the spinal cord, with special reference to catecholamine-containing neurons. Brain Res 221:35–55
8. Brown DL, Guyenet PG (1984) Cardiovascular neurons of brain stem with projections to spinal cord. Am J Physiol 247:R1009–1016
9. Brown DL, Guyenet PG (1985) Electrophysiological study of cardiovascular neurons in the rostral ventrolateral medulla in rats. Circ Res 56:359–369
10. Abbott SB, Stornetta RL, Fortuna MG, Depuy SD, West GH, Harris TE, Guyenet PG (2009) Photostimulation of retrotrapezoid nucleus phox2b-expressing neurons in vivo produces long-lasting activation of breathing in rats. J Neurosci 29:5806–5819
11. Berkowitz RG, Chalmers J, Sun QJ, Pilowsky P (1999) Identification of posterior cricoarytenoid motoneurons in the rat. Ann Otol Rhinol Laryngol 108:1033–1041
12. Pilowsky PM, Jiang C, Lipski J (1990) An intracellular study of respiratory neurons in the rostral ventrolateral medulla of the rat and their relationship to catecholamine-containing neurons. J Comp Neurol 301:604–617
13. Verberne AJ, Stornetta RL, Guyenet PG (1999) Properties of C1 and other ventrolateral medullary neurones with hypothalamic projections in the rat. J Physiol 517(Pt 2): 477–494
14. Sartor DM, Verberne AJM (2003) Phenotypic identification of rat rostroventrolateral medullary presympathetic vasomotor neurons inhibited by exogenous cholecystokinin. J Comp Neurol 465:467–479
15. Sartor DM, Verberne AJ (2010) Gastric leptin: a novel role in cardiovascular regulation. Am J Physiol Heart Circ Physiol 298:H406–414
16. Verberne AJ, Sartor DM (2010) Rostroventrolateral medullary neurons modulate glucose homeostasis in the rat. Am J Physiol Endocrinol Metab 299:E802–807
17. Biscoe TJ, Sampson SR (1970) Field potentials evoked in the brain stem of the cat by stimulation of the carotid sinus, glossopharyngeal, aortic and superior laryngeal nerves. J Physiol 209:341–358
18. Riddell JS, Hadian M (2000) Field potentials generated by group II muscle afferents in the lower-lumbar segments of the feline spinal cord. J Physiol 522(Pt 1):97–108
19. Lipski J (1981) Antidromic activation of neurons as an analytic tool in the study of the central nervous system. J Neurosci Methods 4: 1–32
20. Serra J, Campero M, Ochoa J, Bostock H (1999) Activity-dependent slowing of conduction differentiates functional subtypes of C fibres innervating human skin. J Physiol 515(Pt 3):799–811
21. Gee MD, Lynn B, Cotsell B (1996) Activity-dependent slowing of conduction velocity provides a method for identifying different functional classes of C-fibre in the rat saphenous nerve. Neuroscience 73:667–675
22. Fang X, McMullan S, Lawson SN, Djouhri L (2005) Electrophysiological differences between nociceptive and non-nociceptive dorsal root ganglion neurones in the rat in vivo. J Physiol 565:927–943
23. Morrison SF, Milner TA, Reis DJ (1988) Reticulospinal vasomotor neurons of the rat rostral ventrolateral medulla: relationship to sympathetic nerve activity and the C1 adrenergic cell group. J Neurosci 8:1286–1301
24. Kanjhan R, Lipski J, Kruszewska B, Rong W (1995) A comparative study of pre-sympathetic

and Botzinger neurons in the rostral ventrolateral medulla (RVLM) of the rat. Brain Res 699:19–32

25. McMullan S, Pathmanandavel K, Pilowsky PM, Goodchild AK (2008) Somatic nerve stimulation evokes qualitatively different somatosympathetic responses in the cervical and splanchnic sympathetic nerves in the rat. Brain Res 1217:139–147

26. Lidierth M (2005) Pulser: user-friendly, graphical user-interface based software for controlling stimuli during data acquisition with Spike2 for Windows. J Neurosci Methods 141: 243–250

Chapter 8

Microiontophoretic Study of Individual Neurons During Intracellular Recording

Qi-Jian Sun and Paul M. Pilowsky

Abstract

There is an increasing tendency to combine multidisciplinary methods in studying the central nervous systems. One recently developed technique is the combination of electrophysiological and pharmacological investigations by making intracellular recording during extracellular iontophoresis of drugs to study the effects of neurotransmitters on neuronal activity and connections. In this chapter, we describe this technique in detail.

We show how:

1. Electrodes are made for this type of study.
2. Drug solutions are prepared for current iontophoresis.
3. A microiontophoresis current generator is used to iontophoretically inject drugs onto a neuron while performing intracellular recording.

At the end of this chapter, we provide an example from our recent studies that uses this iontophoretic technique to study effect of counter-intuitive inhibitory effect of blocking $GABA_A$ receptors in the control of post-inspiratory activity of the expiratory laryngeal motoneurons.

Key words: Electrophysiology, Intracellular, Microiontophoresis, Respiratory, Rat, Brainstem, In vivo

1. Introduction

In order to study the central nervous system, it is important to understand the neuronal circuits, where the input is coming from and processed and where the projections go in order to control the various parts of our body. It is also important to know which neurotransmitters are released from neuronal terminals to communicate between these neurons. In general, the effects of potential neurotransmitters on the central nervous system can be studied at three different levels. Studies from the whole animal are conducted by intravenous injection of drugs via the cannulated vein (1, 2).

Paul M. Pilowsky et al. (eds.), *Stimulation and Inhibition of Neurons*, Neuromethods, vol. 78,
DOI 10.1007/978-1-62703-233-9_8, © Springer Science+Business Media, LLC 2013

Studies of the local region are conducted by microinjection of drugs into certain nuclei within the central nervous system (3, 4). Studies at the cellular level are conducted by microiontophoretic injection of drugs directly onto individual neurons (5–7). One great advantage of iontophoretic studies, in comparison to the other two, is its accuracy in applying drugs at cellular level without artefactually affecting the activity of other neurons. Many early iontophoretic studies were carried out using extracellular recording (5, 8). Although extracellular recording is able to show whether an individual neuron is activated, i.e. increase firing rate, it is not known, in cases where the recorded neuron stops firing, whether this silence is the result of increasing inhibitory inputs or withdraw of excitatory inputs. This limitation affects our ability to understand neuronal mechanisms that underlie many important functions.

In one recent iontophoretic study, during extracellular recording (9), we found that activity of expiratory laryngeal motoneurons (ELMs) was silenced during iontophoretic application of gamma-aminobutyric acid (GABA). This is an expected result because we know that GABA is a common inhibitory amino acid neurotransmitter which has a strong inhibitory effect on many, if not all, neurons in the central nervous system (10). However, we also found that, during iontophoretic injection of bicuculline (BIC), a $GABA_A$ receptor antagonist, post-inspiratory activity of the ELM was reduced rather than increased (Fig. 1). This is interesting because blockade of $GABA_A$ receptors should unmask neuronal activity that is inhibited by the GABA-related inhibition; thus, an increased but not decreased neuronal activity should be seen. This is certainly the case in some previous studies (8, 11). The possible neuronal mechanisms underlying our BIC result are not clear. To further address this question, more direct evidence of excitatory or inhibitory inputs received by the ELM need to be obtained, and this requires information from intracellular recordings.

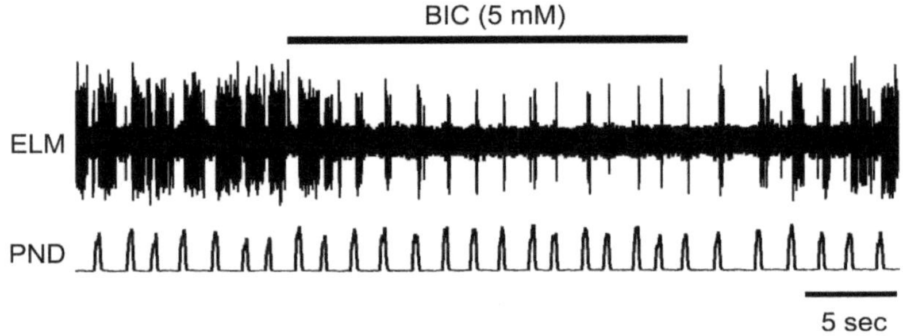

Fig. 1. Single-unit extracellular recording from an expiratory laryngeal motoneuron (*ELM*), showing inactivity during phrenic nerve discharge (*PND*, i.e. 'inspiratory' phase), but activity between the PND's (i.e. 'post-inspiratory' and 'expiratory' phases). Period of bicuculline (*BIC*, 5 mM, 100 nA) iontophoresis is indicated by the *top bar*. Note that BIC iontophoresis strongly reduce number of action potentials of the ELM within each respiratory cycle (reproduced from Sun et al. (9) with permission).

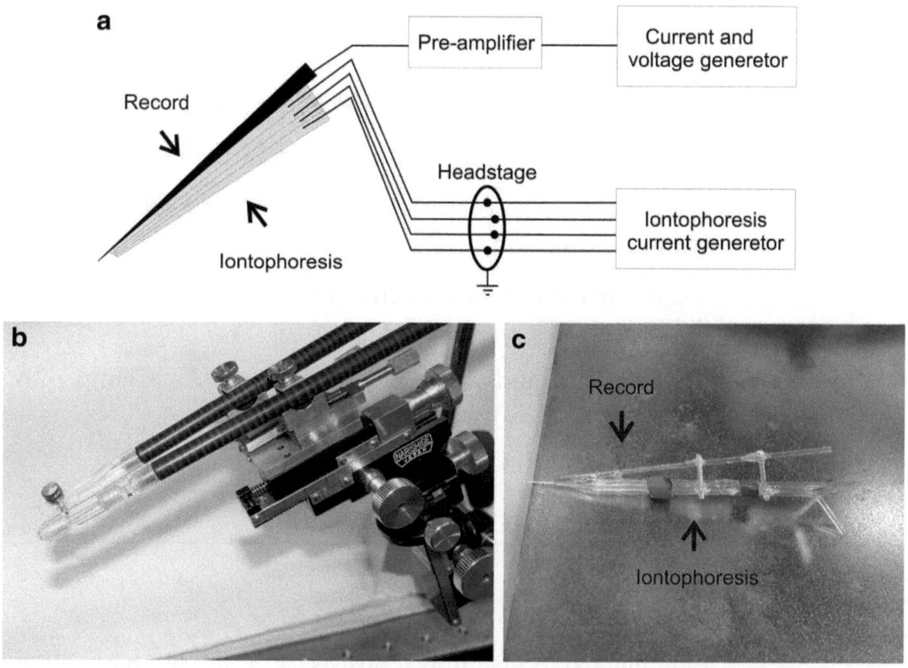

Fig. 2. Arrangement for iontophoretic study during intracellular recording. (**a**) Diagram, showing how an electrode assembly is connected to the current and voltage generator for recording, and the iontophoresis current generator for iontophoresis. (**b**) A photograph, showing how the manipulator is used for electrode assembly. The intracellular recording electrode is mounted onto the *upper holder*, and the five-barrel electrode into the *lower holder*. Positions of the two electrodes can be adjusted separately before assembly. (**c**) A photograph, showing a completed electrode assembly with glass struts to provide additional strength.

In order to conduct microiontophoretic studies during intracellular recording, an electrode assembly (Fig. 2) needs to be made, as there are no such electrodes available for purchase.

2. Materials

Glass capillaries: e.g. GC100F-10, Harvard Apparatus Ltd., UK,

for making both recording and multibarrel electrodes (filament glass is essential for filling electrodes)

Single pipette puller: e.g. P-97, Sutter Instruments Co., USA,

for pulling recording electrodes

Multibarrel pipette puller: e.g. PMP-100, MicroData Instrument Inc., USA,

for pulling multibarrel electrodes

Micromanipulator: e.g. MMD-4, Narishige, Japan,

for placing and holding recording and multibarrel electrodes during assembly

Glue: Araldite (5 min), Selleys Pty Ltd., Australia,

for fabricating a recording and multibarrel electrode assembly

Microiontophoresis current generator: 6400 Advanced, Dagan Corporation, USA,

for generating multi-channelled currents to retain and eject drugs

Current and voltage generator: AxoClamp 2B, Axon Instruments, USA,

for intracellular and extracellular recordings

3. Electrode Assembly and Drug Preparation

Our electrode assembly consists of a single sharp electrode for recording, attached to a multibarrel electrode for drug application and current balancing (Fig. 2a). All electrodes are made from filament-containing glass capillaries (GC100F-10), which have an outside diameter of 1.0 mm and an inside diameter of 0.58 mm. First, a single sharp electrode is pulled using a single pipette puller (P-97, see Table 1 for detailed program settings), with a tip size of <0.5 μm. In order to prepare a five-barrel electrode, we first use heat-shrink tubing (PTFE) to hold five glass barrels together. The multibarrel electrode is then pulled using a multibarrel pipette puller (PMP-100, see Table 2 for detailed program settings). Four of the five barrels are bent 10 mm from the end to an angle of about 30°, using a butane gas burner. The tip of the five-barrel electrode is carefully broken to a total tip diameter of 3–5 μm using forceps under a benchtop dissecting microscope. The single recording and five-barrel drug ejection electrodes are mounted separately on the two micromanipulators (MMD-4, Fig. 2b) before carefully aligning the two electrodes. The fine electrode tip is positioned

Table 1
Puller P97 settings

H(eat)	580
P(ull)	60
V(elocity)	60
D(elay)	150

Table 2
Puller PMP-100 settings

M	20.1		
m	3.5		
Step 1	035	H95	
Step 2	003	H85	ROTA
Step 3	<L	H75	PUL:06
Step 4	002.3	H00	
Step 5	< L	H35	PUL:01c
Step 6	003	H00	COOL

25–50 μm ahead of the tip of the five-barrel electrode. The two electrodes are then glued together using a two parts Araldite (5 min) under the microscope. To strengthen the assembly, the glue is not only applied at very front to connect the tips but also at back with two pieces of glass to form a skeleton and strengthen the assembly (Fig. 2c).

One hour before recording, the intracellular electrode is backfilled (Microfil tubing, WPI) with 1 M $C_6H_5O_7K_3$ (potassium methylsulphate) for intracellular recording (resistance ~50 MΩ). This electrolyte is useful to avoid the risk of extra Cl^- entering the cell and creating reversed inhibitory postsynaptic potentials. Clearly, other electrolyte combinations and concentrations can be used as needed. Three of the five barrels are backfilled with different drugs. In this example, we use a glutamate receptor agonist, D,L-homocysteic acid (DLH, ICN Biomedicals, 0.2 M), GABA (Sigma, 0.5 M) and a $GABA_A$ antagonist, bicuculline methiodide (BIC, Sigma, 5 mM). In order to eject the drugs using current application (i.e. iontophoresis), the drug solutions need to be either positively or negatively charged. This is achieved by adjusting the pH of the drug solution using NaOH (2 M) or HCl (2 M). In this study, we make the DLH solution negatively charged by adjusting its pH to 8, but GABA and BIC solutions positively charged by adjusting their pH to 4. The 4th barrel is backfilled with saline (0.9% NaCl solution) for current balancing, during which a current equal but opposite to the sum of all other channels is applied to achieve the net current from all barrels to be close to zero. Otherwise, the current itself may affect cell membrane excitability. The fifth barrel (the one that is not bent) is used to mount the electrode assembly on an electrode holder. The resistance of each filled barrel ranges between 5 and 50 MΩ, and they are regularly checked during the experiment to ensure that they are not blocked.

4. Micro-iontophoresis Techniques

Fine silver wires (0.010″ Bare, A-M Systems Inc., USA) are inserted into the solution of each electrode. The wires that connect the multibarrels are connected to a multipin headstage of the microiontophoresis current generator (6400 Advanced, Fig. 3). The headstage is placed as close as possible to the electrode holder to minimise any potential sources of electrical noise. The wire that connects the recording electrode is connected to preamplifier of the current and voltage generator (AxoClamp 2B).

Once the electrode assembly is lowered into the brainstem, a retaining current of between 20 and 30 nA is applied to each drug electrode via the microiontophoresis current generator (6400 Advanced). The retaining current minimises diffusional efflux of drug molecules from the tip of the micropipette and is therefore always applied with a polarity opposite to the ionic charge of the drug. For example, retention of positive-charged ions (e.g. BIC) is achieved by applying an anionic current (−20 to −30 nA), making the micropipette solution negative relative to the surrounding medium and thereby drawing cations away from the tip. Drugs are ejected from the micropipette by applying an additional current of opposite polarity between 50 and 150 nA. In the case of the BIC ejection, for example, the current will be switched from negative to positive one. The rate of ejection of a compound is dependent on a number of factors, including the concentration within the micropipette, the strength of the current passed and the dimensions of the electrode tips.

Ejection currents used for GABA and BIC are determined by checking their effects on cell firing, either spontaneous activity or activity induced by a continuous ejection of DLH. Once a consistent

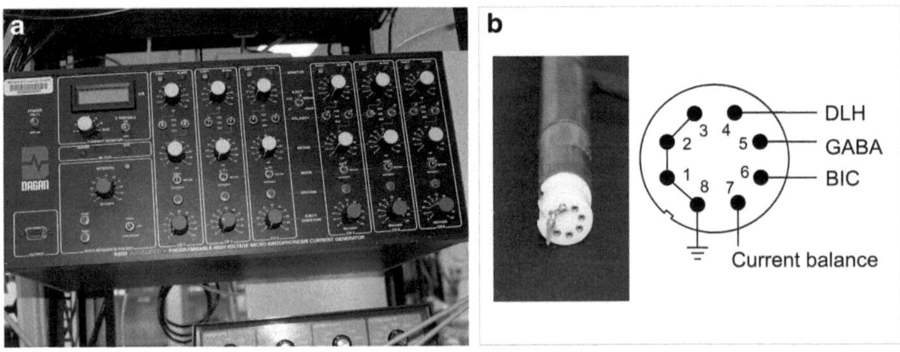

Fig. 3. Microiontophoresis current generator and connections. (**a**) A photograph of front panel of the used microiontophoresis current generator (6400 Advanced). (**b**) A photograph of the headstage used to connect the multibarrel electrode and generator. A cross section of the headstage is shown on the *right* with exact connections for our experiment. Channels 1–3 are not used and connected to ground to reduce current-related noise. Channel 4 is connected to D,L-homocysteic acid (*DLH*)-filled barrel, channel 5 to gamma-aminobutyric acid (*GABA*)-filled barrel and channel 6 to BIC-filled barrel. Channel 7 is connected to saline-filled barrel for current balancing. Channel 8 is ground.

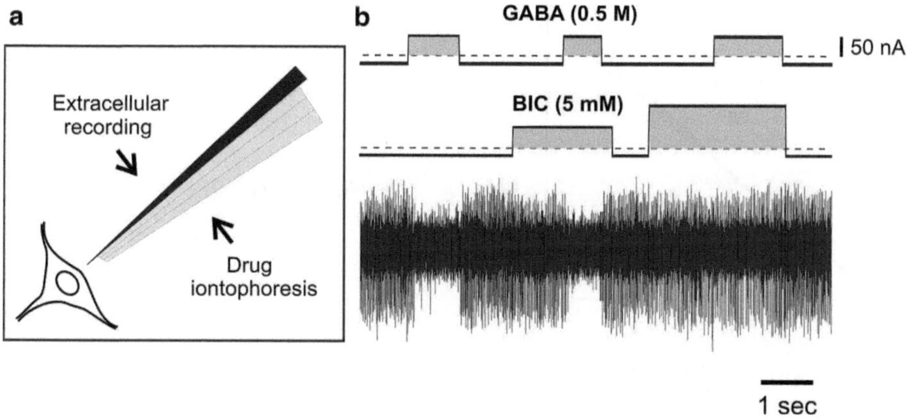

Fig. 4. Extracellular recording and drug iontophoretic application. (**a**) A diagram showing how microiontophoretic study is conducted under extracellular recording. (**b**) Effects of iontophoretic injections of GABA and BIC on a neuron with tonic firing. The fine recording electrode (*black*) is filled with 3 M NaCl (resistance ~10 MΩ). Periods of GABA (0.5 M) iontophoresis are indicated by the *top trace* (*shaded periods*, retaining current −30 nA, ejection current 50 nA). The *dashed lines* in both *top* and *middle traces* indicate the current baseline (zero) when no retaining and ejection current applied. Periods of BIC iontophoresis (5 mM) were indicated by the *middle trace* (*shaded periods*, retaining −30 nA, ejection 50 nA and 100 nA). The *bottom trace* shows that the unit activity was strongly inhibited by GABA iontophoresis. Note that the GABA inhibition is not blocked by the first BIC (lower intensity iontophoresis), but by the second BIC (higher intensity iontophoresis) (reproduced from Sun et al. (9)).

rate of cell firing has been achieved, GABA ejections are manually cycled (1–10 s) with an increasing current until a measurable but submaximal inhibition of the cell firing is observed. One such control experiment during extracellular recording is shown in Fig. 4, in which the GABA application inhibited action potentials of a neuron before and during the 1st but not the 2nd BIC iontophoresis, when the ejection current is increased to 100 nA.

5. Example of Iontophoretic Study During Intracellular Recording

In order to further understand the BIC effect observed during our extracellular study (Fig. 1), intracellular recordings were successfully obtained from five ELMs using the electrode assembly. Before BIC iontophoresis, the ELM hyperpolarized during phrenic nerve discharge (PND, i.e. inspiratory period) and sharply depolarized after the PND (post-inspiratory period), before gradually returning to baseline levels (Fig. 5b). The average resting membrane potential was measured at the end of the expiratory period (−59 ± 13 mV, $n = 5$). One difficulty in making intracellular recordings using the electrode assembly is that the larger multibarrel tip that attaches to the recording electrode will inevitably affect the real, instant speed of the recording electrode's advancing to neurons in the front. This will reduce efficiency of the electrode's penetration into neurons by 50% or more, depending upon the tip

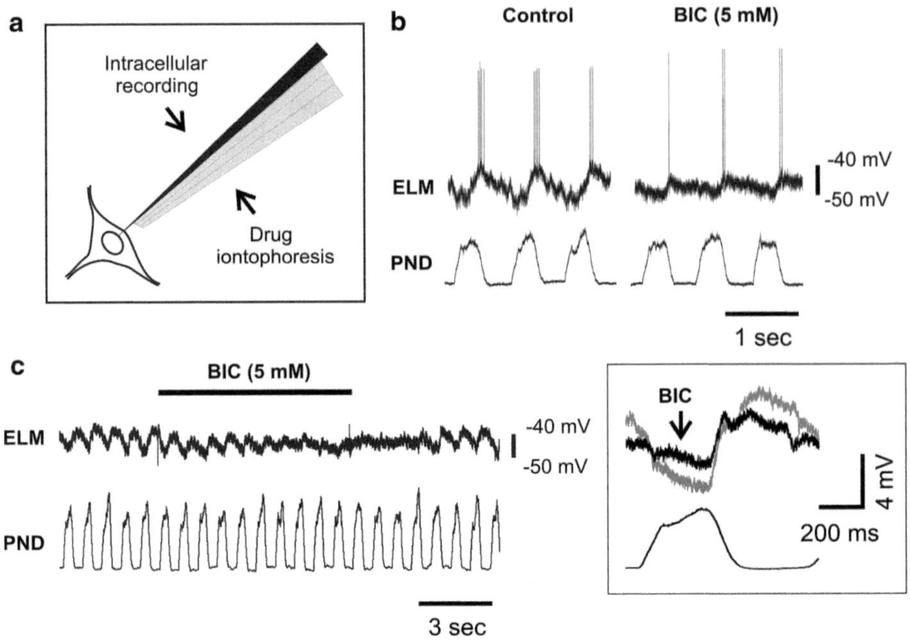

Fig. 5. Effect of BIC iontophoresis on the membrane potentials of two ELMs. (**a**) Diagram of a modified electrode assembly from Fig. 4, in which the glass electrode was replaced by a much finer one for intracellular recording. The recording electrode was filled with 1 M $C_6H_5O_7K_3$ (resistance ~50 MΩ). (**b**) Intracellular recording from one ELM, showing membrane potential changes in relation to the PND before and during BIC iontophoresis. (**c**) Intracellular recording from anther ELM. Note that both its inspiratory hyperpolarisation and post-inspiratory depolarisation are markedly reduced during BIC iontophoresis. On the *right* are membrane potentials averaged from the same ELM, triggered by the beginning of the PND. The averages were made from three respiratory cycles before the BIC (*grey*) and the last three respiratory cycles during the BIC iontophoresis (*black*). The *bottom trace* shows the averaged PND (reproduced from Sun et al. (9)).

size. Accordingly, our technique is more suitable for intracellular recordings from larger motoneurons, such as ELM, when penetrating their membranes and maintaining a recording is easier than with smaller neurons.

During BIC iontophoresis, we found that the inspiratory hyperpolarisation of the ELM was reduced (Fig. 5b). This became more obvious after we compared the averaged membrane potentials triggered from the start of phrenic nerve discharge (Fig. 5c). The averaged ELM hyperpolarization was 2.5 ± 1.4 mV measured before BIC, but 1.5 ± 0.4 mV measured at the end of the BIC iontophoresis: a significant reduction of about 40% ($P < 0.05$). In addition, during post-inspiration, the depolarized membrane potential was also reduced (Fig. 5c). The averaged post-inspiratory depolarization was 3.0 ± 1.3 mV measured before the BIC, but 1.6 ± 0.4 mV measured at the end of the BIC iontophoresis ($P < 0.05$). There was a nearly linear relationship between reductions of hyperpolarized membrane potentials during inspiration and depolarized membrane potentials during post-inspiration after the BIC. Note that, despite these reductions, BIC had little effect on baseline membrane potential as measured with respect to the end of expiration (Fig. 5b,c).

Given the results above (Fig. 5), we propose that BIC application blocks $GABA_A$ inhibition during inspiration, which in turn leads to a smaller depolarisation of the membrane potential during post-inspiration. Earlier work suggests that the inspiratory inhibition of some respiratory neurons is mediated by $GABA_A$ receptors (12, 13). However, our results show for the first time that, in ELM, blockade of this inspiratory inhibition also affects the excitability of the neuron in post-inspiration, providing fresh evidence to support the idea that the post-inspiratory patterning of ELM is generated, at least in part, via a post-inhibitory rebound mechanism (14, 15). $GABA_A$ receptors are not only directly involved in phasic inhibition of the ELM during inspiration but also play an important role in the depolarisation of the membrane potential in post-inspiration.

References

1. Kaczynska K, Szereda-Przestaszewska M (2006) Clonidine-evoked respiratory effects in anaesthetized rats. Exp Physiol 91:269–275
2. Ulger F, Bozkurt A, Bilge SS, Ilkaya F, Dilek A, Bostanci MO, Ciftcioglu E, Guldogus F (2009) The antinociceptive effects of intravenous dexmedetomidine in colorectal distension-induced visceral pain in rats: the role of opioid receptors. Anesth Analg 109:616–622
3. Pilowsky PM, Abbott SB, Burke PG, Farnham MM, Hildreth CM, Kumar NN, Li Q, Lonergan T, McMullan S, Spirovski D, Goodchild AK (2008) Metabotropic neurotransmission and integration of sympathetic nerve activity by the rostral ventrolateral medulla in the rat. Clin Exp Pharmacol Physiol 35:508–511
4. Sun QJ, Bautista TG, Berkowitz RG, Zhao WJ, Pilowsky PM (2011) The temporal relationship between non-respiratory burst activity of expiratory laryngeal motoneurons and phrenic apnoea during stimulation of the superior laryngeal nerve in rat. J Physiol 589: 1819–1830
5. Curtis DR, Duggan AW, Felix D, Johnston GA (1971) Bicuculline, an antagonist of GABA and synaptic inhibition in the spinal cord of the cat. Brain Res 32:69–96
6. Haji A, Pierrefiche O, Foutz AS, Champagnat J, Denavit-Saubie M, Takeda R (1996) Pharmacological properties of peripherally induced postsynaptic potentials in bulbar respiratory neurons of decerebrate cats. Neurosci Lett 211:17–20
7. Berkowitz RG, Sun QJ, Pilowsky PM (2005) Congenital bilateral vocal cord paralysis and the role of glycine. Ann Otol Rhinol Laryngol 114:494–498
8. Dogas Z, Krolo M, Stuth EA, Tonkovic-Capin M, Hopp FA, McCrimmon DR, Zuperku EJ (1998) Differential effects of GABAA receptor antagonists in the control of respiratory neuronal discharge patterns. J Neurophysiol 80:2368–2377
9. Sun QJ, Berkowitz RG, Pilowsky PM (2008) GABA A mediated inhibition and post-inspiratory pattern of laryngeal constrictor motoneurons in rat. Respir Physiol Neurobiol 162:41–47
10. Schmid K, Bohmer G, Gebauer K (1991) GABAA receptor mediated fast synaptic inhibition in the rabbit brain-stem respiratory system. Acta Physiol Scand 142:411–420
11. McCrimmon DR, Zuperku EJ, Hayashi F, Dogas Z, Hinrichsen CF, Stuth EA, Tonkovic-Capin M, Krolo M, Hopp FA (1997) Modulation of the synaptic drive to respiratory premotor and motor neurons. Respir Physiol 110:161–176
12. Onimaru H, Arata A, Homma I (1990) Inhibitory synaptic inputs to the respiratory rhythm generator in the medulla isolated from newborn rats. Pflugers Arch 417:425–432
13. Ramirez JM, Telgkamp P, Elsen FP, Quellmalz UJ, Richter DW (1997) Respiratory rhythm generation in mammals: synaptic and membrane properties. Respir Physiol 110:71–85
14. Barillot JC, Grelot L, Reddad S, Bianchi AL (1990) Discharge patterns of laryngeal motoneurones in the cat: an intracellular study. Brain Res 509:99–106
15. Gauthier P, Hilaire G, Monteau R (1983) Onset and control of expiratory laryngeal discharge: cross-correlation analysis. Respir Physiol 54:67–77

Chapter 9

Neurobiotin Electroporation for Combined Structural and Functional Analysis of Neurons in Developing Mouse Brain Slices

Refik Kanjhan and Mark C. Bellingham

Abstract

Here we describe two Neurobiotin electroporation techniques for combined analysis of neuronal structure and function in 300-μm-thick brain slices during development, using mouse hypoglossal (XII) motoneurons as a model. Individual neurons were studied with patch electrodes either in semi-loose (R ~50–300 MΩ) or in tight-seal (GΩ) configurations. The semi-loose seal allows prolonged recording of spontaneous (if present) or evoked (by voltage steps or extracellular K^+) spiking. In tight seal, the cell membrane is first broken down with voltage pulses (50–100 mV). Resulting pore formation in tight-seal mode allows for the stable recording of spontaneous or evoked postsynaptic potentials or currents and the passage of Neurobiotin without spillover into the adjacent cells. With either configuration, Neurobiotin electroporation is achieved by passing 500-ms pulses at 1 Hz (5–25 mV equivalent of ~300–500 pA) for ~5 min. Labelling quality of recovered cells is similar between the two configurations; however, the recovery rate of labelled neurons is higher with the semi-loose method. The efficiency and reliability of the semi-loose seal makes it particularly suitable for the serial recording and labelling of multiple neurons within the same preparation. Using semi-loose-seal and tight-seal configurations in combination can increase efficiency and data collected per preparation.

Key words: Brain slice, Dendritic morphology, Electroporation, Electrophysiology, Motoneuron, Neurobiotin, Riluzole

1. Introduction

Intracellular sharp-electrode (1–4), whole-cell patch-clamp (5–7), and juxtacellular (8) recording and labelling methods using glass microelectrodes have previously been used for combined analysis of cellular structure and function. Each of these methods has limitations, as outlined below, both in obtaining good electrophysiological recordings and in recovering the axonal, dendritic, and somatic morphology of the recorded neurons.

Paul M. Pilowsky et al. (eds.), *Stimulation and Inhibition of Neurons*, Neuromethods, vol. 78,
DOI 10.1007/978-1-62703-233-9_9, © Springer Science+Business Media, LLC 2013

Eccles and colleagues reported the first intracellular recording of action potentials in neurons of the vertebrate CNS (i.e. cat spinal cord motoneurons) (9). About 25 years later, sharp-electrode methods were used for dye filling and labelling of individual neurons with horseradish peroxidase and Lucifer yellow (LY) (1–3, 10). Smaller molecules, viz. biocytin and Neurobiotin (biotinamide), were subsequently discovered and used for intracellular dye filling (4, 11, 12) and tracing of long neuronal process (13). While sharp-electrode methods produce high-quality dye fills, obtaining and holding stable intracellular recordings is often problematic because the high-resistance tip needed for smaller neurons is so fine, making it fragile and prone to blockage. Consequently, cell recording is often lost before sufficient dye is injected, or the electrode becomes blocked during injection. Therefore, sharp electrodes are rarely used in brain slice preparations (14); instead, whole-cell patch clamping is preferred (7).

Although whole-cell patch-clamp methods (7, 15) overcome many of the electrophysiological problems of sharp-electrode recording, they are laborious and invasive. A tight GΩ seal requires a clean pipette tip, often leading to multiple electrode changes before recording is achieved. In principle, the neuron can be labelled by passive dye diffusion during the patch-clamp recording, but cellular damage, either during this process or on electrode withdrawal, often results in poor quality labelling. Therefore, dye filling using whole-cell patch-clamp recording has a low efficiency and limited reliability for sequentially obtaining electrophysiological and morphological data from multiple cells.

Juxtacellular-labelling methods (8) have been developed to label individual cells with dyes. The juxtacellular method relies on medium- to high-impedance electrodes for extracellular spike recording, by pushing onto the cell without monitoring seal resistance. Therefore, labelling of individual cells with the juxtacellular method requires large-amplitude pulses (10–20 V or 5–50 nA usually for 3–25 min) that can lead to cell damage and/or dye spillover to adjacent cells. It is therefore possible for cells other than the recorded neuron to be labelled, and moreover, the labelled cell(s) may not include the recorded neuron. The juxtacellular method has primarily been used by in vivo experiments (8, 16, 17). Juxtacellular method entirely relies on extracellular spike recording, and therefore, it is usually biased towards labelling spontaneously spiking neurons. This bias may at least partially be overcome by using antidromic and/or orthodromic electrical stimulation to make silent neurons fire.

We have recently developed a semi-loose-seal Neurobiotin electroporation method in whole-mount retina preparations, which largely overcome the limitations of the methods described above (18). Here, we describe modifications to this Neurobiotin electroporation method that can be used with either semi-loose or tight-seal configurations for electrophysiological recordings and reliable single-neuron dye filling in brain slices. As well as overcoming some of the difficulties of each method described previously, this method also increases the reliability of obtaining data from each recording and/or labelling of spontaneously spiking and non-spiking cells in brain slices. Serial and rapid recording and labelling of cells is likely to help provide faster answers to many remaining questions about neuronal structure and function relevant to physiopathological changes at cellular and subcellular levels.

2. Materials

2.1. Extracellular Solutions for Cutting Brain Slices at Different Stages of Mouse Development

High-Mg^{2+} Ringer solution used for cutting brain slices from E17 late embryonic to newborn pups (<P14) contained (in mM) the following: 130 NaCl, 3 KCl, 26 $NaHCO_3$, 1.25 NaH_2PO_4, 5 $MgCl_2$, 1 $CaCl_2$, and 10 d-glucose. However, we have also had successful slicing in embryonic (E17–E18) and newborns (P0–P3) using normal Ringer solution. Thus, use of high-Mg^{2+} solution seems to be more critical for ages older than P3. This may simply be due to a faster drop in the tissue temperature in younger animals due to small body mass, lack of fat, and faster dissection or exposure of brain tissues to the cold solution.

High-sucrose-Mg^{2+} solution used for cutting brain slices from juvenile rats (P14–P28) contained (in mM) the following: 218 sucrose, 3 KCl, 26 $NaHCO_3$, 1.25 NaH_2PO_4, 5 $MgCl_2$, 1 $CaCl_2$, and 10 d-glucose (19). This solution differs from that originally described by Aghajanian and Rasmussen (14), in that the sucrose concentration is lowered (from 252 to 218 mM) to reduce extracellular osmolarity from 355 to 290 mOsm, close to normal extracellular osmolarity. This solution is made fresh for each use (no stock).

Normal Ringer solution used for recording in all ages contained (in mM) the following: 130 NaCl, 3 KCl, 26 $NaHCO_3$, 1.25 NaH_2PO_4, 1 $MgCl_2$, 2 $CaCl_2$, and 10 d-glucose.

All Ringer solutions are continuously bubbled with 95% O_2/5% CO_2 to maintain pH at 7.4. Preparation of 10× salt solution stocks (without glucose and divalent cations) for normal and high-Mg^{2+} Ringer solutions is time saving, as they can be kept in the fridge for up to 3 months at 4°C. The same 10× stock solution, used for both the normal and the high-Mg^{2+} Ringer solutions, contains (in mM) the following: 1,300 NaCl, 30 KCl, 260 $NaHCO_3$, and 12.5 NaH_2PO_4. Glucose is added during diluting out the 10× solution to 1×. $MgCl_2$ (1 M or 5 M stocks) and $CaCl_2$ (2 M stock) are always added in to the diluted 1× solutions, following ~5-min bubbling with 95% O_2/5% CO_2 to prevent precipitation of divalent cations.

2.2. Artificial Intracellular Solutions Used for Filling Electrodes (Pipette Solutions)

Pipette solution for recording action potentials contained 135 mM K-methanesulphonate, 6 mM KCl, 1 mM EGTA (ethylene glycol-bis(2-aminoethyl ether)-*N*,*N*,*N*′,*N*′-tetraacetic acid), 2 mM $MgCl_2$, and 5 mM Na-HEPES (Na-4-2-hydroxyethyl-1-piperazineethanesulfonic acid), 3 mM ATP-Mg^{2+}, and 0.3 mM GTP–Tris (pH 7.25 with KOH, 305 ± 5 mOsm). In voltage-clamp mode, for recording persistent sodium current and excitatory and inhibitory currents, K^+-methanesulphonate in the pipette solution was replaced with Cs^+-methanesulphonate to reduce voltage-gated K^+ currents.

For labelling neurons, 2% Neurobiotin™ (catalogue # SP-1120, Vector Labs, Burlingame, CA, USA) was added to either pipette solution (see above). Neurobiotin solution can be kept in freezer and defrosted multiple times for up to ~4 weeks. The electrodes were pulled with a Narishige (Japan) PC-10 dual-stage microelectrode puller from borosilicate glass capillaries (1.5 mm OD, nonfilamented plain Micro Haematocrit Tubes, Modulohm, Vitrex Medical). The heater level is set to 78 for the #1 and 47 for the #2 heating to obtain outer pipette tip diameter of ~1.5 μm (giving a R ~3 MΩ). Electrode tips were initially dipped in the Neurobiotin solution for 30–60 s and then filled from top-end with ~3 μl of the same solution using a 25-μl Hamilton syringe (Grace Davison Discovery Sciences, Victoria, Australia). Any bubbles are removed by flicking the middle part of the electrode.

3. Methods

3.1. Preparation of Brain Slices

Embryonic pups were surgically obtained following euthanasia of the mother by cervical dislocation. The dams' abdomen was opened up with a large incision, and the uterus containing embryonic pups was removed and then cut through to get the pups out. Embryonic (E17–18) and newborn pups (P0–P3) were anaesthetized by hypothermia for ~3 min, while older mice were deeply anaesthetized by pentobarbital (i.p. 60–80 mg/kg). Heads of animals were dipped in ice-cold Ringer solution for ~10 s, following removal of scalp and underlying bones, and subsequent decerebration rostral to the pons. The preparation was then pinned down with 25-G needles the dorsal side up in a dissection bath (14 cm × 11 cm × 3 cm) filled with ice-cold Ringer slurry (see Note #1; high Mg^{2+} for <P14 and high-sucrose Mg^{2+} for >P14) to a level that just immersed the dorsal surface (Fig. 1a). The occipital bone was removed. The brainstem and the cervical spinal cord were then carefully dissected from the surrounding tissues by a laminectomy and sectioning ventral nerve rootlets with the aid of a Carl Zeiss Jena 212F OPM dissection stereomicroscope. For obtaining high-quality and healthy cells, it is critically important that the medulla oblongata

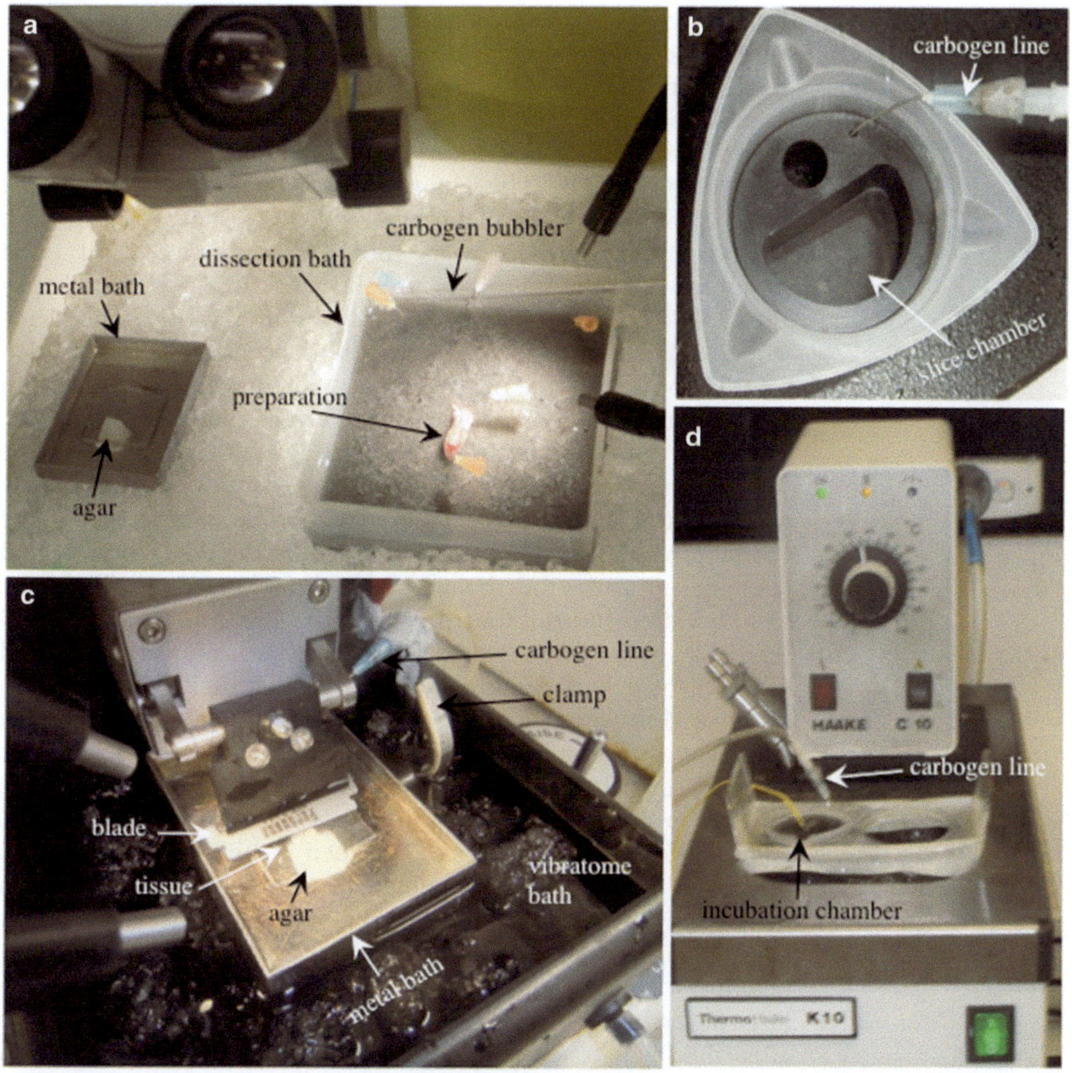

Fig. 1. Pictures of equipment used in preparation of brain slices. (**a**) Picture showing the dissection bath placed on ice with the preparation (E18) pinned in dorsal side up position immersed in ice-cold dissection solution (see Sect. 2.1) that was continuously bubbled with carbogen. The metal bath with superglued agar support was also kept on ice, ready for gluing dissected brainstem in front of the agar. (**b**) The metal bath containing the tissue in ice-cold solution was placed in the vibratome's bath and stabilized with a clamp. The vibratome's bath was then filled with ice-cold water or slurry to maintain temperature of metal bath and tissue at a low level. (**c**) Picture of an incubation chamber used for incubating brain slices prior to recording and labelling. (**d**) Picture of incubation chamber placed in a warm water bath heated to 35°C.

and the upper spinal cord are exposed to cold solution as fast as possible, preferably in less than a minute from the start of dissection. The rostral face (cut at the level of rostral pons) of the tissue block was quickly dried with a piece of filter paper and immediately fixed with a small drop of superglue (Selleys Quick Fix) onto a holding metal bath (55 mm × 75 mm, Fig. 1a), spinal end up and dorsal side of the medulla facing the vibratome blade, while the ventral side rested on an agar block (see Note #2) previously superglued to the metal bath (Fig. 1a, b). The metal bath was then immediately filled with ice-cold extracellular solution (same type as the one used during dissection) and clamped onto the vibratome's larger bath (Fig. 1b). The vibratome's bath was filled with ice-cold water or slurry to keep metal bath and the tissue cool (Fig. 1b). Transverse brainstem sections (300-μm thick) were cut on a vibratome (DSK, frequency set at 9, speed set at 3). Bilateral hypoglossal nuclei can usually be identified with a dissection microscope, as darker spots located ventral to the central canal or the 4th ventricle on either side of the midline. The nucleus is ~1,200-μm long, and therefore, only four sections can be obtained per animal. The most caudal section is collected about 600 μm caudal to the cerebellum where the central canal also positioned immediately dorsal to the centre of the tissue. In the following sections, the central canal moves more dorsally, reaching the dorsal surface and opening out to the fourth ventricle in the third section. The most rostral section has the fourth ventricle with a wider V shape and cerebellar tissue present dorsally. Selected slices are then moved from ice-cold bath solution (high Mg^{2+} or high-sucrose Mg^{2+}) using suction with a Pasteur pipette cut and flame rounded to a diameter of 3–4 mm and incubated in an incubation chamber (Fig. 1c; see below) for 45 (for <P14 mice) to 60 (for >P14 mice) minutes in the same bath solution bubbled with 95% O_2 and 5% CO_2 at 35°C (Fig. 1d). Slices obtained in high-sucrose solution from mice >P14 are additionally incubated in high-Mg^{2+} solution for 30 min. The sections are then moved to normal Ringer solution in a similar incubation chamber and kept at room temperature for at least 30 min prior to start of recording and labelling. Slices will typically produce viable recording for up to 6 h after this final transfer. Bacterial infections can be a significant problem and reduce the life of tissue if the tools, baths, and the tubing in contact with the tissues or the extracellular solutions are not kept sterile. We use treatment with 0.5 M NaOH for 20–30 min, followed by several rinses with plenty of de-ionized water prior to drying in an oven set at 55°C.

The incubation holder is a small plastic beaker, with a custom-made insert that fits in the upper part of the beaker; the insert has a larger hole with stocking material stretch across it for slice storage and a smaller hole into which a gas line emits carbogen bubbles, so that solution flows up out of the small hole and down through the stocking net over the large hole, keeping slices securely pressed against the net (Fig. 1c). Warming up solutions often results in formation of small bubbles (microbubbles) that can prevent sliced tissue's contact with the external solution. These bubbles should be removed by tapping the incubation holder against a flat surface (e.g. table), prior to transfer of slices.

A single brainstem slice was placed in a tissue chamber and stabilized with the aid of a U-shaped metal mesh (10 mm × 10 mm) made from flattened platinum wire (diameter, 1 mm) on a Zeiss Axioskop 2 FS microscope fitted with IR-DIC optics and continuously superfused at 3–4 ml/min with the normal Ringer solution at room temperature (22–24°C) for the duration of experiment, similar to previous studies (18); while the Ringer solution can be heated to 27–32°C to more closely mimic in vivo temperatures, this significantly shortens the time over which healthy neurons can be maintained in brain slices. The tissue and recording electrodes were viewed on a Sony monitor through a 40× water-immersion objective and a 2× relay attachment located in front of the infrared video camera (WAT-902H, Watec Co. Ltd., Japan) mounted above the camera port of the microscope.

3.2. Recording and Labelling with Microelectrodes Filled with Solutions Containing Neurobiotin

Recordings and electroporation (controlling the amplitude and the duration of the voltage pulses) were made with an Axopatch 1D amplifier (Axon Instruments, Foster City, CA, USA). Data were acquired at a sampling rate of 5 or 10 kHz, low-pass filtered at 2 kHz, and stored on a Macintosh computer using Axograph 4.9 software. Hypoglossal motoneurons with large somata were clustered in the brainstem positioned immediately ventrolateral to the 4th ventricle or the rostral end of the central canal. Individual motoneurons were targeted microscopically, and a patch electrode was advanced towards the cell body with the aid of a manipulator (MPC-200, Sutter Instrument Company). No pressure was applied till the electrode was placed immediately above the cell of interest. Under visual guidance from the video monitor, a brief positive internal pressure was then applied via a mouth piece for 5–10 s to keep the microelectrode tip clear during penetration into the tissue. If cleaning debris from the target cell surroundings is required, then an initial electrode filled with extracellular medium can be used prior to introduction of Neurobiotin containing electrode (18). Prolonged (>10 s) and extreme positive pressure (e.g. using a 10- or 20-ml syringe) blowing much Neurobiotin in to the tissue should be avoided to prevent undesired background or non-specific labelling due to leakage of Neurobiotin. Tip of the electrode was gently and rapidly pushed against the soma, and then a continuous suction was applied until a stable seal of preferably >30 MΩ was obtained (see Note #3).

3.2.1. Loose-Seal and Semi-Loose Seal Configurations

XII motoneurons (MNs) usually did not show any spontaneous spiking when recordings were done from intact somatic membranes with electrodes filled with the bath (normal Ringer) solution in loose-seal (<25 MΩ), semi-loose-seal (50–300 MΩ), or tight-seal (>500 MΩ) configurations. If the soma membrane injured, an injury discharge lasting than a minute would be observed. However, when recordings were made from intact membranes with electrodes filled with K^+-based artificial intracellular solution (see Sect. 2.2), XII MNs may display varying levels of spontaneous spiking that can be seen with loose-seal (not shown) and semi-loose-seal (50–300 MΩ) configurations (Fig. 2a). Such spiking likely results from membrane depolarization due to local increases in extracellular K^+ concentrations originating from pipette tip during blowing of pipette solution to clean access to the soma and also leakage of the internal solution from loosely sealed pipette-membrane interface.

Although loose-seal method can be useful for recording extracellular spiking, we find it not so successful for filling neurons. Our attempts to label cells in loose-seal configuration often (~50%) resulted in no filling, filling with low intensity, or filling of more than one cell, possibly due to dye spillover because of low seal resistance allowing a larger gap between the electrode tip and the target cell's membrane.

Semi-loose-seal Neurobiotin electroporation procedure was modified from that described for the retinal ganglion cells (18). Square-wave voltage pulses (0.5 s, 1 Hz) of 5–25 mV that generated current pulses of 300–500 pA were typically applied for 5–6 min (Fig. 2a; see Note # 4). In semi-loose configuration, usually, membrane breaks down easily in the presence of Neurobiotin in the pipette solution (18). Both the baseline shift and the spike amplitude are increased with increasing seal resistance. In the semi-loose seal, the baseline shift was variable ~−20 to −30 mV, often requiring hyperpolarizing DC current injections (−0.1 to −1.0 nA) to bring the membrane potential to a level where the spiking is significantly reduced (−30 to −50 mV). Recording of action potential spikes in this mode allows prolonged recording to test actions of drugs such as riluzole (Fig. 2b). Although this method does not provide an ideal voltage clamp, it is often sufficient to voltage-clamp the cell at a hyperpolarized level that can be stimulated by voltage steps or ramps to more positive potentials (Fig. 2c). Given that most XII MNs are not spontaneously active when recorded with loose-seal method, unless stimulated by an increase in extracellular K^+, these voltage manipulations can be useful in triggering spiking activity (Fig. 2c). Once spiking is initiated, drugs can be tested (Fig. 2c).

The effects of riluzole are hard to test in whole-cell patch-clamp mode; in particular, preparing concentration-response curves are impossible. Firstly, the persistent sodium current present in MNs usually washes out within minutes of rupture, and secondly, spiking never recovers after riluzole treatment in whole-cell mode. In contrast, semi-loose seal method does not wash out the intracellular content and allows spike recording at the termination of riluzole treatment. Therefore, semi-loose seal recordings can provide some vital datapoints where whole-cell patch clamp fails.

Furthermore, because the semi-loose-seal method is less invasive on the neuronal membranes, survival rates of Neurobiotin

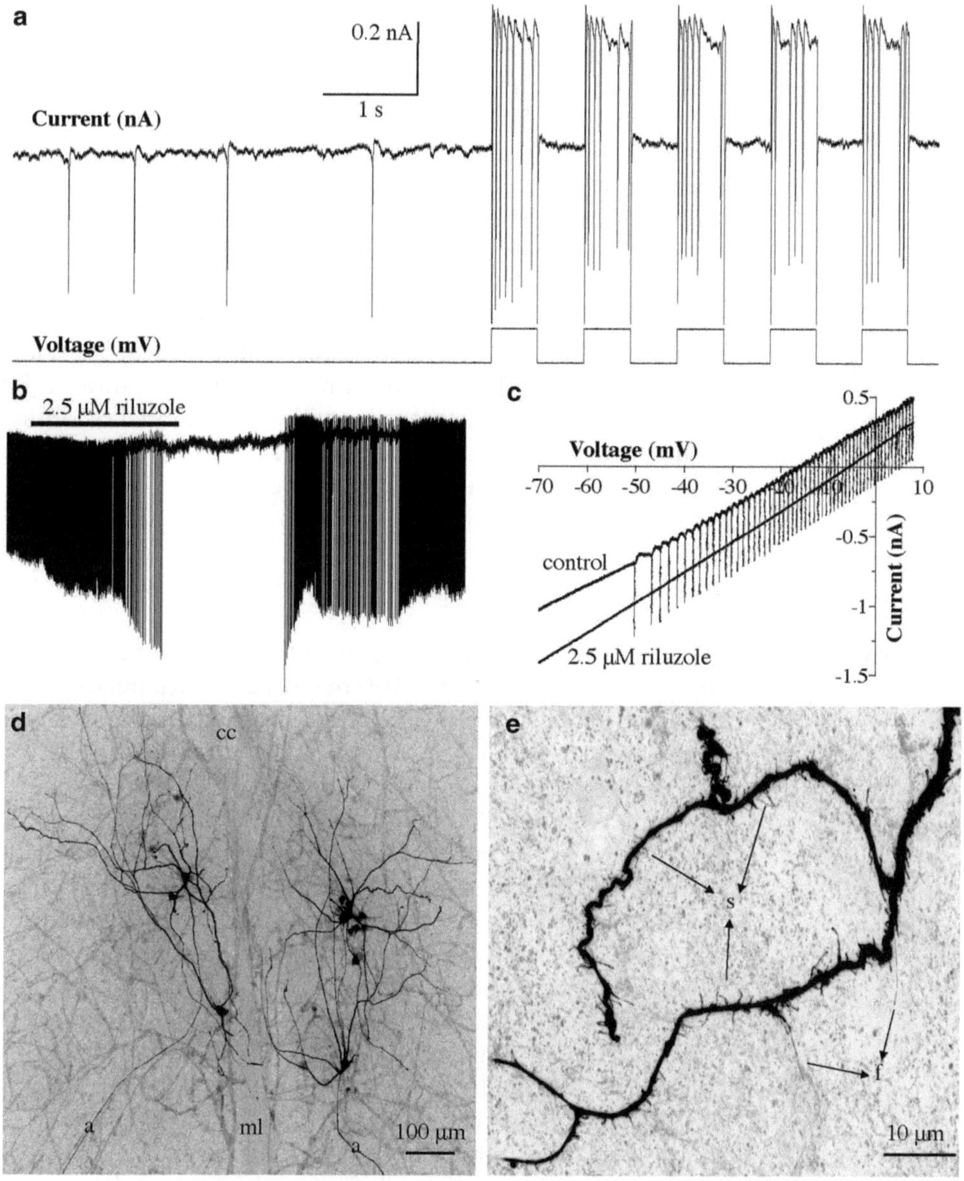

Fig. 2. Semi-loose-seal recording and Neurobiotin electroporation in a hypoglossal motoneuron (XII MN). (**a**) Spike recordings of a XII MN in semi-loose-seal configuration with a patch electrode containing 2% Neurobiotin in a K-methanesulphonate-based artificial intracellular solution. Neurobiotin was electroporated into the cell by application of +15 mV (350 pA) pulses at 1 Hz ($V_h = -50$ mV; $R_{seal} = 50$ MΩ). (**b**) Semi-loose-seal configuration allows prolonged spike recordings desired for testing drug actions, as shown here for riluzole ($V_h = -30$ mV, –0.1 nA hyperpolarizing current applied). (**c**) In cells, voltage-clamped at –30 mV or more negative, spiking activity evoked by a slow (4 s long) voltage ramp from –70 mV to +10 mV can be mostly blocked by bath application of riluzole (2.5 μM). (**d**) A low-magnification confocal image of four XII MNs individually electroporated with Neurobiotin using semi-loose-seal configuration in a P0 mice brainstem slice. The cells are located in the hypoglossal nucleus located beneath the central canal (*cc*), two on the *left* (dorsal and ventral), and two on the *right side* (dorsal and ventral) separated by the midline (*ml*). Axons (*a*) of the labelled cells can be seen extending at the ventrolateral direction to exit the brainstem. Dorsally located MN on the *right side* is tracer coupled to other XII MNs. (**e**) A high-magnification confocal image showing dendritic spines (*s*) and filopodia (*f*) on ventrolateral dendrites of a XII MN.

electroporated and recorded neurons after electrode withdrawal are very high at ~95% compared to whole cell at ~65%, allowing serial recording and labelling of many cells from the same slice preparation (Fig. 2d). This is important when the number of slices is limited (3–4 slices per animal in case of the hypoglossal nucleus). Furthermore, the cells in these slices are usually viable only for 4–6 h; this method can help in obtaining more neuronal labelling per experiment. Semi-loose seals can be made with a lower probability of killing the cell, even in the presence of small obstructions at the pipette interface (18, 20, 21). This property allows the same electrode to be used, if required, for multiple recordings and dye filling. The semi-loose-seal Neurobiotin electroporation technique is simple and fast, potentially allowing the sequential recording and dye filling of multiple cells in a small area of tissue (Fig. 2d). We currently have ~95% success rate in recovering MNs labelled with this method. The quality of labelling is very good, and further quantitative analysis can be carried out at high magnification of dendritic filopodia and spines (Fig. 2e).

3.2.2. Tight-Seal Mode

If a tight seal formed during suction through the pipette, then the cell was stimulated with 50–100 mV square-wave pulses of 500-ms duration at 1 Hz, until a significant drop in the membrane input resistance (e.g. ~80 MΩ), which usually occurred within 10–30 s (Fig. 3a, arrow); this signals a breakdown of the membrane at the pipette tip. It is usually better to try 50-mV steps for up to 30 s; if the membrane does not break, then the voltage stimulation can be increased to 100 mV, as shown in Fig. 3a. Alternatively, increasing amplitude of 1–2-s-long voltage steps at 10-mV increments can be used (18). Either way, as soon as the membrane breaks, the voltage steps are stopped and the synaptic currents can be recorded in desired holding potentials (Fig. 3b; see also end of tracing in Fig. 3a). The currents recorded at −70 mV, close to the reversal potential for Cl^-, will mainly represent excitatory synaptic inputs, while recordings at −15 mV close to cation current reversal potential will represent inhibitory synaptic currents. In the tight-seal configuration, the baseline shift was close to resting membrane potential (−50 to −65 mV), and the spike amplitude was ~80 mV. Since this is a good voltage clamp compared to semi-loose-seal configuration, the membrane currents of interest can be recorded (Fig. 3c). In our case, we were interested in persistent Na^+ currents present in motoneurons, and therefore, we used Cs-based artificial intracellular solution (see Sect. 2.2) to block voltage-gated K^+ channels that disguise the persistent current (Fig. 3c). Bath application of riluzole completely blocks persistent Na^+ current including action current and potential. For recording action potentials in current-clamp mode, the pipette solution must contain K^+-based artificial intracellular solution (see Sect. 2.2), as seen in Fig. 3d. This allows recording of spontaneous and current- or voltage-evoked action potentials at rest or at various holding potentials (Fig. 3d).

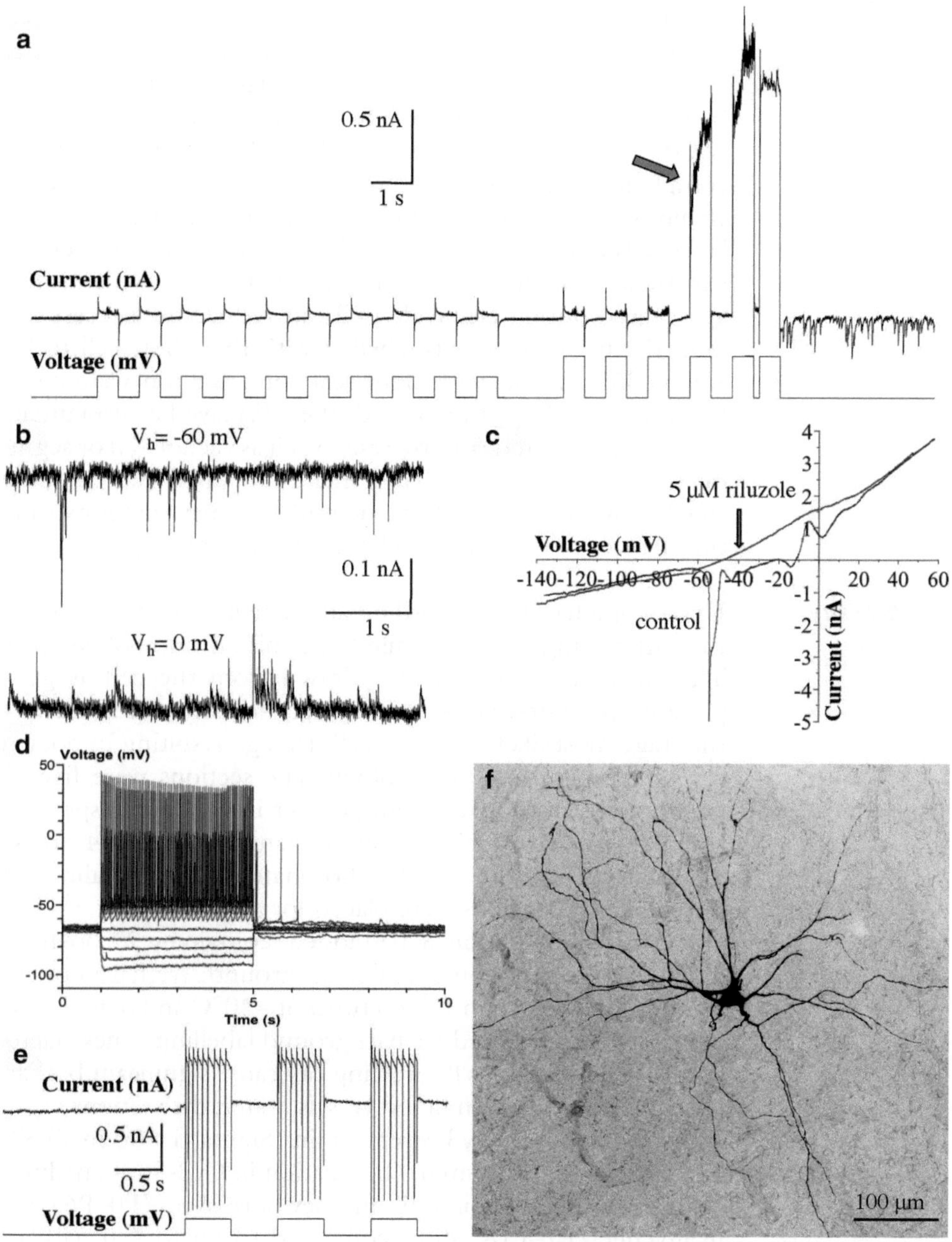

Fig. 3. Recordings and Neurobiotin electroporation in tight-seal mode of hypoglossal motoneurons. (**a**) Example of Neurobiotin electroporation in tight-seal configuration with a patch electrode containing Cs-methanesulphonate-based artificial intracellular solution ($V_h = -60$ mV, $R_{seal} = 1.5$ GΩ). Application of 50-mV pulses to the membrane for 10 s at 1 Hz did not break the membrane (on the *left*). Increasing pulse amplitude to 100 mV rapidly broke the membrane (*arrow*) and allowed recording of synaptic currents, as seen after stopping voltage pulses (on the right). (**b**) Recording spontaneous synaptic currents at −60 mV (*top*) and at 0 mV (*bottom*) holding potentials; reversal of synaptic currents demonstrates recording of transmembrane potential has occurred. (**c**) Current responses evoked by a voltage ramp before and after riluzole application ($V_h = -60$ mV). (**d**) Example of membrane potential recordings in current-clamp mode using K-methanesulphonate-based artificial intracellular solution in the pipette. The membrane potential was held at resting level of −65 mV, and 4-s-long pulses were applied to change membrane potential; larger depolarizing steps elicited repetitive firing of overshooting action potentials. (**e**) Neurobiotin labelling was usually done at the end of a recording session by application of 300–500-pA currents at 1 Hz for 3–5 min. (**f**) Confocal image of a XII MN that was Neurobiotin electroporated in tight-seal mode.

In this mode, the cells can be labelled by diffusion of Neurobiotin through the pores formed during electroporation, given that the recording configuration lasts for at least 20 min. To be sure, we often stimulate a further 3–5 min with current pulses of 300–500-pA amplitude to fill the cells with Neurobiotin at the completion of protocols and recording (Fig. 3e). Whole-cell patch clamp with dye-containing electrodes has previously been used to label cells in brain slices, usually limited to one or two cells. In our experience, the difficulties with tight-seal formation and the damage caused during electrode pull-out reduce the success rate and limit the number of cells labelled with the whole-cell technique. About 35% of cells with tight-seal configuration were either not recovered at all, or if recovered, they displayed a substantial damage to cell soma and/or processes, such as vacuolated or segmented dendrites and/or missing cell soma. Immediate fixation of the slice (see below), following electrode pull-out, may sometimes increase the chances of recovering full neuronal morphology.

3.3. Histochemistry

Upon completion of recording and labelling, sections were left in the bath resting for ~5–10 min to permit diffusion of Neurobiotin. The electrode was then withdrawn from the cell as gently as possible; in tight-seal mode, this is particularly important, as this is the stage most likely to cause cell damage resulting in poor recovery of the neuronal morphology. The sections were fixed in 4% paraformaldehyde made from powder in 0.1 M phosphate buffer (pH 7.4) for 20–30 min and subsequently washed 3–4 times (each 15–20 min) in 0.1 M phosphate-buffered saline (PBS). Subsequently, sections were placed for a minimum of 2 h in PBST (PBS with 0.05% Triton-X 100 added) containing 2% bovine serum albumin to reduce non-specific background. We have noticed that using fixative stored in 1.5-ml tubes at −20°C and defrosted on the day of use also help reduce background labelling. These factors are critical as endogenous fluorescing can cause significant background labelling particularly in blood vessels. Later, the sections were incubated for 2–4 h in Cy3-Streptavidin (Sigma, 1:500 in PBST with 2% bovine serum albumin). Incubation in Cy3-Streptavidin can be reduced to 2 h in embryonic and newborn slices (P0–P6) or to 3 h in juvenile (P7–P18) slices compared to up to 4 h with mature slices (>P18). The sections were viewed with a fluorescence microscope in PBS to ensure that the labelled cells were positioned upwards facing the cover slip, not the glass slide. Some of the sections that contained excellent-filled cells were selected for further double or triple immunolabelling with antibodies to various excitatory (VGLUT-2, PSD-95) and inhibitory (VGAT, gephyrin, GABA-A alpha 1 subunit) synaptic markers (not shown). The sections were individually mounted on uncoated glass slides using a glycerol-based mounting medium, containing 90% glycerol and 10% paraphenylenediamine (10 mg/ml in PBS, pH 8.0). Small glass cover slips (22 mm × 40 mm) were further cut into three smaller pieces with a diamond knife to reduce the amount of pressure they put onto brain slices. Using smaller and lighter cover slips is particularly important in embryonic and newborn slices, as the weight of the cover slip can squeeze and damage fragile sections at younger ages. The slides were then left to settle for ~half an hour; subsequently, the edges of the cover slips were sealed with a nail polish and dried for ~10 min. The sections were then stored in a sealed container in the fridge (safely for up to a year) until they were imaged with a Zeiss LSM 510 Meta confocal microscope. ImageJ64 and Imaris programs can be used for subsequent analysis of neurons.

The tissue was processed to reveal the full morphology of the dye-filled cells with exceptional contrast: the Neurobiotin was confined to the soma and dendrites of the recorded cell, with negligible spillover in the extracellular space (Figs. 2d–e, 3f). Some motoneurons showed tracer coupling to neighbouring motoneurons in late embryonic and early neonatal mice (Fig. 2d, top cell on the right), as described previously for the rat (22). Although we have described labelling of XII MNs in this short section, combination of these two methods that rely on seal resistance measurements can also be used in study of other cell types that can be located under the microscope, such as interneurons, glia, and ependymal cells, alone or in combination to reveal morphological interactions among different cell types (not shown). This is particularly relevant when the sections are obtained from animals that have genetically fluorescent cells, such as mice with GFP-labelled somatostatin or GAD-67-containing interneurons.

Subsequently, morphological properties (dendritic projection, branching and length, dendritic spines, tracer coupling) of filled cells can be analyzed. In addition, filled cells can be further characterized by combining immunocytochemical methods using antibodies against various pre- and postsynaptic markers, ion channels, or receptors. As recovery of neuronal morphology uses a fluorescent marker conjugated to streptavidin, the number of possible subsequent antibody labels is increased. These can be used for (1) determining normal developmental changes in wild-type animals and (2) comparing these findings to genetically modified animals.

4. Notes

1. An ice-cube tray filled with the needed solution (e.g. high-sucrose Mg^{2+}) is frozen in the freezer, and these ice cubes are mixed with the same solution using an electrical kitchen blender to make a slurry immediately prior to the slice preparation.

2. The agar is made by adding 4.5 g agar into the 100-ml Ringer solution, heated to boiling temperature while stirring, then poured into a Petri dish to cool to room temperature. The agar block (~1.2 cm height) then can be stored at upside down position in the same Petri dish at 4°C for ~2 months, and small pieces (~1 cm × 1 cm) are cut with a blade for each experiment. These small pieces of agar are carved into shapes that will provide support to tissue during cutting.
3. Generally we considered seal resistances <25 MΩ as loose seal, 50–300 MΩ as semi-loose seal, and >500 MΩ as tight seal. Although extracellular spiking can be reliably recorded in loose-seal configuration, usually seal resistances of >30 MΩ is required for increasing successful filling of individual neurons without dye spillover that resulted in labelling of surrounding cells or no labelling at all. This is particularly important in the embryonic and newborn stages when motoneurons are densely packed and display dye-coupling (the transfer of dye from originally filled cell via gap junctions to other cells). Loose-seal filling also often resulted in labelling of multiple motoneurons in animals >P14 (not shown), ages in which we do not see dye-coupling with semi-loose-seal or tight-seal fillings. The recovery rate of successfully filled cells without spillover or damage was higher under semi-loose seal (~95%) than with tight seal (~65%) or loose seal (~55%).
4. Biocytin is similar to Neurobiotin and is equally good to use for filling cells, with an advantage that it is of nearly neutral charge and thus can be electroporated either with depolarizing or hyperpolarizing voltage/current pulses. Lucifer yellow can also be used with either method to label cells in a different colour, although it requires stronger hyperpolarizing voltage or current steps of ~100 mV/1 nA to fill cells (18) because it is a larger molecule. Use of hyperpolarizing pulses to fill neurons stabilizes membranes compared to large depolarizing pulses that can destabilize the cell and lead to death. Another disadvantage is that LY signal is low and rapidly fades, therefore, its use, as a permanent marker, often requires further amplification using anti-LY antibodies.

Acknowledgements

We are grateful to David Vaney for his generous support. Matthew Fogarty and Luke Hammond are thanked for their help with confocal imaging. The project was supported by grants to MB from the Australian National Health and Medical Research Council

(569827), the Australian Brain Foundation, and the Motor Neurone Disease Research Institute of Australia.

References

1. Snow PJ, Rose PK, Brown AG (1976) Tracing axons and axon collaterals of spinal neurons using intracellular injection of horseradish peroxidase. Science 191:312–313
2. Brown AG, Rose PK, Snow PJ (1977) The morphology of spinocervical tract neurones revealed by intracellular injection of horseradish peroxidase. J Physiol 270:747–764
3. Brown AG, Fyffe RE (1984) Intracellular staining of mammalian neurons. Academic, London
4. Horikawa K, Armstrong WE (1988) A versatile means of intracellular labeling: injection of biocytin and its detection with avidin conjugates. J Neurosci Methods 25:1–11
5. Imanaga I, Kameyama M, Irisawa H (1987) Cell-to-cell diffusion of fluorescent dyes in paired ventricular cells. Am J Physiol 252:H223–H232
6. Werblin F, Maguire G, Lukasiewicz P, Eliasof S, Wu SM (1988) Neural interactions mediating the detection of motion in the retina of the tiger salamander. Vis Neurosci 1:317–329
7. Edwards FA, Konnerth A, Sakmann B, Takahashi T (1989) A thin slice preparation for patch clamp recordings from neurones of the mammalian central nervous system. Pflugers Arch 414:600–612
8. Pinault D (1996) A novel single-cell staining procedure performed in vivo under electrophysiological control: morpho-functional features of juxtacellularly labeled thalamic cells and other central neurons with biocytin or Neurobiotin. J Neurosci Methods 65:113–136
9. Brock LG, Coombs JS, Eccles JC (1952) The recording of potentials from motoneurons with an intracellular electrode. J Physiol 117:431–460
10. Stewart WW (1978) Functional connections between cells as revealed by dye-coupling with a highly fluorescent naphthalimide tracer. Cell 14:741–759
11. Kita H, Armstrong W (1991) A biotin-containing compound N-(2-aminoethyl)biotinamide for intracellular labeling and neuronal tracing studies: comparison with biocytin. J Neurosci Methods 37:141–150
12. Vaney DI (1991) Many diverse types of retinal neurons show tracer coupling when injected with biocytin or Neurobiotin. Neurosci Lett 125:187–190
13. Lipski J, Zhang X, Kruszewska B, Kanjhan R (1994) Morphological study of long axonal projections of ventral medullary inspiratory neurons in the rat. Brain Res 640:171–184
14. Aghajanian GK, Rasmussen K (1989) Intracellular studies in the facial nucleus illustrating a simple new method for obtaining viable motoneurons in adult rat brain slices. Synapse 3:331–338
15. Hamill OP, Marty A, Neher E, Sakmann B, Sigworth FJ (1981) Improved patch-clamp techniques for high-resolution current recording from cells and cell-free membrane patches. Pflugers Arch 391:85–100
16. Schreihofer AM, Guyenet PG (1997) Identification of C1 presympathetic neurons in rat rostral ventrolateral medulla by juxtacellular labeling in vivo. J Comp Neurol 387:524–536
17. Pilowsky PM, Makeham J (2001) Juxtacellular labeling of identified neurons: kiss the cells and make them dye. J Comp Neurol 433:1–3
18. Kanjhan R, Vaney DI (2008) Semi-loose seal Neurobiotin electroporation for combined structural and functional analysis of neurons. Pflugers Arch 457:561–568
19. Bellingham MC, Berger AJ (1996) Presynaptic depression of excitatory synaptic inputs to rat hypoglossal motoneurons by muscarinic M2 receptors. J Neurophysiol 76:3758–3770
20. Neher E, Sakmann B, Steinbach JH (1978) The extracellular patch clamp: a method for resolving currents through individual open channels in biological membranes. Pflugers Arch 375:219–228
21. Roberts WM, Almers W (1992) Patch voltage clamping with low-resistance seals: loose patch clamp. Methods Enzymol 207:155–176
22. Mazza E, Nunez-Abades PA, Spielmann JM, Cameron WE (1992) Anatomical and electrotonic coupling in developing genioglossal motoneurons of the rat. Brain Res 598: 127–137

Chapter 10

Juxtacellular Neuronal Labelling, Physiological Characterization and Phenotypic Identification of Single Neurons In Vivo

Anthony J.M. Verberne and Ida J. Llewellyn-Smith

Abstract

Characterisation of the phenotype and physiological responsiveness of a neuron and identification of its innervation target(s) are important objectives for neuroscientists. This chapter describes the technique of juxtacellular neuronal labelling and presents procedures for the light and electron microscopic visualisation of labelled neurons and the identification of their neurochemical phenotypes and inputs. For those with little experience in extracellular neuronal recording, a detailed description of the required pieces of equipment and how they are configured is also provided.

Key words: Extracellular, Electrophysiology, Immunohistochemistry, Neurochemistry, Ultrastructure

1. Introduction

Extracellular single-unit recording is an important technique in the neuroscientist's armamentarium. It has helped to identify the physiological properties and projection targets of a broad range of neuronal types in the central nervous system (1). Extracellular single-unit recording combined with antidromic mapping initially served as an important method for identification of the projection targets of single neurons (2). However, this method was largely supplanted by modern anterograde and retrograde neuronal tracing agents as well as by neurotropic viral tracers (3). Nevertheless, extracellular single-unit recording continues to be a useful method for studying the responsiveness of single neurons to sensory, chemical or other types of stimuli. Prior to the development of the juxtacellular neuronal labelling technique by Pinault (4, 5), the phenotype of a recorded neuron could only be identified with certainty by combining intracellular recording and intracellular filling

Paul M. Pilowsky et al. (eds.), *Stimulation and Inhibition of Neurons*, Neuromethods, vol. 78,
DOI 10.1007/978-1-62703-233-9_10, © Springer Science+Business Media, LLC 2013

with a dye such as Lucifer yellow or biocytin combined with subsequent histochemical processing (6, 7). While intracellular recording can provide important information about intracellular currents and potentials that cannot be accessed by extracellular recording, it is difficult and time-consuming, and the yield of filled neurons is low. Consequently, its use for routine experiments on anaesthetised breathing experimental animals is usually not justified. In contrast to intracellular dye filling, it is possible to label several electrophysiologically characterised neurons per day with the juxtacellular method. Furthermore, juxtacellular neuronal labelling appears to be a generally applicable technique and has been used to label a diverse array of brain neurons. These include neurons in the rostral and caudal ventrolateral medulla, the nucleus of the solitary tract, retrotrapezoid nucleus, Barrington's nucleus, locus coeruleus, motor cortex and horizontal limb of the diagonal band of Broca (8–15). Juxtacellular neuronal labelling has also been applied in brain regions with high neuronal density and compactness, such as the locus coeruleus (16) or the nucleus of the solitary tract (15), another indication of the method's utility.

Juxtacellular neuronal labelling involves extracellular single-unit recording with glass microelectrodes filled with a solution containing a biotin derivative, such as biotinamide ((5-((3aS,4S,6aR)-2-oxohexahydro-1H-thieno(3,4-d)imidazol-4-yl)pentanamide) also referred to as (*N*-(2-aminoethyl)biotinamide; Neurobiotin) or biocytin (*N*ε-biotinyl-L-lysine). This chapter outlines the basic technique of juxtacellular neuronal labelling, its advantages and disadvantages, the equipment required for the technique and the procedures that are used to detect labelled neurons and to identify the neurochemical phenotypes of the neurons and the axons that innervate them.

The precise mechanism underlying juxtacellular neuronal labelling is not completely understood, and it has recently been suggested that juxtacellular labelling is similar to electroporation (17). Nevertheless, for over two decades, extracellular application of DC current, either as pulses or continuously, has been known to alter the discharge of a recorded neuron (18). In practice, application of positive current (microelectrode tip positive) excites neurons, while negative current inhibits neuronal discharge. Pulses of positive DC current are usually used to obtain juxtacellular neuronal labelling. Labelling can also be achieved using continuous current application, but the electrode tip is more likely to block and becomes unusable during this procedure. It is also possible to identify the innervation targets of extracellularly recorded neurons by antidromic stimulation and the collision test (2). When neurons are silent, it is not possible to perform a collision test. However, application of a small amount of juxtacellular DC current is often sufficient to induce spiking activity, allowing the completion of this characterization.

Evidence supporting the conclusion that a juxtacellularly labelled neuron is identical to the neuron monitored during extracellular recording is substantial: (1) a labelled neuron is only found if the recorded unit is entrained by the juxtacellular current pulses, (2) application of juxtacellular current pulses in the absence of action potentials does not result in a labelled neuron and (3) application of current pulses large enough to impair the integrity of the neuronal membrane after completion of the filling procedure yields a labelled but damaged cellular profile. There are several safeguards that, if strictly adhered to, ensure that only the recorded neuron is labelled:

1. We strongly recommend using an electrode impedance meter to measure the impedance of a biotinamide-filled electrode 'in vitro' prior to insertion into the brain. Optimum electrode impedance ensures that the spike amplitude of the recorded neuron is in excess of 1 mV. This spike amplitude means that the signal-to-noise ratio is excellent and makes it easy to confirm that only a single neuron is being recorded. Measuring impedance also saves time because electrodes whose impedance is too low or too high can be rejected before they are inserted into the brain.
2. It is critical to monitor the discharge of the neuron constantly during application of juxtacellular current pulses so that it is possible to confirm in real time that only a single neuron is being recorded. We use an oscilloscope to monitor extracellular neuronal discharge while attempting entrainment. We also recommend recording each entrainment using data acquisition software such as Spike 2 (Cambridge Electronic Design, Cambridge, UK).
3. Entrainment of neuronal activity to the juxtacellular current pulses is a prerequisite for successful labelling. Entrainment is indicated by a reduced probability of neuronal firing during the 'off' phase of the pulse train and a marked increase in the probability of firing during the 'on' phase (Fig. 1). The degree of filling depends on the duration of the entrainment period as well as the intensity of the applied current pulses. In our experience, neurons are often labelled after even very brief periods (5–10 s) of entrainment to the juxtacellular stimulation, but only their cell bodies and most proximal dendrites are filled. Extensive labelling of dendritic trees usually requires up to 10 min of entrainment.
4. Labelling a second neuron in close proximity to the first labelled cell should never be attempted. This ensures that a labelled neuron can be identified unambiguously in a tissue section. If it is desirable to label a second neuron and the local neuroanatomy permits discrimination, we advise moving to a site several hundred micrometres distant. Alternatively, a second

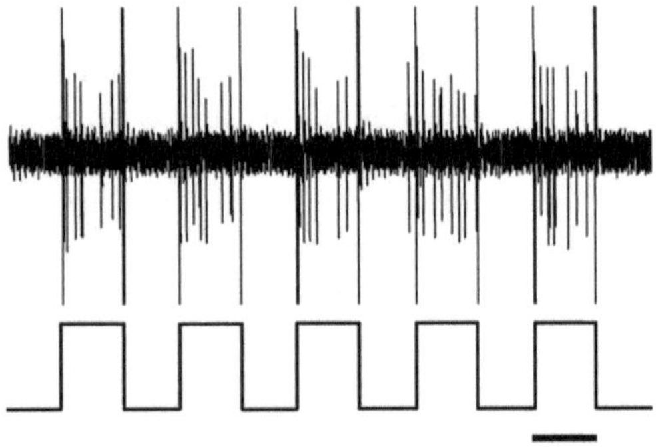

Fig. 1. Entrainment of neuronal discharge to juxtacellularly applied current pulses. Note the absence of spiking activity during the 'off' phase (lowest point) of the pulse train. The pulse width is 200 ms at a rate of 2.5 Hz.

neuron can be identified and labelled on the contralateral side of the brain.

5. If the neuron can still be recorded after juxtacellular labelling, it is prudent to establish that antidromic stimulation elicits an antidromic spike with a latency identical to that obtained while the filled neuron was being characterised.

2. Materials

2.1. Equipment

1. Microelectrode puller (e.g. Narashige PE-22).
2. Intracellular electrometer amplifier (e.g. Intra 767, World Precision Instruments; Intracellular Electrometer, Model 3100, A-M Systems; Electrometer Intracellular Amplifier IE-210, Warner Instruments; Intracellular amplifier AM3100, AD Instruments).
3. Bandpass amplifier (e.g. Xcell3, Frederick Haer).
4. Analogue or digital storage oscilloscope.
5. Microcomputer-based data acquisition system (e.g. Cambridge Electronic Design Power 1401 and Spike 2 software).
6. Analogue or digital stimulator or PIC microcontroller-based pulse generator (see Fig. 2). To control the intensity of the current pulses, it is usually necessary to alter the amplitude of the pulses provided by the stimulator. Therefore, if the amplitude of the pulses from the digital stimulator cannot be varied (standard transistor–transistor logic (TTL) pulses), we recommend using the PIC microcontroller design depicted in Fig. 2. This device could be constructed by any university electronics workshop. The BASIC code used to program the PIC microcontroller is shown in Fig. 3.

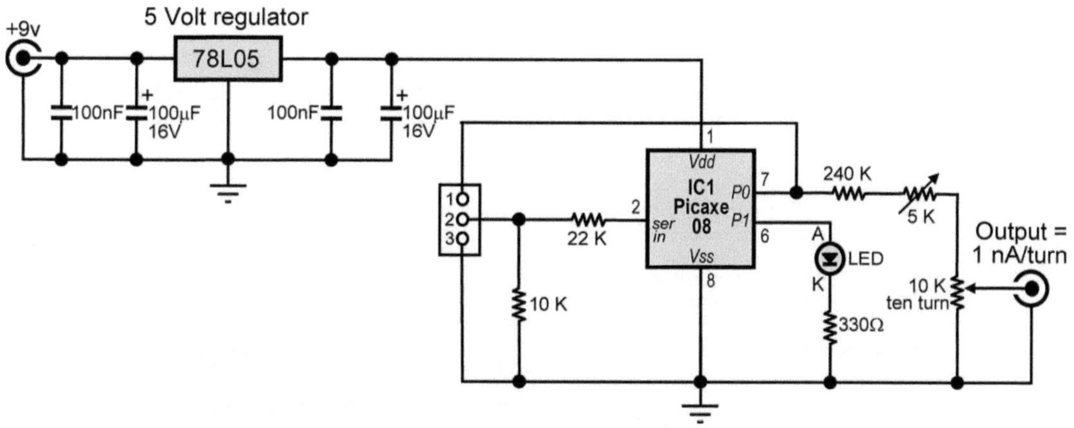

Fig. 2. A pulse generator circuit for juxtacellular neuronal labelling. The device is based on a PIC microcontroller (PICAXE 08M; available from a range of electronic component suppliers or online, e.g. at http://www.techsupplies.co.uk) and has been designed so that the intensity of the current pulses is easy to set. The circuit is powered by a 9 V DC plug pack. The PICAXE chip is programmed using BASIC code that raises the voltage on pins 6 and 7 of the microcontroller chip to 5 V for 200 ms. At the end of the 200 ms pulse, the voltage drops to 0 V for another 200 ms, which is equivalent to a pulse train of 2.5 Hz with a 50% duty cycle. The pulse at pin 6 controls a light-emitting diode (LED) which signals operation of the pulse generator, while the pulse at pin 7 is connected to a voltage divider circuit.

```
************************************************************************
PICAXE-08 Pulse Generator (This BASIC code generates 200 ms pulses at 2.5 Hz.)
main:
high 0                                  'make pin 7 high
high 1                                  'make pin 6 high to turn LED on
pause 200                               'wait 200 ms
low 0                                   'make pin 7 low
low 1                                   'make pin 6 low to turn LED of
pause 200                               'wait 200 ms
gotomain                                'repeat
************************************************************************
```

Fig. 3. BASIC code for programming the PICAXE-based pulse generator circuit. The programming software and complete instructions for programming the PICAXE chip are available at http://www.rev-ed.co.uk/picaxe/.

7. Microelectrode impedance meter.
8. Audio amplifier and loudspeaker (e.g. Grass AM-10).

2.2. Configuration of the Recording Apparatus for Juxtacellular Neuronal Labelling (Fig. 4)

To perform juxtacellular neuronal labelling successfully, the apparatus must be configured correctly. This section is aimed at those who are new to the technique and will help them to configure their recording apparatus.

Amplifiers that are used for intracellular recording generally have the capacity to inject current intracellularly via the recording microelectrode. In general, amplifiers that are used for extracellular recording do not have this capability. Successful juxtacellular labelling involves application of pulsatile positive current (electrode tip positive) in order to entrain the recorded neuron. To achieve this, an analogue stimulator with a variable output voltage can be used as a

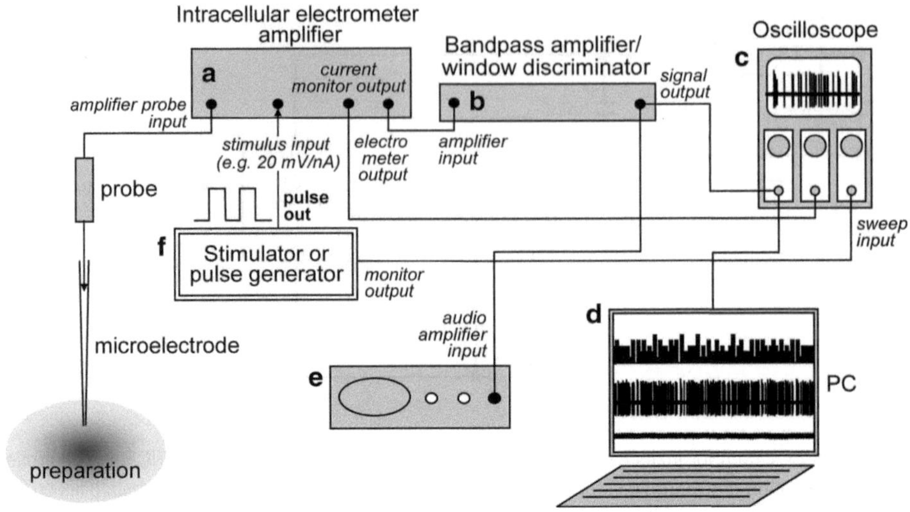

Fig. 4. Equipment required for juxtacellular neuronal labelling. An intracellular electrometer amplifier (**a**) capable of delivering current pulses is an essential element. The intracellular amplifier is coupled to a bandpass amplifier (**b**) whose output is monitored using an oscilloscope (**c**), a computerised data acquisition system (**d**) and an audio monitor (**e**). Positive current pulses are applied juxtacellularly via the amplifier probe connected to the recording microelectrode. Pulses are generated by the pulse generator (**f**). The pulse generator can be a conventional laboratory stimulator or a dedicated pulse generator as described in this chapter. During application of the juxtacellular current, the sweep of the oscilloscope should be synchronised with the pulses so that entrainment of the recorded neuron can be constantly monitored. The sweep can be synchronised by connecting the 'monitor' output of the pulse generator to the external input of the oscilloscope time base. The time base should be set to 100 ms/division. It is also prudent to monitor the pulsed output of the electrometer to detect changes in electrode impedance that may result from blockage of the electrode.

source of timed pulses. Although a range of stimulation parameters are possible, most investigators use pulses of 200-ms duration and a 50% duty cycle, which corresponds to a pulse rate of 2.5 Hz.

We have designed a simple pulse generator that can be coupled to the intracellular electrometer to provide the extracellular current pulses that are required for juxtacellular labelling (Fig. 2). This device can be used instead of a digital or analogue stimulator and has the advantage that the intensity of the pulses in nanoamperes can be dialled up using a ten-turn potentiometer. In addition, this pulse generator obviates the need for a dedicated multiple-output physiological stimulator. The circuit is based around a simple PIC microcontroller chip and is easy to construct. In order to function correctly, the chip must be programmed with the basic code in Fig. 3. The program editor may be obtained from the Revolution Education Web site at http://www.rev-ed.co.uk/picaxe. The PICAXE chip is programmed through the PC USB port using a special USB cable available from suppliers of the PICAXE-08M chip. Note that the output of this device has been designed to provide 20 mV per turn of the controller (equivalent to 1 nA) to suit the Intra 767 electrometer (World Precision Instruments, Sarasota, FL, USA). The device could be readily adapted to suit other intracellular amplifiers.

2.3. Microelectrode Preparation

Microelectrodes are pulled from borosilicate glass capillaries (1.5–2.0 mm OD) and are filled with a solution containing 1.5–5% biotinamide (*N*-2-aminoethyl biotinamide hydrobromide, Invitrogen, Eugene, OR, USA; see Sect. 5 for recipe) or Neurobiotin (Vector Laboratories, Burlingame, CA, USA) or biocytin (Sigma, St. Louis, MO, USA). Satisfactory labelling of the soma and proximal dendrites can be achieved using 1.5% biotinamide. Higher concentrations will result in more complete filling so that distal dendrites and the axon are labelled. Tetramethyl rhodamine conjugated to biocytin is also effective but is more expensive than biotinamide. Because it is inherently fluorescent, biocytin–tetramethyl rhodamine does not require processing of the labelled neuron using a streptavidin conjugate (see Sect. 3).

We strongly recommend paying careful attention to electrode impedance prior to insertion of the electrode into the brain. An electrode whose impedance is too low (<12–14 MΩ) as determined by an impedance meter in vitro is unsuitable because the electrode is unlikely to develop current pulses at the electrode tip that have sufficient intensity to entrain the neuron. Checking the impedance before using it also helps to identify electrodes that are blocked or whose impedance is too high (>25 MΩ).

2.4. Physiological Characterisation

At the very least, juxtacellular neuronal labelling permits precise identification of the location of a physiologically characterised neuron. At its best, it can provide substantial morphological detail, precise location and, in combination with immunohistochemistry or *in situ* hybridization, information about the neurochemical phenotype of the filled neuron. Physiological characterisation procedures will depend on the nature of the population of neurons under investigation. If possible, antidromic stimulation should be used to identify a projection target of the recorded neuron. Confirmation of antidromic activation should be made using the usual three criteria, i.e. constant latency spike, high frequency following and, most importantly, the collision test (2). The first criterion involves the demonstration that each stimulation pulse produces a spike with invariant latency (T_1). Since synaptic activation of a neuron can also sometimes occur with an invariant latency, this criterion is necessary but insufficient for determining whether a stimulation-evoked spike is antidromic in nature. High frequency following involves stimulation with a pair of pulses with a set interpulse interval. Antidromic spikes elicited by a pair of pulses with an interpulse interval of 5 ms correspond to an instantaneous stimulation frequency of 200 Hz. Synaptic activation usually fails at these frequencies because the chemical transmission process is relatively slow and acts like a low-pass filter. However, even here there are exceptions. Therefore, the most important criterion for confirmation of antidromic activation is satisfaction of the collision test. In practice, the test is made by application of a test stimulus that is triggered by the occurrence of a spontaneous spike at a delay that is greater than the putative antidromic latency (i.e. $>T_1$). Under these circumstances, a constant latency spike should be observed. Gradual reduction of the interval between the spontaneous spike and the delay set at the electrical stimulator should eventually lead to failure of the putative antidromic spike, which is termed 'collision'. Collision occurs at a stimulation interval that corresponds to $T_1 + R$ where R = the refractory period of the neuron in milliseconds.

Once a neuron has been physiologically characterised, juxtacellular labelling can be attempted. As the name of the technique implies, juxtacellular neuronal labelling is most likely to succeed if the electrode tip is very close to the neuron under investigation. The distance between the tip and the neuron can be judged by the spike amplitude, which should be >1 mV.

2.5. Entrainment of the Neuron

Once a neuron is isolated and characterised physiologically, a decision is made as to whether it is suitable for labelling. Successful labelling depends on the spike size, which is an indication of proximity between the neuron and the tip of the recording microelectrode, and on whether only a single unit is being recorded. Current pulses are then applied at the lowest intensity (<1 nA). If the PIC microcontroller unit is used, the intensity of the current pulses can be controlled with the 10-turn potentiometer.

It is important that the recorded neuron is continuously monitored so that entrainment can be observed. Recognition of entrainment is vital for successful juxtacellular labelling. We recommend the use of an oscilloscope to monitor this process. As described in the legend to Fig. 4, the oscilloscope time base should be set to 100 ms/division where each sweep is triggered by the pulse source (e.g. monitor output in Fig. 4). This arrangement allows entrainment to be readily identified because the investigator can observe the recorded spikes in the 'on' and 'off' phases of the juxtacellular current pulses, as depicted in Fig. 1. The intensity of the current pulses is increased gradually until entrainment occurs. Entrainment is often preceded by an increase in the level of baseline noise of the recording. The duration of the period of entrainment will determine the amount of biotinamide that is taken up by the recorded neuron. Entrainment of more than one neuron can also be recognised using this method. The output of the bridge circuit of the amplifier should also be monitored simultaneously. This allows the investigator to identify the 'on' and 'off' phases of the applied juxtacellular pulses as well as to detect an increase in electrode impedance as a result of blockage of the electrode. Severe blockage of the electrode may result in a grossly distorted (non-square wave) appearance of the applied current pulses.

We strongly advise investigators *not* to attempt to label more than one neuron in the same nucleus or brain region. Labelling more than one neuron in a nucleus or region can lead to ambiguity when the neurons are being reconstructed. When two neurons are labelled near each other, their dendrites can overlap so closely that it is not possible to accurately identify the neuron of origin of each dendrite. However, a second neuron may be labelled on the contralateral side of the brain or on the same side of the brain as long as the two neurons are separated by several hundred micrometers and lie in different areas that can be readily discriminated anatomically.

3. Identification of the Juxtacellularly Labelled Neuron

Identification of a juxtacellularly labelled neuron requires perfusion to preserve brain structure and to fix the biotinamide or biocytin inside the labelled and characterised neuron. Perfusion should be delayed until the biocytin or biotinamide has had a chance to diffuse throughout the last neuron that was juxtacellularly labelled. We usually wait one hour between the last fill and perfusion.

3.1. Perfusing the Rat

A perfusion pump, pressure-controlled perfusion apparatus or gravity-fed perfusion system connected to a 16-G blunt catheter or 18-G blunt needle can be used for perfusing the rat once one or more neurons in the brain have been juxtacellularly labelled.

While deeply anaesthetised, the rat is placed on a tray in a fume hood, and the abdomen is opened. To improve the flow of fixative to the brain, the gut plus renal arteries and veins or the subdiaphragmatic abdominal aorta can be clamped with haemostats. The xiphoid process is grasped with a haemostat; the diaphragm is opened with sharp scissors, and the ribs are cut far laterally on both sides to provide good visualisation of the heart. The sternum and cut ribs are pulled back towards the head with the haemostat and fastened in place with masking tape. The heart is cleared of connections to the chest wall so that the ascending aorta can be seen easily. The perfusion catheter or needle is inserted through the left ventricular wall into the left ventricle and up into the ascending aorta so that 1–2 mm of the tip is visible in the aorta. One millilitre of heparin (1,000 IU/ml) is injected into the heart followed by 250–500 ml of phosphate-buffered saline (PBS) or Dulbecco's Modified Eagles' Medium/Ham's F12 tissue culture medium (Sigma) to flush all of the blood from the vasculature. Once the PBS or tissue culture medium no longer contains blood and the liver has cleared, the rat is perfused with 500 ml or 1 L of fixative at a flow rate of approximately 50 ml/min. For light microscopy, the fixative is 4% formaldehyde in 0.1 M phosphate buffer, pH 7.4. A glutaraldehyde-containing fixative (e.g. 4% formaldehyde + 0.3% glutaraldehyde or 2.5% glutaraldehyde) in 0.1 M phosphate buffer, pH 7.4, is used for electron microscopy. Brains should be postfixed in the same fixative as used for perfusion. Brains for light microscopy should be left in 4% formaldehyde for 3–5 days at room temperature on a shaker. Brains for electron microscopy should be postfixed for several hours to overnight at either room temperature or 4°C. After postfixation, the brains should be washed in several changes of 0.1 M phosphate buffer, pH 7.4.

Brains for light microscopic studies on juxtacellularly labelled neurons can be stored for extended periods at 4°C in 0.1 M phosphate buffer, pH 7.4, containing 0.05% sodium azide without affecting the ability to detect the juxtacellularly labelled neuron. Brains for electron microscopy should be processed immediately after postfixation to prevent excessive cross-linking of the tissue by glutaraldehyde.

3.2. Sectioning the Brain

In order to recover a juxtacellularly labelled neuron, it is necessary to stain all of the sections through the brain region containing the filled neuron. Either vibratome or cryostat sections are suitable. No pretreatment is required if the perfused brain is to be sectioned on a vibratome. If a brain is to be cut on a cryostat, it should be infiltrated with 20% sucrose overnight and then 30% sucrose for at least 24 h before sectioning. We prepare our sucrose solutions for infiltration in half-strength Tris phosphate-buffered saline (TPBS; 10 mM Tris, 0.9% NaCl, 0.05% thimerosal in 10 mM phosphate buffer, pH 7.4; see Sect. 5 for recipe). Thimerosal, the powerful preservative in TPBS, prevents bacterial contamination during the process of sucrose infiltration.

If the juxtacellularly labelled neuron is to be reconstructed, the sections need to be mounted in serial order. We routinely cut brain blocks containing juxtacellularly labelled neurons into 30-μm sections and stain the sections free-floating. In our experience, it is reasonably easy to mount sections of this thickness in serial order if the block containing the labelled neuron has been cut into a 1:4 series. We do not recommend using free-floating sections <20 μm in thickness because they are difficult to handle.

3.3. Light Microscopic Identification of the Labelled Neuron

The simplest procedure for identifying a labelled neuron within a tissue section is to use streptavidin conjugated to a fluorescent tag. Almost as easy is identifying the neuron with a conjugate of avidin or streptavidin with horseradish peroxidase (HRP) or alkaline phosphatase (AP) and an enzyme reaction that produces a coloured reaction product. In the first procedure, sections of the perfused brain region containing the labelled neuron are simply washed and incubated with streptavidin conjugated to a fluorescent tag, followed by washing and mounting onto unsubbed glass slides using a water-soluble mountant that preserves the fluorescent signal (e.g. 11, 19). In the second procedure, the sections are washed, incubated with an avidin-HRP (11, 19) or avidin-AP conjugate, washed, reacted to reveal the HRP or AP and then mounted onto subbed slides. HRP-reacted sections can be coverslipped with a standard mounting medium, such as DePeX. Most AP reaction products are water soluble, so water-based mountants are required.

3.3.1. Fluorescence Detection

1. Cut 30-μm sections on a vibratome or cryostat into 0.1 M phosphate buffer, pH 7.4.
2. Rinse for 3×5 or 10 min in Tris–saline (TS; see Sect. 5 for recipe) or TPBS.
3. Incubate in streptavidin-Cy3 (Amersham or Jackson ImmunoResearch) diluted 1:1,000 in TS or TPBS containing 0.1–0.3% Triton X-100 for 2 h.
4. Rinse 3×5 or 10 min in TS or TPBS.
5. Transfer into water containing 0.4% Triton X-100.
6. Mount onto uncoated slides. Before sections are completely dry, coverslip with a water-soluble mounting medium, such as buffered glycerol, Vectashield or other commercial anti-fade mounting medium.
7. Store the slides at 4°C before and after viewing.

3.3.2. HRP Detection

1. Cut 30-μm sections on a vibratome or cryostat into 0.1 M phosphate buffer, pH 7.4.
2. Treat with 1% H_2O_2 in 30% methanol for 30 min to block endogenous peroxidase activity.
3. Wash 3×10 min on a shaker in TPBS containing 0.3% Triton X-100.
4. Incubate in a 1:1,500 dilution of ExtrAvidin® – Peroxidase (Sigma) in TPBS containing 0.3% Triton X-100 for 4–24 h at room temperature on a shaker.
5. Wash 3×10 min on a shaker in TPBS.
6. Use a nickel-intensified diaminobenzidine (DAB) reaction (20) to visualise the juxtacellularly labelled neuron.
7. Wash twice quickly in TPBS to stop the reaction.
8. Wash 1×5 min on a shaker in TPBS.
9. Transfer into water containing 0.4% Triton X-100.
10. Mount in serial order onto slides coated with chrome alum gelatine; dry and coverslip with a conventional mounting medium, such as DePeX or Permaslip (Alban Scientific, St. Louis, MO, USA).

3.3.3. Alkaline Phosphatase Detection

1. Cut 30-μm sections on a vibratome or cryostat into 0.1 M phosphate buffer, pH 7.4.
2. Incubate sections in streptavidin-AP (Vector) diluted 1:1,000 in TS containing 0.1% Triton X-100 for 2 h on a shaker.
3. Wash sections 3×5 min on a shaker in TS.

4. Pre-incubate in nitro blue tetrazolium (NBT; 0.1 M NaCl/50 mM $MgCl_2$ in 0.1 M Tris, pH 9.5) for 10 min on a shaker.
5. Remove NBT.
6. Add 5 ml of a solution containing 450 μg/ml NBT and 175 μg/ml BICP (5-bromo-4-chloro-3-indolyl phosphate 4-toluidine salt) plus 1 drop of levamisole (0.125 M; Vector) to inhibit endogenous AP.
7. React on a shaker at room temperature for about 6 min.
8. Rinse 3 × 5 or 10 min on a shaker in TS.
9. Mount onto gelatine-coated-subbed slides from TS and coverslip with Vectashield.

3.4. Electron Microscopic Identification of the Labelled Neuron (Fig. 5)

The HRP detection procedure above can be modified to study the ultrastructure of juxtacellularly labelled neurons (21). Instead of being perfused with formaldehyde, the animals are perfused with a fixative that contains glutaraldehyde, such as 4% formaldehyde + 0.3% glutaraldehyde or 2.5% glutaraldehyde. Triton X-100 is omitted from diluents and washes in order to retain membranes, and the tissue is permeabilised with 50% ethanol (22). Incubation times are increased to allow immunoreagents to penetrate through the more heavily fixed tissue.

Neuronal antigens in axons that synapse on the juxtacellularly labelled neuron can be visualised after the juxtacellularly labelled neuron has been localised. Visualising inputs requires another round of avidin–biotin–peroxidase staining using a primary antibody against the antigen of interest and a reaction product that has a different ultrastructural appearance to that used for detecting the juxtacellularly labelled neuron (e.g. 20, 21).

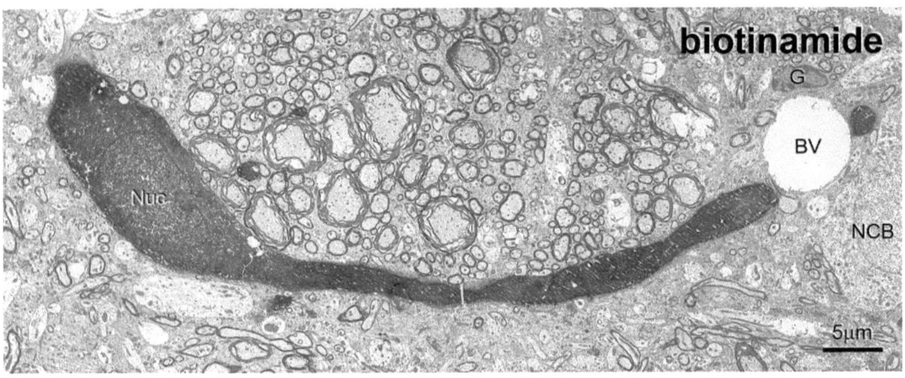

Fig. 5. Revealing the ultrastructure of a juxtacellularly labelled neuron. Electron micrograph showing a juxtacellularly labelled bulbospinal barosensitive neuron in the rat rostral ventrolateral medulla. The neuron was filled with biotinamide; the rat was perfused with 2.5% glutaraldehyde, and biotinamide was detected with a nickel-intensified DAB reaction, which produces an amorphous electron dense reaction product. *BV* blood vessel, *G* glial cell, *NCB* nerve cell body, *Nuc* nucleus.

1. Cut 50-μm or 70-μm sections on a vibratome into 0.1 M phosphate buffer, pH 7.4.
2. Expose to 50% ethanol in distilled water for 30 min to 3 hr.
   - The time in ethanol will depend on the fixative and the quality of the perfusion. More heavily fixed tissue will require longer treatment with 50% ethanol. Do a test to determine what time in ethanol allows the antibodies and avidin-HRP complex to penetrate all the way through a 50-μm or 70-μm vibratome section.
3. Wash 2 times quickly in 0.1 M phosphate buffer, rotating the sections vigorously by hand until they sink.
4. Incubate in a 1:1,500 dilution of ExtrAvidin® - Peroxidase (Sigma) in TPBS *without* Triton X-100 for 3–7 days at room temperature on a shaker.
5. Wash 3 × 30 min in TPBS.
6. Use a nickel-intensified DAB reaction (20) to visualise the juxtacellularly labelled neuron.
7. Wash twice quickly in TPBS to stop the reaction.
8. Rinse 1 × 5 min in TPBS.
9. Wash 3 × 10 min in 0.1 M phosphate buffer, pH 7.4.
   - Sections may be stored overnight at 4°C after this step.
10. Osmicate in 0.5% osmium tetroxide in 0.1 M phosphate buffer, pH 7.4, for 1 hr at room temperature on a shaker.
11. Wash 4 × 5 min in distilled water.
12. Stain with 1% aqueous uranyl acetate for 30 min.
    - Sections should be kept flat during this step.
13. Wash 2 × 5 min in distilled water.
14. Dehydrate through 70, 90 and 95% acetone.
    - Sections should be kept flat during these steps and during the first wash in 100% acetone.
15. Wash 4 × 5 min in 100% acetone.
16. Wash 2 × 5 min in propylene oxide.
17. Infiltrate for 30 min–2 hr with 1:1 propylene oxide: Durcupan™ ACM resin (Sigma-Aldrich) at room temperature.
18. Infiltrate for 10 min at 60°C then overnight at room temperature in Durcupan resin.
19. Embed on clean glass slides under coverslips cut from sheets of Aclar® film (Electron Microscopy Sciences, Hatfield, PA, USA).
20. Polymerise for 48 hr at 60°C.
21. Remove the Aclar™ coverslip and place the slide with the polymerised sections on a hot plate at 55–60°C.

22. Using a razor blade or scalpel, cut around the area of the embedded section that contains the juxtacellularly labelled neuron.
23. Use a razor blade to carefully lift the piece of the section containing the labelled neuron off the slide.
24. Re-embed the labelled neuron in a small drop of fresh resin on a flat blank resin block under a clean glass coverslip.
25. Polymerise for 48 hr at 60°C.
26. Remove the coverslip and trim the block.
27. Cut ultrathin sections and stain with Reynold's lead citrate.

4. Examining the Phenotype of and Inputs to the Labelled Neuron

Immunofluorescent staining should be used to assess the neurochemical phenotype of a juxtacellularly labelled neuron (e.g. 19, 23). Either immunofluorescent or immunoperoxidase staining can be used to identify the neurochemistry of axons that innervate a labelled neuron (e.g. 21). Since sections stained with immunoperoxidase do not degrade with time as sections stained with immunofluorescence do, we prefer immunoperoxidase staining for the assessment of inputs.

4.1. Assessing the Phenotype of the Labelled Neuron (Fig. 6)

1. Cut 30-mm sections on a vibratome or cryostat into 0.1 M phosphate buffer, pH 7.4.
2. Wash 3 × 10 min in TPBS containing 0.3% Triton X-100.
3. Expose to 10% normal horse serum (NHS) or normal donkey serum in TPBS-Triton for 30 min on a shaker to block non-specific antibody binding.
   - Do not wash after this blocking step.
4. Incubate in 1:250 streptavidin-Alexa 488 (Jackson ImmunoResearch) plus the optimal dilution of primary antibody (determined previously by titration) in TPBS-Triton containing 10% NHS for 1–3 days in the dark on a shaker at room temperature.
5. Wash 3 × 10 min in TPBS.
6. Incubate in a Cy3-labelled secondary antibody directed against the immunoglobulin (IgG) of the species in which the primary antibody was raised (e.g. Cy3-anti-rabbit IgG if the primary antibody was raised in rabbit) and diluted with TPBS-Triton for 16–24 hr in the dark on a shaker at room temperature.
   - We recommend secondary antibodies (whole IgG) raised in donkey from Jackson ImmunoResearch. The donkey secondary antibodies for multiple labelling, which show minimal cross reactivity to a range of species, give the best results. With these secondary antibodies, the blocking

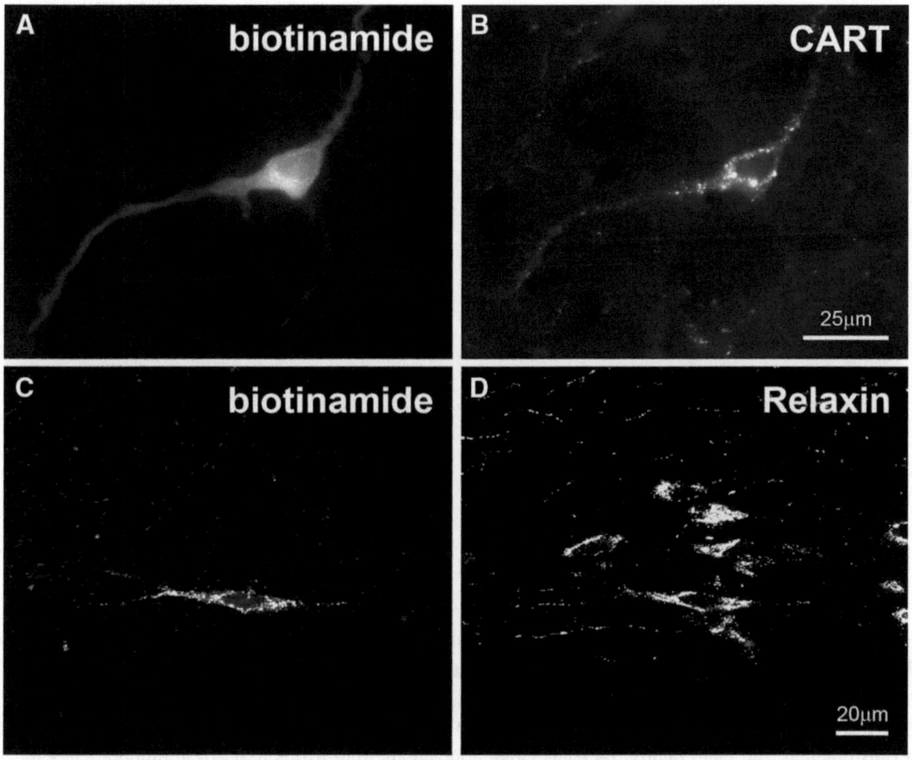

Fig. 6. Identifying the neurochemical phenotypes of juxtacellularly labelled neurons. (**A, B**) Fluorescence micrographs showing a juxtacellularly labelled bulbospinal barosensitive neuron in the rat rostral ventrolateral medulla that is immunoreactive for cocaine and amphetamine-regulated transcript (*CART*). The neuron was filled with biotinamide, and the rat was perfused with 4% formaldehyde. Biotinamide was detected with Alexa 488-streptavidin and CART immunoreactivity, with a rabbit anti-CART antibody and Cy3 donkey anti-rabbit IgG. (**C, D**) Confocal micrographs showing a juxtacellularly labelled relaxin-3-immunoreactive neuron in the rat nucleus incertus. The neuron was filled with biotinamide, and the rat was perfused with 4% paraformaldehyde. Biotinamide was detected with Dylight 488-streptavidin and relaxin-3-immunoreactivity with a rabbit anti-relaxin-3 antibody and Alexa-594 goat anti-rabbit IgG.

serum should be either NHS (Invitrogen) or normal donkey serum (Jackson ImmunoResearch).

7. Wash 3×10 min in TPBS in the dark on a shaker at room temperature.
8. Transfer into water containing 0.4% Triton X-100.
9. Mount onto uncoated slides. Before sections are completely dry, coverslip with a water-soluble mounting medium, such as buffered glycerol, Vectashield or other commercial anti-fade mounting medium.
10. Store the slides at 4°C before and after viewing.

4.2. Assessing the Neurochemistry of Inputs to the Labelled Neuron (Fig. 7)

1. Cut 30-mm sections on a vibratome or cryostat into 0.1 M phosphate buffer, pH 7.4.
2. Treat with 1% H_2O_2 in 30% methanol for 30 min to block endogenous peroxidase activity.

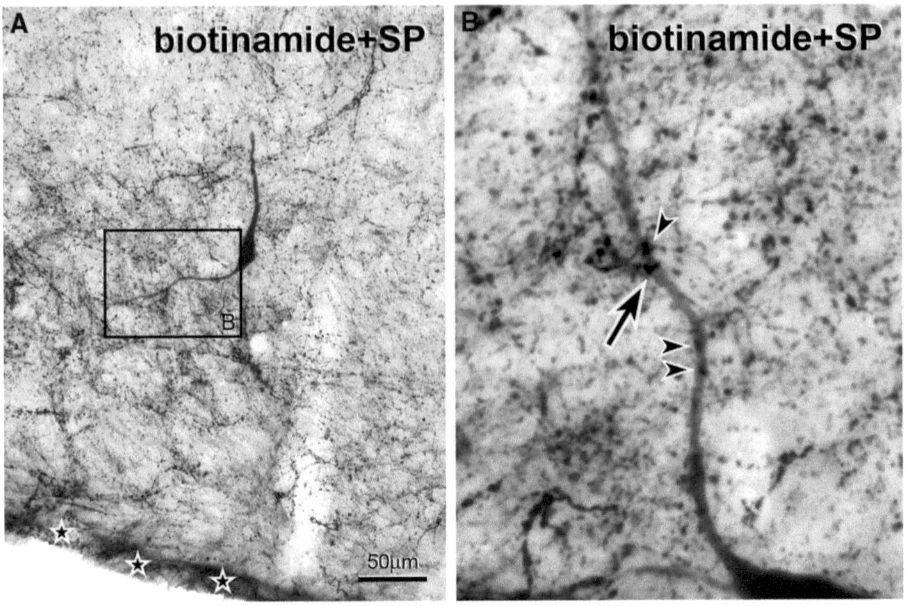

Fig. 7. Identifying the neurochemistry of inputs to a juxtacellularly labelled neuron. Bright-field micrographs showing a juxtacellularly labelled bulbospinal barosensitive neuron in the rat rostral ventrolateral medulla. The neuron was filled with biotinamide, and the rat was perfused with 4% formaldehyde. Biotinamide was detected with ExtrAvidin-HRP, and an imidazole intensified DAB reaction. Substance P (SP)-immunoreactivity was visualised with a rabbit anti-SP antibody and a nickel-intensified DAB reaction. The filled neuron receives close appositions from axons that are immunoreactive for SP. (**A**) The labelled neuron lies close to the ventral surface of the medulla (*stars*). The dendrite in *box B* is shown at higher magnification in (**B**). (**B**) An SP-immunoreactive varicosity (*arrow*) forms a close apposition on the dendrite of the juxtacellularly labelled neuron. Other SP-positive varicosities (*arrowheads*) also closely appose the dendrite but are not in the plane of focus.

3. Wash 3 × 10 min in TPBS containing 0.3% Triton X-100.
4. Expose to 10% NHS or normal donkey serum in TPBS-Triton for 30 min on a shaker to block non-specific antibody binding.
   - Do not wash after this blocking step.
5. Incubate in 1:250 ExtrAvidin® – Peroxidase (Sigma) plus the optimal dilution of primary antibody (determined previously by titration), which is directed against a neurochemical in an input, in TPBS-Triton containing 10% NHS for 1–3 days on a shaker at room temperature.
   - Because the biotin in serum binds avidin conjugates, a higher concentration of ExtrAvidin-HRP is required when serum is present in the diluent.
6. Wash 3 × 10 min in TPBS.
7. Visualise the juxtacellularly labelled neuron with an imidazole-intensified DAB reaction (20).
8. Wash twice quickly in TPBS to stop the reaction.
9. Wash 3 × 10 min in TPBS.
10. Incubate in a biotinylated secondary antibody directed against the immunoglobulin of the species in which the primary antibody was

raised (e.g. biotinylated anti-rabbit IgG if the primary antibody was raised in rabbit) and diluted with TPBS-Triton containing 1% NHS overnight on a shaker at room temperature.

11. Wash 3 × 10 min in TPBS.
12. Incubate in 1:1,500 ExtrAvidin® – Peroxidase in TPBS-Triton for 4–6 hr on a shaker at room temperature.
13. Wash 3 × 10 min in TPBS.
14. Visualise the neurochemical in the input with a nickel-intensified DAB reaction (20).
15. Wash twice quickly in TPBS to stop the reaction.
16. Wash 1 × 5 min in TPBS.
17. Transfer into water containing 0.4% Triton X-100.
18. Mount immediately in serial order onto chrome alum gelatine-coated slides; dry and coverslip with a conventional mounting medium, such as DePeX or Permaslip (Alban Scientific).
    - If the sections are stored after the nickel-intensified DAB reaction, the reaction product may turn from black to dark brown.

5. Solutions and Reagents (Table 1)

Biotinamide (2% *N*-(2-aminoethyl) biotinamide, hydrobromide in 0.5 M sodium acetate).

N-(2-aminoethyl) biotinamide	25 mg
Sodium acetate (0.5 M)	1.25 ml

TS (100 mM Tris, 0.9% NaCl, pH 7.4)

	1 L	2 L
Trizma base	12.11 g	24.22 g
NaCl	9 g	18 g
Distilled water	1 L	2 L
pH to 7.4 with concentrated HCl		

TPBS (10 mM Tris, 0.9% NaCl, 0.05% thimerosal in 10 mM phosphate buffer, pH 7.4).

	1 L	2 L
Trizma base	1.211 g	2.422 g
NaCl	9 g	18 g
Thimerosal	0.5 g	1 g
0.4 M Phosphate buffer, pH 7.4	25 ml	50 ml
Distilled water	800 ml	1.6 L
pH to 7.4 with concentrated HCl. This takes about	0.5 ml	1 ml
Add distilled water to make up to	1 L	2 L

Table 1
Recommended reagents

Supplier	Reagent	Catalogue #	Unit size
Invitrogen http://www.invitrogen.com	*N*-(2-aminoethyl) biotinamide, hydrobromide	A1593	25 mg
	Gibco Horse Serum heat-inactivated	26050-070	100 ml
	Streptavidin-Cy3	434315	1 mg
Jackson ImmunoResearch http://www.jacksonimmuno.com	Normal Donkey Serum	017-000-121	10 ml
	Fluorescent or biotinylated secondary antibodies (whole IgG) for multiple labelling raised in donkey	Various	Usually 0.5 ml
	Streptavidin-Cy3	016-160-084	1 mg
	Streptavidin-Alexa 488	016-540-084	1 mg
Sigma-Aldrich http://www.sigmaaldrich.com	Biocytin hydrochloride	B1758	25 mg, 50 mg, 100 mg
	DMEM/F-12 Ham tissue culture medium	D8900	10 × 1 L bottles
	Thimerosal	T5125	10 g, 25 g
	ExtrAvidin®-Peroxidase	E2886	0.2 ml, 1 ml
	Glucose oxidase	G0543	10 KU
	Durcupan™ ACM resin	44611, 44612, 44613, 44614	100 ml each
Vector Laboratories http://www.vectorlabs.com	Neurobiotin tracer	SP-1120	50 mg
	Streptavidin-alkaline phosphatase	SA-5100	1 ml
	Levamisole solution	SP-5000	18 ml
	Vectashield	H-1000	10 ml

Acknowledgments

The work reported here was supported by the National Health and Medical Research Council of Australia (Project Grants 229907, 480414, 350207). The juxtacellularly labelled neuron in Fig. 5 was filled by Dr. Ann Schreihofer in the laboratory of Dr. Patrice Guyenet. We thank them both for allowing us to use this image. The juxtacellularly labelled relaxin-3-positive neuron in Fig. 6 (panels c and d) was filled by Dr. Sherie Ma in the laboratory of Associate Professor Anthony Verberne.

References

1. Humphrey DR, Schmidt EM (1990) Extracellular single-unit recording methods. In: Boulton AA, Baker GB, Vanderwolf CH (eds) Neuromethods. Neurophysiological techniques: application to neural systems. Humana Press, Clifton, NJ, pp 1–64
2. Lipski J (1981) Antidromic activation of neurones as an analytical tool in the study of the central nervous system. J Neurosci Methods 4:1–32
3. Duque A, Zaborszky L (2006) Juxtacellular labeling of individual neurons in vivo: from electrophysiology to synaptology. In: Zaborszky L, Wouterlood FG, Lanciego JL (eds) Neuroanatomical tract-tracing. 3. Molecules, neurons and systems. Springer, New York, pp 197–236
4. Pinault D (1994) Golgi-like labeling of a single neuron recorded extracellularly. Neurosci Lett 170:255–260
5. Pinault D (1996) A novel single-cell staining procedure performed in vivo under electrophysiological control: morpho-functional features of juxtacellularly labeled thalamic cells and other central neurons with biocytin or neurobiotin. J Neurosci Methods 65:113–136
6. Kitai ST, Penny GR, Chang HT (1989) Intracellular labeling and immunocytochemistry. In: Heimer L, Zaborsky L (eds) Neuroanatomical tract-tracing methods. 2. Recent progress. Plenum Press, New York, pp 173–199
7. Armstrong WE (1992) Intracellular recording and labelling with biocytin-filled electrodes. In: Kettenmann H, Grantyn R (eds) Practical electrophysiological methods. A guide for in vitro studies in vertebrate neurobiology. Wiley-Liss Inc., New York, pp 304–308
8. Schreihofer AM, Guyenet PG (2003) Baro-activated neurons with pulse-modulated activity in the rat caudal ventrolateral medulla express GAD67 mRNA. J Neurophysiol 89:1265–1277
9. Schreihofer AM, Guyenet PG (1997) Identification of C1 presympathetic neurons in rat rostral ventrolateral medulla by juxtacellular labeling in vivo. J Comp Neurol 387:524–536
10. Rouzade-Dominguez ML, Pernar L, Beck S, Valentino RJ (2003) Convergent responses of Barrington's nucleus neurons to pelvic visceral stimuli in the rat: a juxtacellular labelling study. Eur J Neurosci 18:3325–3334
11. Verberne AJM, Stornetta RL, Guyenet PG (1999) Properties of C1 and other ventrolateral medullary neurones with hypothalamic projections in the rat. J Physiol (Lond) 517:477–494
12. Kirouac GJ, Pittman QJ (1999) Identification of barosensitive neurons in the mediobasal forebrain using juxtacellular labeling. Am J Physiol 276:R1766–R1771
13. Schreihofer AM, Stornetta RL, Guyenet PG (1999) Evidence for glycinergic respiratory neurons: Botzinger neurons express mRNA for glycinergic transporter 2. J Comp Neurol 407:583–597
14. Houweling AR, Doron G, Voigt BC, Herfst LJ, Brecht M (2010) Nanostimulation: manipulation of single neuron activity by juxtacellular current injection. J Neurophysiol 103:1696–1704
15. Jones GA, Llewellyn-Smith IJ, Jordan D (2002) Physiological, pharmacological, and immunohistochemical characterisation of juxtacellularly labelled neurones in rat nucleus tractus solitarius. Auton Neurosci 98:12–16
16. Bangasser DA, Zhang X, Garachh V, Hanhauser E, Valentino RJ (2011) Sexual dimorphism in locus coeruleus dendritic morphology: a structural basis for sex differences in emotional arousal. Physiol Behav 103:342–351
17. Houweling AR, Brecht M (2008) Behavioural report of single neuron stimulation in somatosensory cortex. Nature 451:65–68

18. Andrew RD, Fagan M (1990) A technique for controlling the membrane potential of neurons during unit recording. J Neurosci Methods 33:55–60
19. Sartor DM, Verberne AJM (2003) Phenotypic identification of rat rostroventrolateral medullary presympathetic neurons inhibited by exogenous cholecystokinin. J Comp Neurol 465:467–479
20. Llewellyn-Smith IJ, Dicarlo SE, Collins HL, Keast JR (2005) Enkephalin-immunoreactive interneurons extensively innervate sympathetic preganglionic neurons regulating the pelvic viscera. J Comp Neurol 488:278–289
21. Llewellyn-Smith IJ, Schreihofer AM, Guyenet PG (2001) Distribution and amino acid content of enkephalin-immunoreactive inputs onto juxtacellularly labelled bulbospinal barosensitive neurons in rat rostral ventrolateral medulla. Neuroscience 108:307–322
22. Llewellyn-Smith IJ, Minson JB (1992) Complete penetration of antibodies into vibratome sections after glutaraldehyde fixation and ethanol treatment: light and electron microscopy for neuropeptides. J Histochem Cytochem 40:1741–1749
23. Burman KJ, Sartor DM, Verberne AJM, Llewellyn-Smith IJ (2004) Cocaine- and amphetamine-regulated transcript in catecholamine and non-catecholamine bulbospinal, barosensitive neurons of the rat rostral ventrolateral medulla. J Comp Neurol 476:19–31

Part IV

Methods for Analysis

Chapter 11

Analysis of Sympathetic Nerve Activity

Rohit Ramchandra

Abstract

A powerful method to study the role of the autonomic nervous system in preservation of homeostasis is to directly record sympathetic nerve activity (SNA) to various organs. This chapter will introduce methods to analyze the multi-fiber sympathetic nerve activity that is differentially recorded between two electrodes. The pros and cons of using spike counting using a threshold versus rectification and integration of the raw signal will be discussed. The analysis of the amplitude and frequency of sympathetic bursts will be introduced, and examples from the literature where this may be important will be highlighted. The next section will focus on the analysis of baroreflex control of SNA and introduce methods to compare baroreflex curves within animals and between groups of animals. This chapter will end with a discussion on the drawbacks of trying to directly compare SNA between two different groups of animals.

Key words: Sympathetic nerve activity, Rectification and integration, Background noise, Noradrenaline

1. Introduction

Sympathetic nerve fibers innervate internal organs and play an important role in the homeostasis of various bodily functions including body temperature, energy metabolism, and gastrointestinal function. The sympathetic nervous system is traditionally described as one arm of the autonomic nervous system and is classified as the effector system for a "fight or flight" response. Over the last three decades, there has been an increasing body of evidence supporting the concept that sympathetic nerve activity (SNA) to different organs can be differentially regulated depending on the homeostatic challenges experienced. The finding that SNA to different organs is increased in a host of cardiovascular diseases and is detrimental to cardiovascular outcome has led to a renewed interest in trying to understand the mechanisms that control SNA. There are several indirect measures of SNA, and it is important to note that all indirect measures have limitations.

Paul M. Pilowsky et al. (eds.), *Stimulation and Inhibition of Neurons*, Neuromethods, vol. 78,
DOI 10.1007/978-1-62703-233-9_11, © Springer Science+Business Media, LLC 2013

A detailed analysis of the pros and cons of indirect measures of SNA is beyond the scope of this chapter. Suffice to say, direct recordings of SNA offer a stronger method to establish the mechanisms that control SNA. While direct recordings of SNA are invasive and can be difficult to master, such methods enable one to test important hypotheses about the control of the sympathetic nervous system. A detailed Methods section of how to achieve good sympathetic recordings is not the focus of this chapter. The objective of this chapter is to focus on the *analysis* of the SNA signal recorded in both the conscious as well as anesthetized conditions. This chapter will introduce methods to analyze the raw signal that is differentially recorded between two electrodes.

2. Materials for Analyzing SNA

The tools required for the analysis of sympathetic nerve activity are:

1. A data acquisition and analysis system. This can include the PowerLab system from AD Instruments, Spike 2 from Cambridge Electronic Design, or MatLab from MathWorks.
2. The data will need to be processed in a spreadsheet file such as Microsoft Excel.
3. For generation of baroreflex curves, you will need a graphic program that can plot equations. This can include SigmaPlot or GraphPad Prism.

The assumption at the outset is that you have a sympathetic nerve fiber bundle from where you have recorded a differential signal. The most common method to record SNA is to apply band-pass filters with a high-pass filter that varies from 50 to 300 Hz and a low pass between 1 and 5 kHz (1). It is important to ensure that the band-pass filters applied to the signal enable the best signal to noise ratio of the sympathetic bursts. The filtering can also be applied post acquisition of the signal (at 0–5 kHz), but it is important to ensure the quality of the signal using the appropriate filters. The signal typically appears as positive and negative deflections centered around zero, i.e., as an AC signal (Fig. 1). The deflections in a whole fiber recording are bursts of sympathetic nerve activity that are entrained to the cardiac cycle. In anesthetized preparations where baroreceptor modulation is suppressed, this entrainment may be diminished, and a clear respiratory modulation of SNA is observed. The signal to noise ratio determines the average deflection of a single burst relative to the noise level. It is important to listen to the sympathetic signal to ensure that small bursts of sympathetic activity are not being lost in the noise. Problems with the signal can include EMGs and in some cases an ECG signal (Fig. 2 and see Notes (Sect. 4.1)).

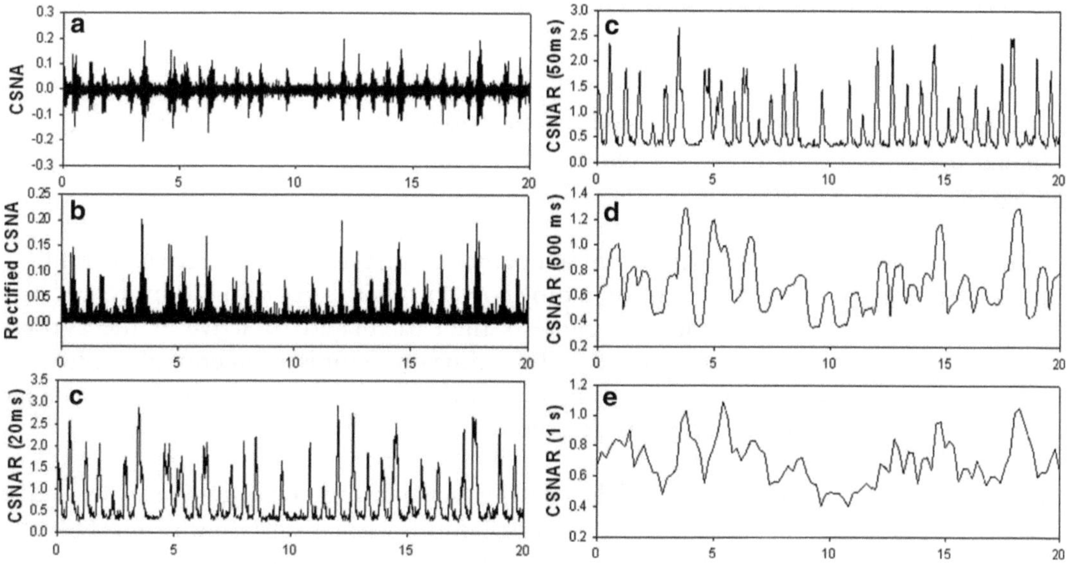

Fig. 1. Denotes a raw sympathetic nerve activity signal (panel **a**), the same signal rectified (panel **b**), and the rectified signal integrated over four different time constants (20, 50, 500, 1,000 ms in panels **c**–**f**, respectively).

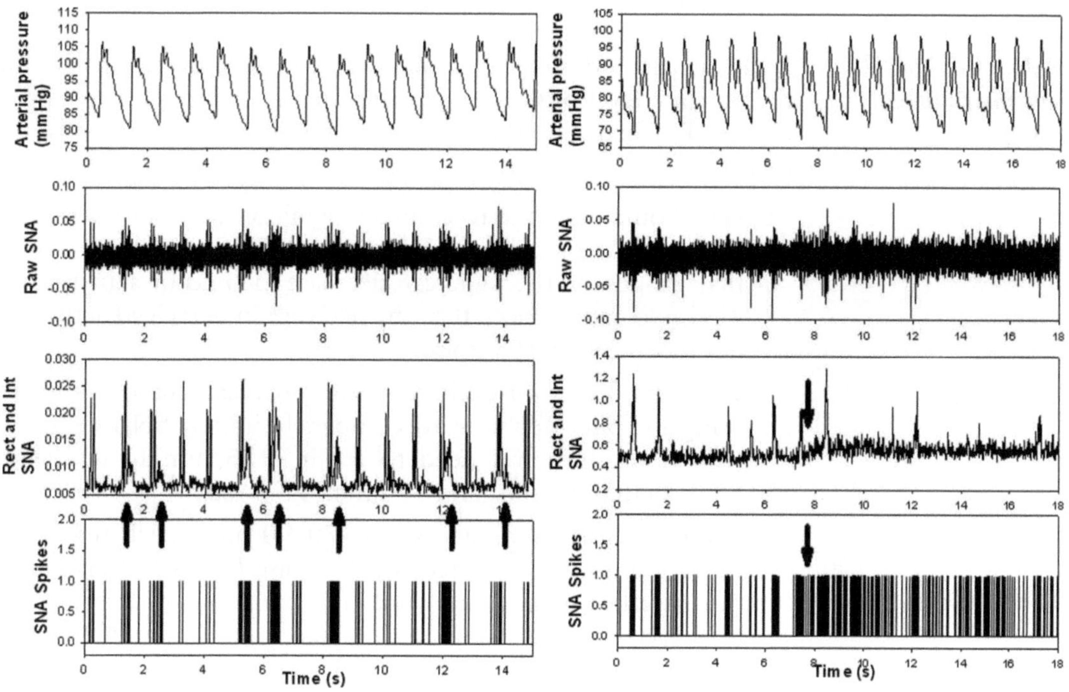

Fig. 2. Examples of ECG (*left panel*) and EMG (*right panel*) in SNA signal. *Arrows* in the *left panel* denotes real sympathetic bursts, and the *arrow* in the *right panel* denotes appearance of EMG and resultant shift in the baseline of the integrated signal.

3. Methods

3.1. Methods to Analyze the Raw Signal

In order to analyze the baseline level of sympathetic drive, two methods are commonly used:

3.1.1. Rectification and Integration of the Signal

Typically, the raw sympathetic signal is full-wave rectified and then integrated using a time constant that varies from 20 ms to 1 s. The area under the curve is then calculated to arrive at the average discharge intensity. If the time constant is small enough, individual bursts can be visualized. If the time constant is large, the average level of SNA is determined (Fig. 1).

3.1.2. Setting a Threshold and Count of Individual Spikes

This method relies on the fact that the raw activity is expressed as deviations from baseline. In this case, a threshold is selected, and instances of the signal crossing the threshold are counted. This can be done using programs such as Spike 2. The threshold should be set marginally over the noise level to enable individual spikes to be counted (Fig. 3). Ideally, the threshold should be set so that during periods of no nerve activity, there is no crossing of threshold and hence no spikes counted.

3.2. Zero Level of Sympathetic Nerve Activity

Both of the methods mentioned above require determination of the baseline level of noise when there is no activity present in the nerves. To determine the zero level of SNA, several methods can be used:

1. Ganglionic blockade to completely inhibit SNA allows one to determine the level noise of the signal. It is important to note that ganglionic blockade can decrease locomotor activity, and it is important to note that any decrease in perceived activity is not locomotor EMG noise.
2. Using baroreceptor excitation, e.g., increased blood pressure using i.v. phenylephrine to completely inhibit SNA (Fig. 4). (Blood pressure will need to reach ~105 mmHg in sheep, 85 mmHg in rabbits, and 180 mmHg in rats.)
3. If the baseline heart rate is low (70–90 bpm) such that individual bursts can be detected, the noise level between bursts can be taken as baseline noise. For conscious rodents where baseline heart rate may be high such that individual bursts are not discernable, using nasopharyngeal stimulation to excite SNA allows one to determine the noise level as detailed in these publications (1, 2).
4. For anesthetized experiments, the level of noise following death can also be used as the noise level.

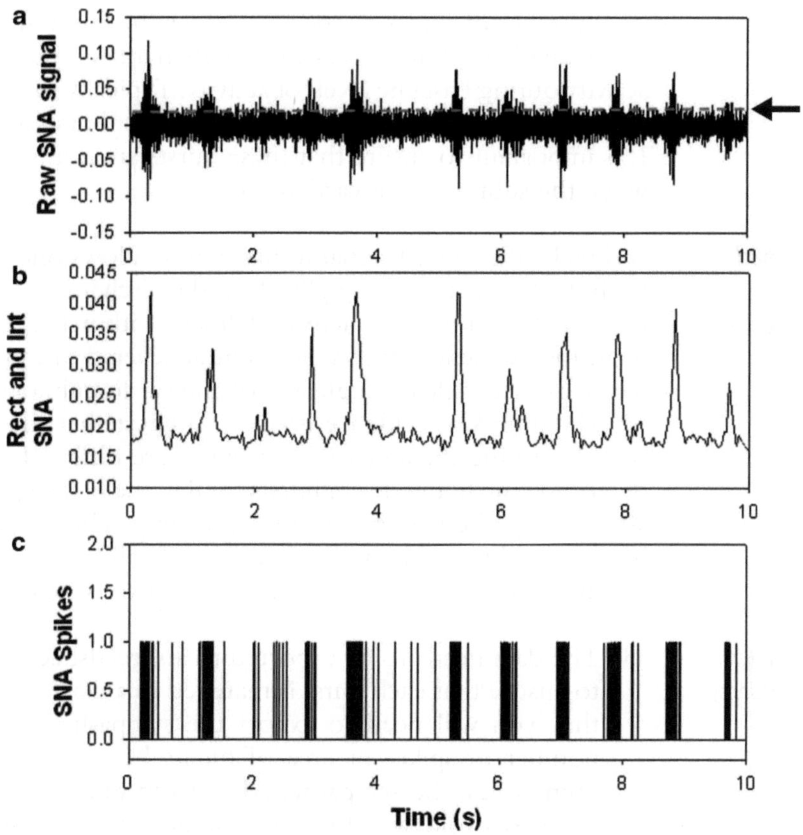

Fig. 3. Raw sympathetic nerve activity examined as a rectified and integrated signal (panel **b**; time constant is 20 ms) or using a threshold to count spikes crossing the threshold (panel **c**).

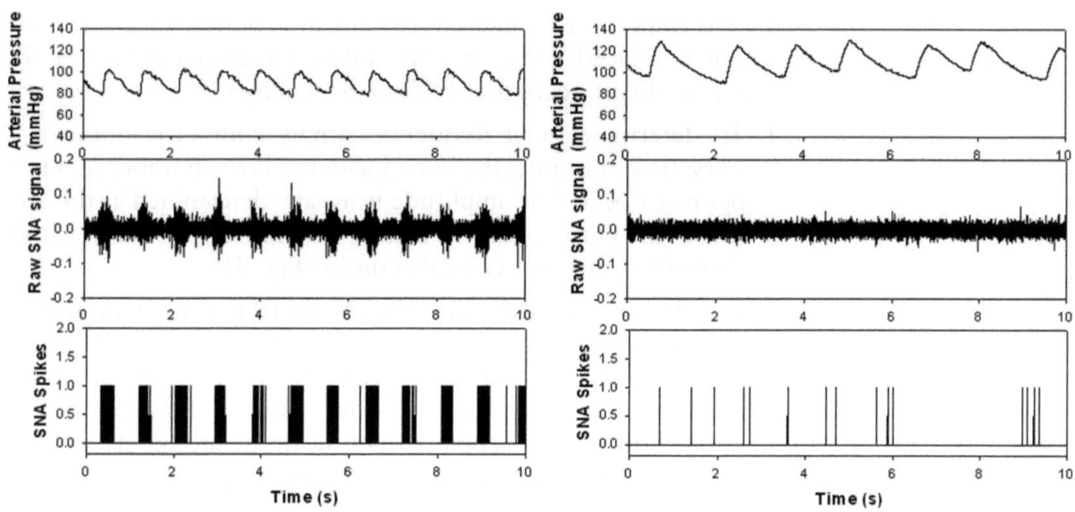

Fig. 4. Raw sympathetic nerve activity during baseline (*left panel*) and after phenylephrine to increase blood pressure (*right panel*). Note the inhibition of sympathetic nerve activity at high blood pressures. The resulting spikes above threshold can be used as the noise level.

The number of spikes or the area under the curve when there are no sympathetic bursts present can then be subtracted from the activity during baseline levels of activity. Typically the nerve recording needs to be visualized so that the smallest bursts are picked out. It is important to ensure that these bursts are not lost in the noise when the subtraction is carried out.

3.3. Amplitude and Frequency of Sympathetic Signals

A signal with a good signal to noise ratio allows one to determine both the frequency and amplitude of the discharges. The amplitude reflects the number of activated fibers within the cardiac cycle, while the frequency reflects their central generation and entrainment to the cardiac cycle through arterial and cardiopulmonary receptors (3, 4). Abundant evidence exists to suggest that the mechanisms that determine the amplitude of bursts are different to those that determine the frequency of bursts (5–8). The first step is to identify all the sympathetic bursts in your recording. The common method is to use a threshold voltage/number of spikes or a rate of rise of the integrated signal or in some cases a mixture of the two (1).

3.3.1. Determining Burst Amplitude and Frequency

1. The data needs to be exported to a spreadsheet file. You need to ensure that each burst is matched to a diastolic nadir. To do this, you will need to export the sympathetic nerve data as number of spikes or area of bursts between diastolic peaks. Sometimes, the sympathetic burst can begin before the nadir of the diastolic waveform, so you need to make sure that each burst fits within two diastolic troughs.
2. Alternatively, if your data have a very quick heart rate (>150 bpm), you can export the peak of the rectified and integrated SNA signal.
3. It is important to identify the smallest bursts, so a threshold can be placed and that only spikes/integrated area over the threshold are recognized as bursts of activity.
4. To determine burst frequency, simply count the number of bursts per unit time; the burst incidence is the number of bursts per heart beat. The amplitude is usually determined as the area under the curve of each burst or the number of spikes within diastolic nadirs that cross threshold (Fig. 5).
5. It is also important to note that the signal to noise ratio usually deteriorates over time and one needs to be aware of this (see Notes (Sects. 4.2 and 4.3)). While determining average burst amplitude, it is important to only consider those bursts that cross threshold. A simple averaging of all the bursts will not provide an accurate answer (see Notes (Sect. 4.4)).

Very few animal studies have reported on the amplitude and frequency of bursts. This is partly because the majority of studies were carried out in rodents where the high heart rate means

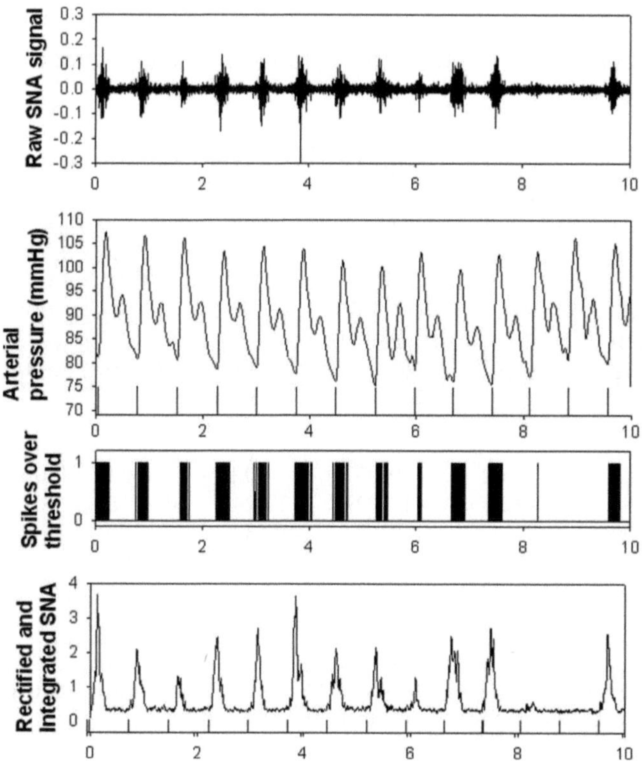

Diastolic nadir	Rect and Int SNA values	Spikes over threshold
0.03	0.5475	19
0.76	0.4762	15
1.53	0.3551	12
2.27	0.4933	20
3.01	0.4884	16
3.75	0.6062	23
4.49	0.4896	19
5.23	0.4757	14
5.98	0.3257	7
6.69	0.5814	24
7.41	0.516	17
8.11	0.2685	1
8.83	0.2504	0
9.56	0.4658	16

Fig. 5. Analysis of the burst amplitude using two different methods is depicted in this figure. The panels represent from *top* to *bottom*, the raw SNA signal, arterial pressure trace, SNA analyzed as spikes above threshold, and SNA analyzed as a rectified and integrated signal of the raw trace. The column on the *right* depicts the time of the diastolic nadir and the corresponding burst size expressed as area under the curve or spikes above threshold.

that during periods of high activity, it is difficult to delineate one burst from the other. In this situation, determining the exact number of bursts can get difficult. One way is to rectify and integrate the raw signal and determine the peak of the rectified and integrated signal. The peak of the rectified and integrated signal is also a good indicator of the amplitude of the burst.

3.4. Rectification and Integration versus Spike Counting

In order to determine if there is a difference in analyzing SNA signals by rectification and integration of the entire signal vs. threshold counting, we subjected SNA signals to both analyses to determine if there was a difference between the two. As detailed in the example, the conclusion is that we did not observe a difference in the statistical significance reached at the end of the experiment, although the percentage change using either method differed significantly.

Real-life example: Cardiac SNA was recorded in conscious sheep before and following volume expansion using 500 ml of gelofusine, a plasma expander. The data was then analyzed two different ways, and the mean results are shown in Table 1. These were then

Table 1
Changes in cardiac sympathetic nerve activity following volume expansion using gelofusine

Time	Animal 1	Animal 2	Animal 3	Animal 4	Animal 5	Animal 6	Mean	SEM
Using rectified and integrated area under the curve								
Control	0.00	-1.40	0.02	-0.12	0.23	0.54	-0.12	0.05
100 ml in	-6.91	-3.05	-7.16	15.37	-16.26	-11.66	-4.94	2.02
200 ml in	-5.47	-10.73	-6.32	9.32	-23.76	-7.08	-7.34	3.00
300 ml in	-4.77	-8.39	-23.40	-2.15	-21.05	-12.70	-12.08	4.93
400 ml in	-7.62	-13.24	-30.52	-6.15	-20.06	-18.33	-15.99	6.53
One-way anova *p* value 0.041								
Using spike counting								
Control	0.00	-0.02	0.07	0.02	-0.05	0.02	0.01	0.00
100 ml in	-11.64	-11.79	-35.54	33.84	-42.72	-29.24	-16.18	6.61
200 ml in	-15.55	-26.63	-27.26	22.95	-49.62	-15.64	-18.62	7.60
300 ml in	-18.05	-30.57	-45.63	-5.19	-57.69	-20.90	-29.67	12.11
400 ml in	-24.40	-29.60	-62.04	-13.22	-66.04	-43.36	-39.78	16.24
One-way anova *p* value 0.032								

The top panels represent the CSNA data analyzed using rectification and integration of the raw CSNA signal, and the bottom columns represent the same data analyzed as spikes above a threshold. Note that although the percentage changes are different, the statistical significance using a one-way anova is the same for both methods

subjected to statistical analysis using a one-way anova. It is important to note here that using either method resulted in a significant change in SNA. This indicates that either method to analyze SNA is acceptable, but it is important to be consistent throughout.

3.5. Analyzing Baroreflex Curves

One common feature of sympathetic nerves is that they are under tight control by the arterial baroreceptors. Baroreflex control of SNA is usually investigated by decreasing and increasing blood pressure to reflexly influence SNA. This can be done using vasopressor and vasodilatory agents or by using cuffs around the descending vena cava and aorta. The method below will detail the construction of a baroreflex curve.

Step by step methods:

1. For the basis of this example, we will assume that the baroreflex curve raw data is generated using either infusions of vasoactive drugs or using cuff electrodes to increase and decrease blood pressure and thereby SNA.
2. The data needs to be analyzed as either a rectified or integrated output or using a threshold to count spikes. It is best to start

with the data generated during phenylephrine infusion as this will enable you to pick the threshold to determine the zero baseline. You will need to determine the burst area or number of spikes between each diastolic nadir.

3. The generated data is then exported as a text file to allow manipulations in a graphical program. If you are only interested in mean sympathetic activity, the data can be exported as spikes or area of SNA per unit time. If you are interested in burst amplitude, you need to export the data per unit heart beat to ensure that you can compare burst sizes at different time points. To do this, you need to identify the diastolic nadirs and calculate the area between each diastolic nadir (see Notes (Sect. 4.5)).

4. If you have a quick baseline heart rate and large changes in heart rate (as in rodents), you can plot the raw data. If working with larger animals with slow heart rates, you may have to reduce the number of points that are plotted. To do this, sort your data from lowest to highest diastolic blood pressures. Then calculate the average of every 5 data points so that you now have a new list of diastolic blood pressures and analyzed SNA values.

5. Plot the data as mean sympathetic activity vs. diastolic blood pressure (Fig. 6). It is important to determine here if there is a plateau of SNA during decreases in blood pressure. If there is no plateau of the data, it will be difficult to determine the precise point at which SNA plateaus (see notes (Sect. 4.6)). Usually, the data should plateau at the lower end (i.e., no bursts when the pressure is too high).

6. A 4- or 5-parameter sigmoidal fit is used to fit the data accurately. This can be done in any statistical or graphical program (SigmaPlot or GraphPad Prism is a good one). This can then be used to determine the parameters of the fit of the curve. The parameter equation is detailed below:

$$y = y_0 + \frac{a}{1 + e^{b(x - x_0)}}$$

where y_0 = lower plateau, a = range of the baroreflex curve, b = gain of the baroreflex curve, $x_0 = BP_{50}$ of the curve, the blood pressure when SNA is 50 % of the maximum.

In the examples given here, we have used a 4-parameter sigmoidal curve fit. As discussed by Ricketts and Head (9), a 5-parameter sigmoidal fit may be better in terms of goodness of fit, but in my experience, the difference is marginal.

7. It is important to inspect the raw data to ensure that the fitted curve is not dependent on one or two outliers. If they are, recheck the data to ensure that there is not mistake and replot.

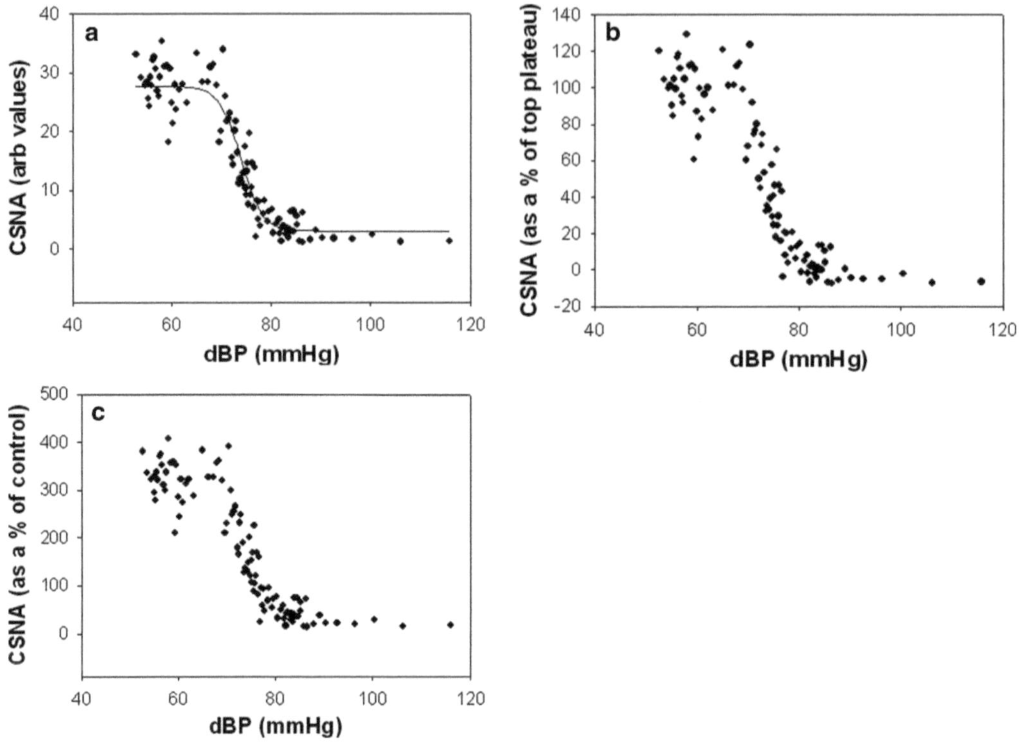

Fig. 6. Baroreflex curves drawn from one animal. Panel (**a**) represents the raw data drawn as a scatter plot. Panel (**b**) represents the data redrawn using the top plateau as 100 %. Panel (**c**) represents the data redrawn using the control level as 100 %.

8. The drawn baroreflex curve reflects changes in SNA using raw baseline activity. You can now either redraw the curve using the baseline levels as 100 % or using the top plateau (derived from the sigmoidal parameters) as 100 % (Fig. 6).

3.6. Comparing Measures of SNA

3.6.1. Within Animal Comparisons

As long as the signal to noise does not deteriorate, the comparison can be made as percentage change from control. If comparisons are made between days, it is important to ensure that there is no change in the signal to noise ratio. Head and colleagues (2) have shown that the maximum responses to nasopharyngeal stimulation remain stable in an animal over time and that any degradation of the signal is associated with degradation of the maximum response (however, see notes (Sect. 4.7)). This then allows one to compare baseline levels of activity between different days. This technique has been used successfully by different groups to compare levels of activity between days. For comparison of baroreflex curves in a within animal comparison, use the control data as 100 %. This allows one to observe real changes in gain as well as the upper and lower plateaus.

3.6.2. Between Animal Comparisons

The best way to compare sympathetic activity between groups is to compare the number of bursts per unit heart beat (or per unit time

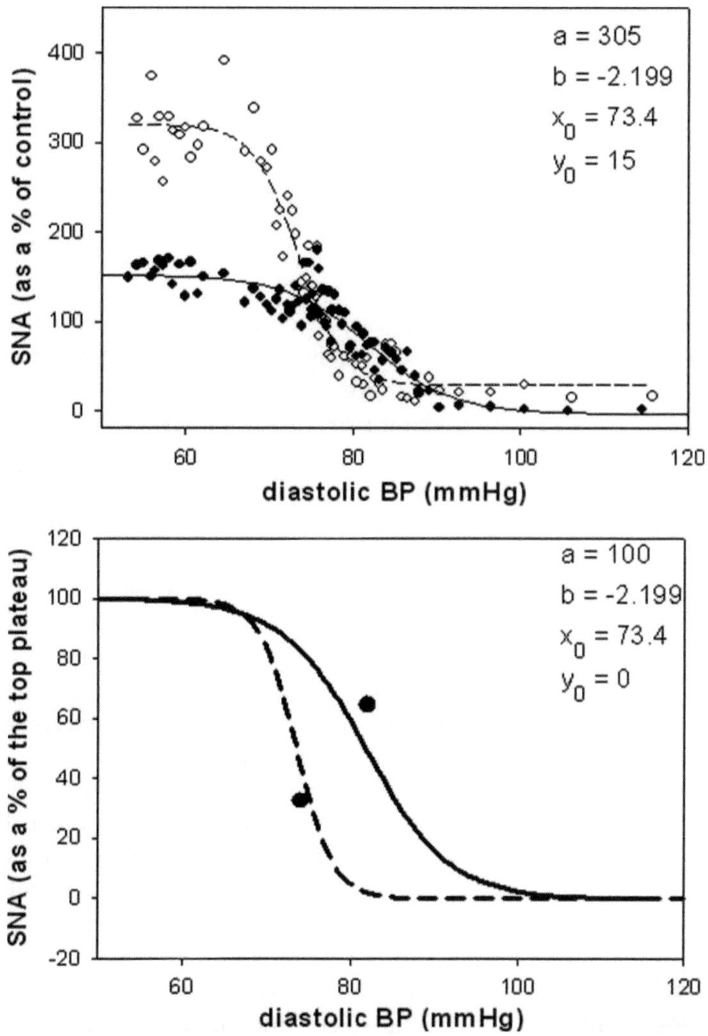

Fig. 7. *Top panel* represents the raw data for nerve activity recorded in two different organs (heart and kidney) within the same animal. A four-parameter sigmoidal fit is applied to the data, and the inset represents the values for the fit for cardiac SNA (represented by *open circles*). The *bottom panel* represents the same sigmoidal fit redrawn as a percentage of the top plateau. The only variables that can be compared between the two nerve activities are the gain and the BP_{50}

if there is no change in heart rate). This method unfortunately also has an assumption. It is not known if the same burst releases the same amount of noradrenaline into the postsynaptic cleft. This is an assumption that is difficult to accurately test at present. While other methods are available and are used by many researchers, it is important to consider their limitations and assumptions.

For comparisons between nerves in the same animal, the conservative approach is to use the top plateau (with sodium nitroprusside) as 100 %. This will only allow one to determine changes in the gain and the threshold and saturation points of the curve (Fig. 7).

If the data are constructed relative to control as 100 %, it is important to understand that changes in the signal to noise ratio can influence your results. The considerations when comparing SNA between different groups of animals remain the same.

4. Notes

4.1. Dealing with ECG and EMG in the SNA Signal

The presence of EMG in the signal can be very annoying. If the preparation is an anesthetized preparation, one should spend some time trying to eliminate EMG at the source. This can be done by making sure the amplifier is well grounded. If you cannot, or you are working in a conscious preparation, if the EMG interference is limited to small time scales, you can try and work around the periods of EMG. If the EMG is present in a lot of the data, one option is to try setting your threshold over your EMG. If you can still detect bursts, you may be able to use the data. If the amplitude of the EMG and the SNA signal is similar, it may be time to start on another preparation.

ECG in the signal trace is slightly easier to eliminate. In this case, try changing your filters (decrease the low-frequency cutoff) to increase the amplitude of the ECG signal. The ECG will occur at a different time point in the cardiac cycle to the sympathetic burst. This means that you can select the ECG trace and delete the ECG off your trace. Alternatively, you can count the area under the curve or the number of spikes for each ECG blip and subtract the same number of spikes along with your estimation of the noise.

4.2. Changes in the Noise Level

It is important to ensure that the noise level of the signal does not change over time since a larger level of noise will result in an increase in the rectified and integrated signal amplitude. One way to watch for changes in baseline noise is to make sure your integrated time constant is small and allows identification of individual bursts. If your integrated time constant is small, you should be able to detect changes in baseline noise more easily (Fig. 2). If your signal to noise ratio is high, you can change the threshold to only pick the bursts when there is EMG.

4.3. Changes in Heart Rate and Changes in Burst Amplitude

It is important to keep in mind that a change in heart rate will influence the burst amplitude. An increased heart rate or decreased heart period will decrease burst amplitude via baroreceptor mechanisms. This makes meaningful interpretations of changes in burst amplitude difficult. If your intervention results in large changes in heart rate, it is likely that baroreceptor inhibition will play a major role in determining the size of the burst.

4.4. Burst Amplitude when There Are Quiet Periods

It may be tempting to just add the average burst amplitudes of all bursts and divide by the number of heart beats to arrive at a mean burst amplitude. However, this only reflects the mean nerve activity level. It is important to identify which heart beats have a burst and eliminate the baseline levels of activity from the heart beats where there is no SNA. To do this, you need to find the smallest burst and corresponding spikes/area for the burst. The next step is to make sure that any heart beats which have less than this number is considered a heart beat with no burst. Once you eliminate all of these heart beats, you can then calculate the average of the burst amplitudes of the remaining heart beats to arrive at a mean burst amplitude.

4.5. Bursts Within Diastolic Nadirs

In order to accurately determine sympathetic burst amplitude, it is important to determine the start and the finish of a burst. If the diastolic nadirs are used to determine burst area/number of spikes, it is important to note that in some cases, the sympathetic burst may start before the diastolic pressure reaches a nadir. In this case, during your analysis, you may have to shift the arterial pressure curve or the analyzed SNA signal to ensure that the bursts are contained within diastolic nadirs.

4.6. Plateaus of Baroreflex Curves

It is important to ensure that baroreceptor loading completely shuts down sympathetic nerve activity. To achieve this, it is important to ensure that blood pressure is elevated long enough for you to determine if a plateau exists. If plotting your raw data does not reveal a plateau in SNA in the majority of your animals, the only parameter that can be accurately measured is the gain of the baroreflex curve. In addition, if baseline heart rate is very slow, then you need to produce slower changes in heart rate to ensure that you have enough points to produce a baroreflex curve.

4.7. Comparisons Using Nasopharyngeal Smoke or Baroreceptor Unloading

When comparing SNA within animals, comparing the responses relative to baseline activity which is normalized to 100 % is the best method as long as there is no change in the signal to noise ratio over time. If there is a change in the signal to noise ratio or if SNA is being compared between animals, another method is to determine the maximum response of the nerve activity and then relate the baseline as a percentage of the maximum activity. Nasopharyngeal stimulation is used as an alternative to baroreceptor unloading, and some reports suggest that this is the most robust method to compare nerve activities. The assumption is that the maximum activity using either method is the same between groups, but this is not always the case (10). One recommendation is that both raw data and normalized data be shown to reduce ambiguity.

5. Concluding Remarks

This chapter has provided an overview of the methods used to analyze multi-fiber recordings of sympathetic nerve activity. Examining changes in sympathetic burst frequency and amplitude adds another dimension to understanding control of the sympathetic nervous system, and readers are especially encouraged to consider publishing data on changes in these two components. While some aspects may seem time-consuming, the analysis of sympathetic activity is relatively straightforward, and the reader is encouraged to seek advice from others in the field. Some of the data analysis can be automated on data acquisition programs, and scripts in Spike 2 that automate part of the analysis can be made available by emailing the author.

Acknowledgements

Work in the lab has been funded by the National Health and Medical Research Council and the National Heart Foundation of Australia. I would like acknowledge the helpful comments from Dr. Lindsea Booth in the preparation of this chapter.

References

1. Guild SJ, Barrett CJ, McBryde FD et al (2010) Quantifying sympathetic nerve activity: problems, pitfalls and the need for standardization. Exp Physiol 95:41–50
2. Burke SL, Head GA (2003) Method for in vivo calibration of renal sympathetic nerve activity in rabbits. J Neurosci Methods 127:63–74
3. Malpas SC (1998) The rhythmicity of sympathetic nerve activity. Prog Neurobiol 56:65–96
4. Malpas SC, Ninomiya I (1992) A new approach to analysis of synchronized sympathetic nerve activity. Am J Physiol Heart Circ Physiol 263:H1311–1317
5. Keller DM, Cui J, Davis SL, Low DA, Crandall CG (2006) Heat stress enhances arterial baroreflex control of muscle sympathetic nerve activity via increased sensitivity of burst gating, not burst area, in humans. J Physiol 573:445–451
6. Malpas SC, Bendle RD, Head GA, Ricketts JH (1996) Frequency and amplitude of sympathetic discharges by baroreflexes during hypoxia in conscious rabbits. Am J Physiol Heart Circ Physiol 271:H2563–2574
7. Malpas SC, Ninomiya I (1992) Effect of asphyxia on the frequency and amplitude modulation of synchronized renal nerve activity in the cat. J Autonom Nerv Sys 40:199–205
8. Ramchandra R, Hood SG, Frithiof R, May CN (2009) Discharge properties of cardiac and renal sympathetic nerves and their impaired responses to changes in blood volume in heart failure. Am J Physiol Regul Comp Integr Physiol 297:R665–674
9. Ricketts JH, Head GA (1999) A five-parameter logistic equation for investigating asymmetry of curvature in baroreflex studies. Am J Physiol Regul Comp Integr Physiol 277:R441–454
10. Mousa TM, Gao L, Cornish KG, Zucker IH (2005) Effects of angiotensin II on autonomic components of nasopharyngeal stimulation in male conscious rabbits. J Appl Physiol 98: 1607–1611

Chapter 12

Insight into Autonomic Nervous System Control of Heart Rate in the Rat Using Analysis of Heart Rate Variability and Baroreflex Sensitivity

Cara M. Hildreth, Ann K. Goodchild, and Jacqueline K. Phillips

Abstract

Heart rate variability and baroreflex sensitivity are two readily accessible methods available to assess control of heart rate by the autonomic nervous system. Heart rate variability can provide information regarding neural (parasympathetic, sympathetic, reflex) and nonneural (hormones, thermoregulation) control of heart rate. Baroreflex sensitivity provides information on how effective the autonomic nervous system is in rapidly changing heart rate in response to changes in blood pressure. Methods are presented on how to calculate heart rate variability using frequency domain analysis and how to calculate baroreflex sensitivity using the sequence method and α-coefficient method of analysis. Data is presented from a variety of rat strains (control, hypertensive, kidney disease, depressed) to indicate the utility of these methods for comparing autonomic regulation of the heart in both anaesthetised animals via intra-arterial recordings and conscious animals using telemetry.

Key words: Heart rate variability, Baroreflex sensitivity, Frequency domain analysis, Sequence method, α-Coefficient, Radiotelemetry

1. Introduction

Heart rate is under the dual influence of the parasympathetic (vagal) and sympathetic components of the autonomic nervous system. The efferent cholinergic parasympathetic pre- and postganglionic neurons in the vagus nerve act to slow heart rate. Cholinergic sympathetic preganglionic neurons and noradrenergic postganglionic neurons increase heart rate. The balance of these two inputs determines the resting level of heart rate. This background 'tone', or relative amount of vagal and/or sympathetic neural input to the heart, is altered on a continual basis by respiration and cardiovascular reflexes in balance with the body's homeostatic requirements. For example, parasympathetic inputs to the

Paul M. Pilowsky et al. (eds.), *Stimulation and Inhibition of Neurons*, Neuromethods, vol. 78,
DOI 10.1007/978-1-62703-233-9_12,

heart alter with respiration to produce respiratory sinus arrhythmia, i.e. the inspiratory-related increase in heart rate and expiratory-related decrease in heart rate, while cardiovascular reflexes, such as the baroreceptor reflex, change both parasympathetic and sympathetic neural input to the heart, altering cardiac output in response to changes in blood pressure.

An imbalance between vagal and sympathetic inputs to the heart, principally too little vagal and too much sympathetic tone, and an inability to modulate these autonomic functions in response to respiration and/or cardiovascular reflexes occur in a number of disease states, including hypertension, kidney disease and depression (1–5). Imbalances of this type are closely associated with an increased risk of cardiac death (6). Given this strong association between aberrant neural control of the heart and risk of cardiac mortality, it is necessary to be able to assess, in both the research and clinical environments, how effectively the parasympathetic and sympathetic nervous systems control heart rate, i.e. what is the level of tone, how much respiratory-related control of heart rate is present and how well do cardiovascular reflexes change heart rate.

Autonomic inputs to the heart, their respiratory and reflex modulation as well as nonneural inputs, such as circulating hormones, temperature and the physical act of breathing (i.e. the compression of the heart through inspiration), produce a healthy waxing and waning of heart rate termed heart rate variability (HRV) (7, 8). In a normal individual, variability in heart rate indicates the ability of the heart to respond appropriately to changes in the external and internal environment. If the system is unhealthy, then the heart begins to beat like a metronome, and little variability is observed. HRV can be readily measured in both the clinical and laboratory setting using computational methods that examine variations in the inter-beat interval between one heart beat and the next using either time domain, frequency domain or non-linear methods of analysis (8). Baroreceptor reflex-mediated changes in heart rate (the increase in heart rate that opposes a decrease in blood pressure and vice versa, in order to bring blood pressure back to normal), which comprise a significant proportion of HRV, can be exclusively examined using computational methods that assess spontaneous or induced changes in blood pressure and the resultant change in heart rate (9).

Time domain analysis is the simplest mathematical approach to assess HRV and is achieved by creating Poincare plots: geometrical analysis of the histogram of all inter-beat intervals or statistical analysis of the standard deviation of the inter-beat interval. The latter is the most commonly used method, and parameters measured include the following: the standard deviation of all normal R–R intervals (SDNN); standard deviation of the average R–R interval in all 5-min segments (SDANN); square root of the mean squared differences of successive R–R intervals (RMSDD), which provides a short-term estimation of HRV and reflects vagal control

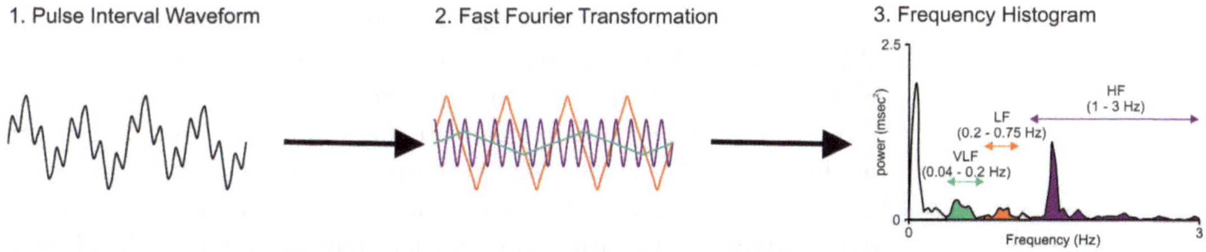

Fig. 1. Schematic illustration of the method to calculate heart rate variability using frequency domain analysis in the rat. A pulse interval waveform is: (1) broken down into its simpler waveform components using fast Fourier transformation; (2) the amount that each waveform contributes to the overall waveform is then plotted using a frequency histogram; (3) in the rat, very low-frequency (VLF) waveforms are defined as waveforms oscillating between 0.04 and 0.2 Hz. Low-frequency (LF) waveforms are defined as waveforms oscillating between 0.2 and 0.75 Hz. High-frequency (HF) waveforms are defined as waveforms oscillating between 1 and 3 Hz.

of heart rate and the proportion of adjacent R–R intervals that are more than 50 ms apart (pNN50). SDNN provides an estimate of overall HRV, is inversely related to resting heart rate, and decreases as respiratory rate increases. SDANN provides an estimate of long-term changes in heart rate, and RMSDD and pNN50 estimates correlate with high-frequency changes in heart rate and are an index of vagal control of heart rate (8, 10, 11).

Frequency domain analysis, or spectral analysis, of heart rate separates the inter-beat interval into its different frequency components (10, 12). When the inter-beat interval, measured using either pulse or R–R intervals, is broken into its different frequency components, through fast Fourier transformation, three main power spectral densities are observed as indicated in Fig. 1: very low frequency (VLF; <0.04 Hz humans, 0.04–0.2 Hz rats), low frequency (LF; 0.04–0.15 Hz humans, 0.2–0.75 Hz rats, 0.15–0.6 Hz mice) and high frequency (HF; 0.15–0.4 Hz humans, 1–3 Hz rats, 2.5–5 Hz mice) (8, 13, 14). The different frequency ranges between species reflect differences in resting heart rate. As parasympathetic, sympathetic and nonneural inputs to the heart oscillate at different frequencies, the different power spectral densities can reflect differing aspects of heart rate control.

HRV estimates can provide information on different aspects of heart rate control, both neural and nonneural. Total HRV power reflects the overall ability of heart rate to vary in response to both neural and nonneural influences. Very low-frequency power is purported to reflect hormonal and thermoregulatory influences on the heart (7, 8). Low-frequency power reflects both sympathetic and vagal influences on heart rate as well as baroreceptor-mediated changes in heart rate (7, 8). High-frequency power is a general index of vagal control of heart rate, reflecting tonic vagal control of heart rate and the reflex and respiratory modulation of vagal inputs to the heart (7, 8).

Baroreceptor-evoked changes in heart rate can be induced in the laboratory setting through a variety of methods. In humans, carotid sinus massage, the Valsalva manoeuvre, head-up tilting and administration of vasoactive agents can be used (9). In rats, electrical stimulation of the carotid sinus nerve, occlusion of the common carotid artery and intravenous injections of vasoactive drugs can be used (9). These methods involve direct activation of the baroreflex; however, the baroreflex is spontaneously activated under normal conditions in order to maintain resting blood pressure. Spontaneous or autogenic baroreceptor function can be assessed using a number of methods including the sequence method, R–R interval–systolic blood pressure cross-correlation and squared ratio of R–R interval/systolic blood pressure spectral powers at LF and HF (α-coefficient) (9).

Baroreceptor function assessed using either *induced* or *autogenic* changes in blood pressure can indicate how effectively baroreceptor activation is able to change heart rate—baroreflex sensitivity (BRS). Induced changes in blood pressure provide additional measures of baroreflex function such as how much activation of the baroreflex alters heart rate and over what range of blood pressure the baroreflex works. However, inducing baroreceptor reflex changes in heart rate in humans in a clinical setting, or conscious animals in a laboratory setting, can be invasive and therefore limits the feasibility of this technique. This can be overcome by using methods for assessing BRS using autogenic changes in blood pressure. Two techniques that provide comparable estimates of BRS are the sequence method and α-coefficient.

The rationale for the sequence method is that spontaneous fluctuations in blood pressure are opposed by a baroreflex-mediated change in heart rate. The sequence method therefore uses an algorithm that seeks out patterns of three or more consecutive heart beats that are characterised by +blood pressure/–heart rate or –blood pressure/+heart rate patterns (see Fig. 2). It is based on strict criteria in which a change in blood pressure up to 1 mmHg in humans and >0.5 mmHg in rats is opposed by a change in heart rate (13, 15). The slope of the linear regression between +blood pressure/–heart rate or –blood pressure/+heart rate patterns is a measure of BRS. BRS calculated with the sequence method correlates with the BRS calculated using the 'Oxford method' (16, 17), where induced changes in blood pressure are achieved with phenylephrine (a vasoconstrictor) and sodium nitroprusside (a vasodilator). It is important to note that the sequence method provides a measure of the gain of the BRS around the set point of arterial blood pressure and, unlike the Oxford method, does not provide information regarding the operating range of the baroreceptor reflex.

The α-coefficient is the square root of the ratio of the power spectral density of heart rate and systolic blood pressure and is calculated by comparing LF (αLF) or HF (αHF) power spectral densities of heart rate with systolic blood pressure (see Fig. 3). The rationale for this method of BRS estimation is that heart rate and

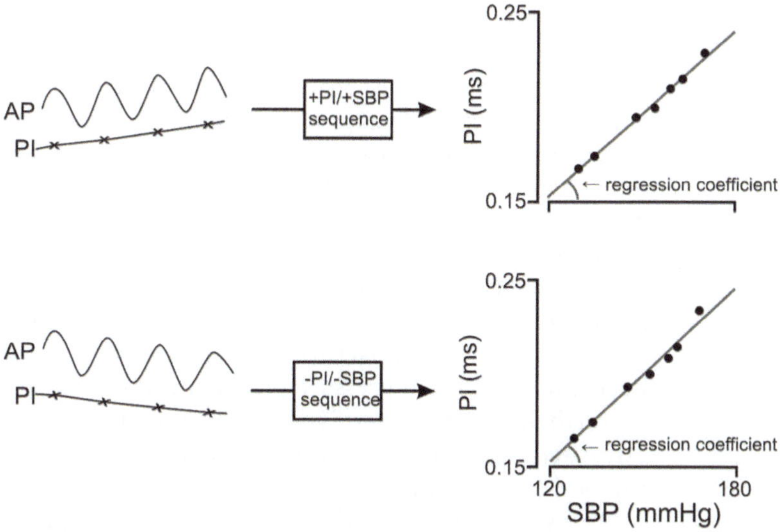

Fig. 2. Schematic illustration of the sequence method of baroreflex analysis. The algorithm searches systolic blood pressure (*SBP*) and pulse interval (*PI*; the reciprocal of heart rate) in the time domain for sequences characterised by an increase or decrease in PI and SBP. The linear regression between changes in SBP and changes in PI is calculated, and the regression coefficient is used as an estimate of baroreflex sensitivity. (AP: arterial pressure). Adapted from Parati et al. (9).

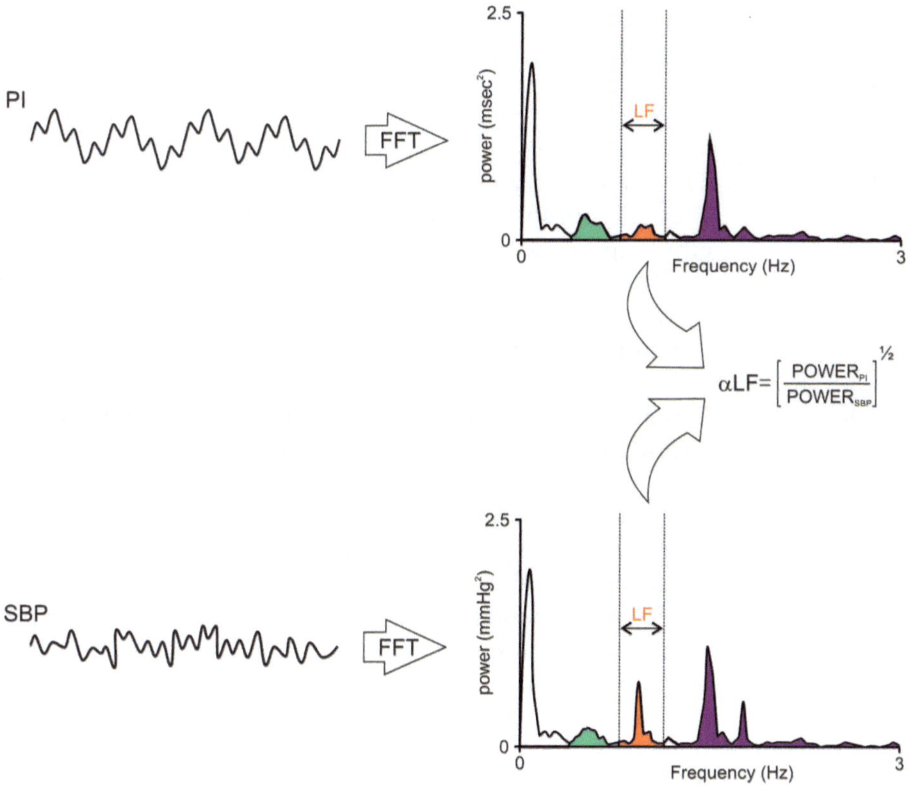

Fig. 3. Schematic illustration of the method to calculate baroreflex sensitivity (*BRS*) using cross spectral analysis of the pulse interval (*PI*) and systolic blood pressure (*SBP*) power spectrum. Power in either high-frequency (>0.15 Hz) or low-frequency (LF, 0.04–0.15 Hz) peaks of PI and SBP power spectrum is compared to give an estimate of BRS. (FFT: fast Fourier transformation). Adapted from Parati et al. (9).

systolic blood pressure power spectrums are highly related (or coherent) in the LF and HF range. The coupling between LF blood pressure and heart rate power spectra is highly dependent upon baroreflex inputs (18, 19). Cutting the nerves that process baroreceptor-related information decreases the incidence of segments that are coherent between blood pressure and heart rate power spectra in the LF range (18). The percentage of coherent segments in the HF range is not affected by cutting these nerves, indicating that other factors, perhaps respiratory-related, couple heart rate and blood pressure power spectra in the HF range (18). Nevertheless, estimates of both αLF and αHF correlate with BRS estimated using bolus injections of phenylephrine and the sequence method (15, 20, 21).

This chapter will provide a detailed overview of how researchers can apply frequency domain analysis of HRV and BRS, estimated using the sequence method and the α-coefficient in the LF range, in rodent models of disease to provide information as to the status of autonomic control of cardiovascular function. The methods provided are primarily concerned with data collection and analysis, and a guide to interpretation of the data provided.

2. Methods

2.1. Animals

The choice of animal, and strain, will be dependent upon the scientific question under consideration. Appropriate institutional ethics approval must be obtained prior to any experimentation involving animals. The size and age of the rat will be determined by the question asked and the design feasibility of the experiment [see note 1 (Sect. 3.1)).

2.2. Equipment

Different set-ups are required for collecting information from anaesthetised versus conscious animals.

2.2.1. Anaesthetised Preparations

The minimum requirements are the capacity to maintain stable anaesthesia with oxygen supplementation if required and access an intravenous (IV) line for administration of IV fluids and anaesthetic supplementation (as and when required) and implantation of an arterial catheter for acquisition of arterial pressure waveforms and subsequent calculation of heart rate from the pulse interval.

Intra-arterial catheters need to be connected to a pressure transducer and then through a bridge amplifier to an analogue to digital signal converter (e.g. Power 1401, Cambridge Electronic Design, Cambridge, UK, or PowerLab, ADInstruments, Colorado, USA) connected to a computer. ECG leads can be attached to obtain a more accurate estimate of heart rate from the R–R interval [see note 2 (Sect. 3.2)]. ECG leads need to be connected to a bioamplifier and then to an analogue to digital signal converter connected to a computer. To record either blood pressure or ECG signal, a data acquisition programme needs to be installed on the computer appropriate to the analogue to digital signal converter used (e.g. Spike 2 for a Power 1401 or LabChart for a PowerLab). A thermometer for recording rectal temperature is required, in conjunction with a thermostatically controlled electric blanket and infrared heating source (if required), to maintain core body temperature at 37 ± 0.5 °C throughout the procedure.

2.2.2. Conscious Preparations Involving Radiotelemetry

Two commercially available telemetry systems currently exist. One is available from Data Sciences International (DSI) (MN, USA) and the other, Telemetry Research (Auckland, New Zealand). For a DSI system, you will need a receiver (RPC-1 for telemetry in rodents), an ambient pressure monitor (APR-1), a data exchange matrix and a telemetry probe (e.g. PA-C40 or PA-C10) to measure blood pressure in rodents; see note 1 (Sect. 3.1) about different uses. Further information of different telemetry probes available can be found at http://www.datasci.com. The data matrix connects directly to a computer. In order to acquire data using a DSI system, you need to install their data acquisition software programme onto the computer (Dataquest A.R.T). You can use this software package to analyse the data and/or export the data to Excel (Microsoft®). Alternatively, you can purchase additional licences to analyse acquired data in third-party software programmes such as Spike 2.

For a Telemetry Research system, you require a SmartPad (TR180, battery charging pad & receiver), Configurator (TR190), an analogue to digital converter, data acquisition software programme and telemetry probe (e.g. TRM54P probe) to measure blood pressure in rodents. Information regarding other available telemetry probes can be found at http://www.millar.com/products/telemetry. The telemetry research system does not stipulate which analogue to digital converter or data acquisition software programme needs to be used and which custom-built software programmes are compatible with this system.

2.3. Protocols for Data Collection

1. Protocol for HRV, systolic blood pressure variability (SBPV, needed to calculate BRS using the α-coefficient method) and BRS data collection from anaesthetised animals.
   (a) For non-recovery procedures, animals are anaesthetised with ethyl carbamate (urethane 1.3 g/kg i.p.). Depth of anaesthesia should be assessed regularly using reflex responses to tactile (corneal stroking) and noxious (hind-paw pinch) stimuli. Additional doses of urethane (0.13 g/kg IV) are administered as required.
   (b) The right femoral artery and vein are cannulated to record arterial blood pressure and administer drugs, respectively.

(c) After an initial period of 20 min for stabilisation, data for basal cardiovascular parameters [blood pressure and heart rate] are acquired over a 5-min period. HRV, SBPV (needed to calculate αLF) and BRS will be determined from an 80-s segment during this period where blood pressure and pulse interval are stationary [see note 3 (Sect. 3.3)].

(d) All data is subsequently analysed offline.

2. Protocol for HRV, SBPV (needed to calculate BRS using the α-coefficient method) and BRS data collection from conscious animals.

   (a) Animals are anaesthetised and maintained under isoflurane anaesthesia to facilitate rapid postoperative recovery. Appropriate sterile surgical approach and pain relief for telemetry probe implantation and recovery are critical to the success of the procedure.

   (b) Telemetry transmitter catheters can be implanted into the aorta either via the femoral artery or directly into the aorta [see note 4 (Sect. 3.4)]. The type and size of transmitter will depend on the body weight of the animal [see note 1 (Sect. 3.1)]. If the probe possesses a suture rib (DSI probes only), it can be sutured to the abdominal muscle. Muscle incisions are sutured closed, and skin incisions are stapled closed. Animals need to be monitored closely postoperatively for signs of pain and distress and postsurgical complications [see note 5 (Sect. 3.5)].

   (c) After a 7–10-day recovery period [see note 6 (Sect. 3.6)], arterial pressure is recorded continuously over a 5-min period every 15 min over the day–night cycle for 24 h in order to determine blood pressure levels over 24 h [see note 7 (Sect. 3.7)]. From this data, stationary 80-s periods from a 2-h period during the middle of the day and during the middle of the night are used to determine HRV, SBPV and BRS thereby avoiding the impact of circadian rhythms [see note 8 (Sect. 3.8)].

   (d) All data are subsequently analysed offline.

2.4. Data Analysis

1. Heart Rate Variability

   Pulse interval can be calculated from the arterial pressure waveform or R–R interval calculated from the ECG waveform if available [see note 2 (Sect. 3.2)]. The methods provided to calculate HRV use Spike 2 software. The pulse interval or R–R interval waveform is then resampled at 10 Hz and converted from seconds to milliseconds, and the channel offsets to zero by applying a DC remove process to the waveform. The resampled pulse interval waveform is then processed using fast

Fourier transformation, and the resulting frequency histogram is viewed in a Hanning window with a bin size of 256 to show frequencies from 0 to 5 Hz in 128 bins with a resolution of 0.04 Hz. The frequency histogram produced is then copied as text into Excel. The sum of the HRV power between 0.04–3 Hz, 0.04–0.2 Hz, 0.25–0.75 Hz and 1–3 Hz is then calculated in order to estimate total, VLF, LF and HF power, respectively. The process used to calculate HRV is illustrated in Fig. 4.

2. Systolic Blood Pressure Variability
For the purpose of illustration, the method to calculate SBPV is provided in order to be able to calculate the α-coefficient (see below). The methods provided to calculate SBPV use Spike 2 software and are similar in approach to that used to calculate HRV. Systolic blood pressure is calculated from the arterial pressure waveform as the value of blood pressure at the peak of every arterial pressure waveform. The systolic blood pressure waveform is then converted to a virtual channel and resampled at 10 Hz, and the channel offsets to zero by applying a DC remove process to the waveform. The virtual systolic blood pressure waveform is then processed using fast Fourier transformation, and the resulting frequency histogram is viewed in a Hanning window with a bin size of 256 to show frequencies from 0 to 5 Hz in 128 bins with a resolution of 0.04 Hz. The frequency histogram produced is then exported as text into Excel. The sum of the HRV power between 0.04–3 Hz, 0.04–0.2 Hz, 0.25–0.75 Hz and 1–3 Hz is calculated in order to estimate total, VLF, LF and HF power, respectively.

3. Baroreflex Sensitivity
    (a) *Spontaneous BRS*: sBRS is calculated using the sequence method from the pulse interval and the arterial pressure waveform (13, 22]. A script needs to be written that is specific for the data acquisition/analysis software package used. The script needs to be provided with a blood pressure waveform and a pulse interval or R–R interval waveform. The script then needs to be able to search through these two waveforms and detect when an increase in blood pressure of at least 0.5 mmHg has occurred three times in a row, simultaneous with three successive increases in pulse interval or R–R interval. Each one of these lengthening sequences within the blood pressure recording needs to be identified, and the linear regression of all of these sequences needs to be calculated. The script then needs to search through the blood pressure waveform again to identify when a decrease in blood pressure of at least 0.5 mmHg has occurred three times in a row simultaneous with three successive shortenings in pulse interval or R–R interval. Again, the script needs to identify each one of these

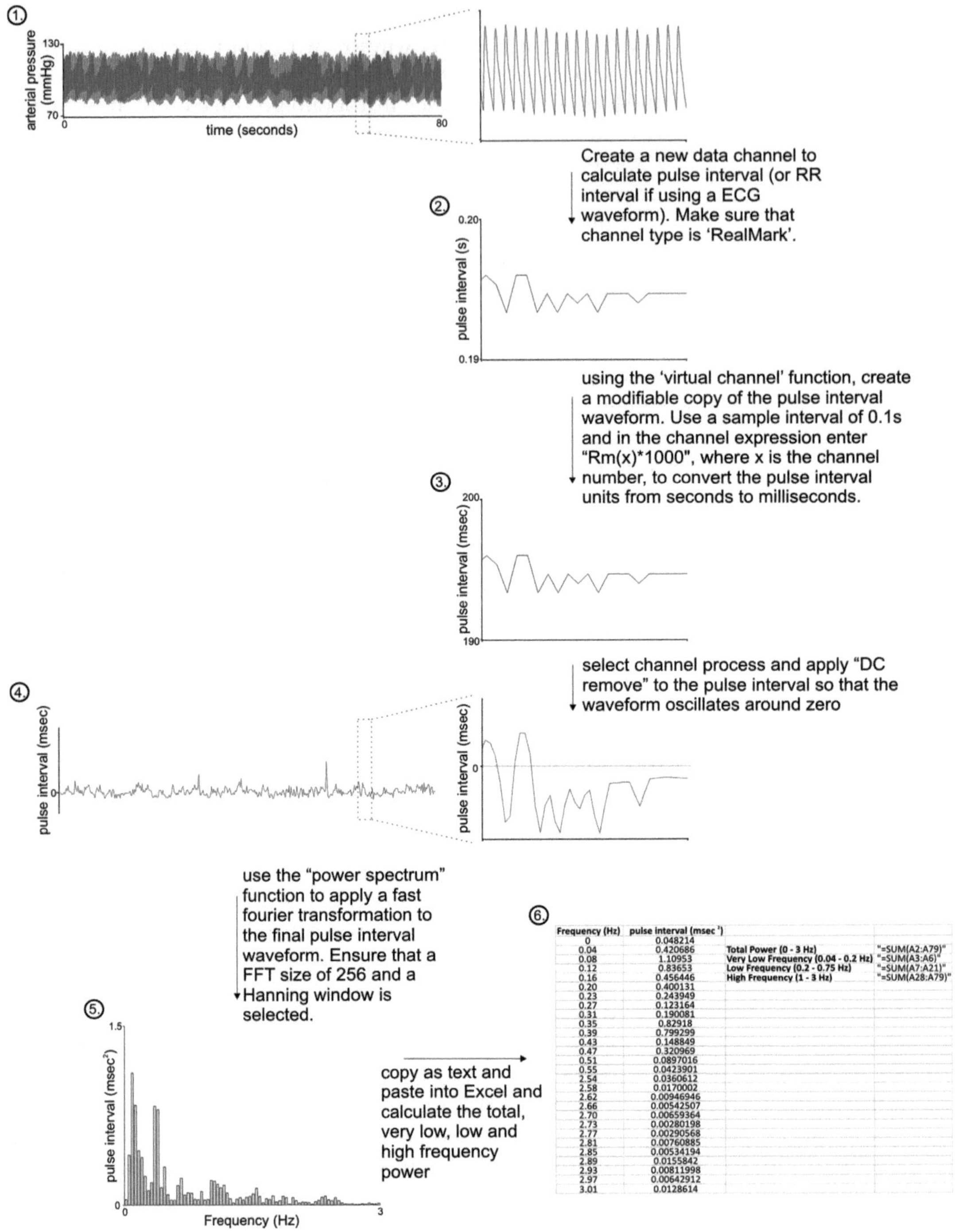

Fig. 4. Schematic illustration of the process used to calculate heart rate variability using a blood pressure recording in Spike 2. From the blood pressure recording (1), the pulse interval is derived (2). A virtual copy of the pulse interval channel is made and the units converted from seconds to milliseconds (3). The pulse interval channel is then offset to zero (4), and a fast Fourier transformation is applied to create a frequency histogram (5). The frequency histogram is copied into excel, and the amount of power within the total- (0.04–3 Hz), very low- (0.04–0.2 Hz), low- (0.2–0.75 Hz) and high (1–3 Hz)-frequency ranges is calculated (6).

shortening sequences and calculate the linear regression of all sequences (22). sBRS is then calculated as the average regression of all lengthening and shortening sequences for each animal. Alternatively, programmes such as HemoLab can be downloaded that can calculate sBRS using the sequence method (see: http://www.haraldstauss.com/HemoLab).

(b) α-*Coefficient*: LF and HF power spectral densities of HRV and SBPV are estimated. The square root of the ratio between LF and HF of HRV and SBPV is calculated. αLF and αHF, respectively, are calculated as the average of all estimates for each animal.

3. Notes

3.1. Animal Age and Size

For conscious animal data collection using telemetry, the body weight of the animal will affect the choice of telemetry probe used for implantation, with the recommended minimum body weight when using the commercially available telemetry systems supplied by Data Sciences International (USA) being 175 g for PA-C40 and 17 g for PA-C10 probes (both probes record blood pressure using a fluid filled catheter) and 200 g for TRM54P probes supplied by Telemetry Research (New Zealand) (probe measures blood pressure using a solid state Millar pressure sensor). For experiments conducted under anaesthetised conditions, the weight and age of the animal does not affect the ability to conduct the study, and as such should be relative to the study in question.

3.2. Determination of Heart Rate from Blood Pressure or ECG Waveform

Heart rate can be calculated from either the pulse interval (i.e. the time between the peak of one arterial pulse wave and the peak of the next) or the R–R interval (i.e. the time between one R wave on the ECG waveform and the next). As the peak of the R wave is sharper than the peak of the pulse wave, detecting the peak of the R wave can be achieved more accurately leading to less variability in the R–R interval compared with pulse interval and therefore a more accurate calculation of heart rate. This is important for experiments assessing HRV, as any variability in estimating the pulse interval or R–R interval can lead to overestimations of HRV. If you can only obtain a measurement of arterial pressure (e.g. if you have a telemetry probe that only records blood pressure), you can overcome this by taking the first derivative of the blood pressure waveform. This results in a sharp peak at the steepest point of the pressure waveform. Two peaks will normally be identified—a positive peak at the rising phase of the pressure waveform (i.e. between diastole and systole) and a negative peak at the falling phase of the pressure waveform (i.e. between systole and diastole). Heart rate can be calculated from the time between one positive peak and the next. This is the method that the Data Sciences acquisition package uses to calculate heart rate from a blood pressure waveform. The most important point, however, is to be consistent and choose a method to calculate heart rate and do not vary between subjects.

3.3. Choosing Data to Be Used for Analysis

Given the nature of the data analysis—to determine the amount of variability ordinarily present—choosing segments of data where PI or SBP continually change can introduce variability into the data and therefore result in overestimations of HRV and/or SBPV. For conscious animals, this natural movement can occur with changes in activity, breathing and external stimuli such as noise. To avoid this, expand the y-axis of the resampled systolic blood pressure and pulse interval waveforms. Place a horizontal line through the data at $y=0$. Choose data segments that generally track along the horizontal line (a little deviation from the horizontal line is normal); avoid using data that constantly deviates from the horizontal line. For telemetry systems that provide an indication of the animals movement (i.e. DSI), avoid using segments where the animal was overly active.

3.4. Method of Telemetry Probe Placement

Telemetry transmitter catheters can be implanted directly into the descending aorta or into the femoral artery and then fed up into the descending aorta. Both approaches produce excellent blood pressure signals; however, both have considerations. Directly inserting the catheter into the descending aorta needs to be done quickly, as blood flow to the lower extremities is temporarily occluded during the procedure (less than 5 min is optimal), and following insertion, blood flow beneath the catheter towards the iliac bifurcation needs to be ensured to guarantee adequate perfusion of the lower extremities. Failure to do so will result in tissue hypoxia and subsequent damage to the lower limbs. Implanting the catheter via the femoral artery and feeding it into descending aorta, while not time dependent, can also result in poor tissue perfusion due to inadequate collateral blood flow. This is of greater risk in younger animals (bodyweight less than 170 g), and in these animals, the direct aortic approach is recommended.

3.5. Potential Surgical Complications

The most common surgical complications that occur following implantation of a telemetry probe are wound dehiscence, exposure of the transmitter body and poor blood flow to the lower limbs [see note 4 (Sect. 3.4)]. Wound dehiscence can occur when the skin wound is not closed adequately or there is poor wound healing. Sterile surgical technique is critical, and the use of surgical skin staples as opposed to sutures provides greater protection against this problem. Postoperative infections should be treated with the appropriate antibiotic cover. Exposure of the transmitter body, or a prolapsed probe, can occur when the abdominal muscles are not sutured together adequately, although care should be taken not to overtighten muscle sutures as this may cause ischaemia. For probes that have a suture rib (DSI probes only), this can be avoided by ensuring that one suture is placed in between each suture rib in order to strengthen the wound.

3.6. Recovery Period Prior to Telemetry Data Collection

For studies using telemetry to assess autonomic function as described here, it is imperative that animals recover for an adequate amount of time prior to data collection. Failure to do so will affect the quality of data obtained. Markers of appropriate surgical recovery include body weight gain, increased food and water intake, wound healing and a return of circadian rhythms in heart rate, blood pressure, temperature and activity. A study conducted by Leon et al. (23) indicated that this process could take about 7 days in the rat. In our experience, the recovery time is also heavily influenced by the choice of anaesthetic, with the return of circadian rhythms taking longer when barbiturate anaesthetics, such as sodium pentobarbitone, are used (recovery time close to 10 days) compared with inhalational anaesthetics such as isoflurane (recovery time often less than 7 days).

3.7. How Much Telemetry Data to Collect?

Copious amounts of data can be recorded with radiotelemetry—often more than is manageable and, for that matter, can be feasibly analysed. A recent study by Guild and colleagues demonstrated that recording data for 10 s every 10 min was sufficient to determine the resting level of heart rate or blood pressure over a 24-h period (24). However, this is not a sufficiently long enough period of time in which to calculate HRV and BRS as the number of pulse intervals in a 10 s period for a rat would only be ~30. In order to calculate HRV and BRS accurately, at least 80 s of data is required. Furthermore, as stationary data can be difficult to acquire in a conscious animal, it is advisable to record more than the 80-s period of data needed. We recommend recording 5 min worth of data every 15 min in order to obtain enough segments of data in which to calculate HRV and BRS.

3.8. Minimising the Impact of Circadian Rhythms

In order to avoid the influences of circadian rhythm (i.e. the nocturnal rise in heart rate and blood pressure) on HRV and BRS, we calculate HRV during the middle of the day (1100–1300 h) and during the middle of the night (2300–0100 h). This 2-h window should provide a sufficient number of segments from each of the 5-min period recordings every 15 min to calculate HRV and BRS on in each animal (~8 segments per animal during the day and during the night).

4. Results

4.1. Typical Results

As an example of the utility of HRV and BRS as a method to assess neural control of the heart, the following data is presented to address two questions: (1) What is the effect of disease on HRV and BRS? (2) What is the effect of anaesthesia on HRV and BRS? The animal models used to provide this data are the spontaneously hypertensive rat (SHR), a model of essential hypertension (25); the Lewis polycystic kidney (LPK), an animal model of kidney disease (26); the Flinders sensitive line (FSL), an animal model of depression (27) and their respective genetic controls; the Wistar Kyoto (WKY), Lewis and Sprague Dawley (SD) rats.

All experiments were approved by Macquarie University's Animal Ethics Committee and conducted in accordance with the Australian Code of Practice for the Care and Use of Animals for Scientific Purposes. Experiments conducted on Lewis and LPK rats were also approved by Murdoch University Animal Ethics Committee. Male SD ($n=12$), Lewis ($n=14$), SHR ($n=14$), LPK ($n=18$), FSL ($n=11$) and WKY ($n=12$) rats were used.

All data are expressed as mean ± SEM. A student's *t*-test was used to compare HRV and BRS estimates in each model of disease compared with its genetic control. A one-way ANOVA was used to compare HRV and BRS estimates within each strain under anaesthetised versus conscious (day and night) conditions.

Total, VLF, LF and HF HRV estimates taken under both anaesthetised conditions and in conscious animals using telemetry, during the day and at night, in the various rat strains are shown in Fig. 5. BRS estimates calculated using the sequence method and α-coefficient in the various rat strains under anaesthetised and conscious conditions are shown in Fig. 6.

HRV and BRS in Disease: In the LPK model of kidney disease, total, VLF, LF and HF power were reduced, compared with the control Lewis strain under anaesthetised conditions (Fig. 5). Total and HF power were also reduced in the FSL model of depression, under anaesthetised conditions when compared to its control (Fig. 5a, d). Under conscious conditions, HF power was also reduced in the FSL compared with the SD, its genetic control (Fig. 5d). HRV did not differ between the hypertensive SHR and its genetic normotensive control, the WKY (Fig. 5). Together, this implies that in the LPK, but only under anaesthetised conditions, total HRV, hormonal and/or thermoregulatory (VLF), vagal and sympathetic and respiratory and reflex modulation of heart rate (LF) are altered. Vagal and respiratory modulation of heart rate (HF) is also reduced in the FSL, under both anaesthetised and conscious conditions.

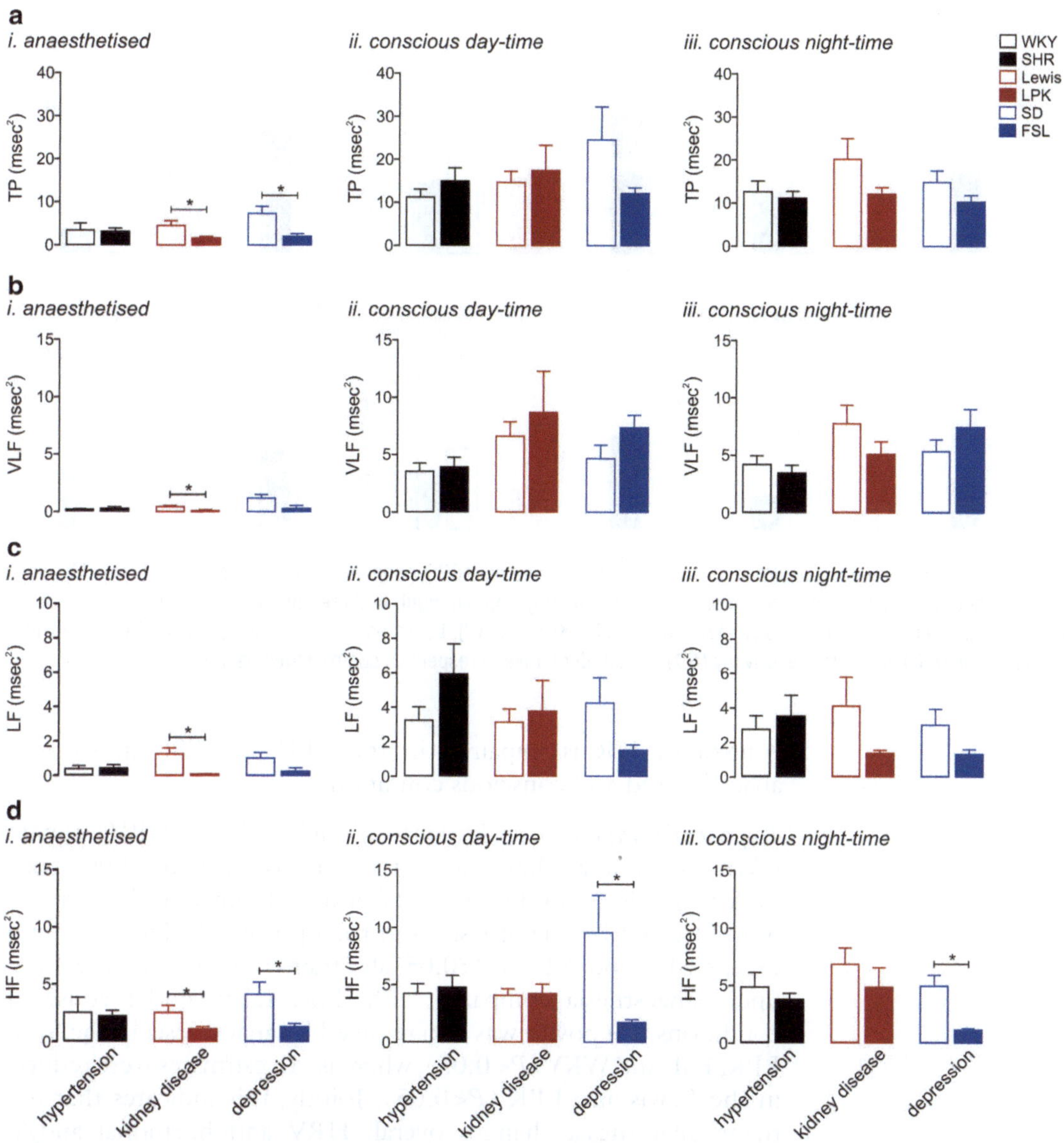

Fig. 5. Heart rate variability estimates calculated using frequency domain analysis, under (i) anaesthetised and conscious conditions during the (ii) day and (iii) night, in three animal models and their respective controls: spontaneously hypertensive (*SHR*) and Wistar Kyoto (*WKY*), Lewis polycystic kidney (*LPK*) and Lewis, Flinders sensitive line (*FSL*) and Sprague-Dawley (*SD*). Total heart rate variability power (*TP*) estimates are shown in panel (**a**). Very low-frequency (*VLF*) estimates are shown in panel (**b**). Low-frequency (*LF*) estimates are shown in panel (**c**), and high-frequency (*HF*) estimates are shown in panel (**d**). **P*< 0.05 relative to respective control strain as indicated.

BRS, estimated using the sequence method (Fig. 6a) and α-coefficient (Fig. 6b), was reduced in both the LPK and FSL compared with their genetic controls under anaesthetised conditions, but not in the SHR. Under conscious conditions, BRS calculated using the sequence method was reduced at night in both the LPK and FSL. These results indicate that baroreceptor reflex

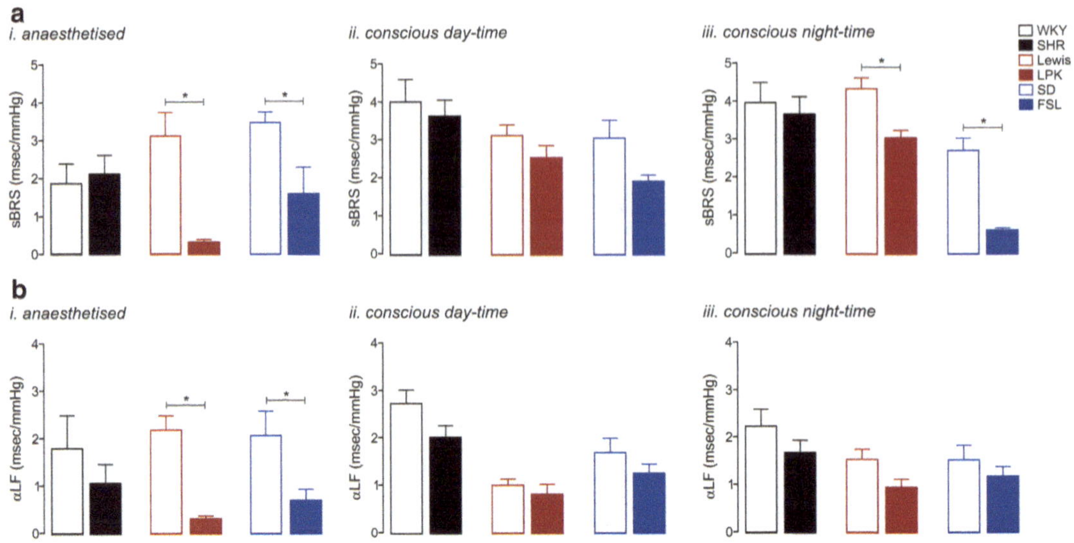

Fig. 6. Baroreflex estimates calculated using the sequence method (sBRS, panel **a**) and α-coefficient (αLF, panel **b**), under (i) anaesthetised and conscious conditions during the (ii) day and (iii) night, in three animal models and their respective controls: spontaneously hypertensive (*SHR*) and Wistar Kyoto (*WKY*), Lewis polycystic kidney (*LPK*) and Lewis, Flinders sensitive line (*FSL*) and Sprague-Dawley (*SD*). $^{*}P<0.05$ relative to respective control strain as indicated.

control of HR is impaired in the LPK and FSL under both anaesthetised and conscious conditions.

Impact of Anaesthesia on Estimates of HRV and BRS: HRV estimates taken under anaesthesia were typically lower than those taken under conscious conditions. Total and VLF power estimates were most affected by anaesthesia, and total power ($P<0.05$ all strains except SD) and VLF ($P<0.05$ all strains) estimates were lower under anaesthesia compared with those taken under conscious conditions. LF power was also reduced by anaesthesia in the SHR, LPK, FSL and WKY ($P<0.05$), whereas HF estimates were reduced in the Lewis and LPK ($P<0.05$). Jointly, this indicates that urethane anaesthesia changes overall HRV and hormonal and/or thermoregulatory control of heart rate and has the capacity to affect vagal, sympathetic and baroreflex modulation of heart rate. Notably, abnormalities observed under anaesthesia in the LPK strain were absent in the conscious state, but this was not apparent in the SHR. This suggests that some disease states, such as kidney disease, may be more sensitive to anaesthesia, with urethane preferentially affecting the neural circuits controlling heart rate, resulting in a reduction in HRV.

BRS estimates were less affected by anaesthesia. Estimates calculated using the sequence method were reduced in the LPK and WKY ($P<0.05$), and estimates calculated using the α-coefficient method were reduced in the LPK and the Lewis under anaesthesia ($P<0.05$). This implies that anaesthesia does not have a substantial effect on the ability of the parasympathetic and sympathetic nervous systems to change heart rate in response to an alteration in blood pressure, as opposed to the widespread effects seen on HRV.

4.2. General Data Interpretation

4.2.1. Heart Rate Variability

VLF power is suggested to reflect hormonal and/or thermoregulatory control of heart rate (7, 8). However, a reduction in VLF power does not necessarily mean that hormonal and/or thermoregulatory control of heart rate is reduced. For example, the VLF component reflects, amongst other hormonal systems, the influence of the renin–angiotensin system on the heart, and inhibition of this system with angiotensin-converting enzyme inhibitors increases VLF (28, 29) so that a reduction in VLF power may reflect excessive hormonal control of heart rate. Therefore, unless specific information is known about hormonal and/or thermoregulatory systems, interpretation of VLF should be limited, especially as some regard the VLF component as having questionable physiological significance with respect to heart rate control (8).

Changes in LF power need to be considered in light of the generally held view that oscillations in LF power are due to both parasympathetic and sympathetic components of the autonomic nervous system as well as baroreflex control of heart rate (7, 8). It is not a clear-cut situation. In rodents, administration of the muscarinic receptor antagonist atropine abolishes LF power; however, administration of the beta receptor antagonist atenolol to the same rodent can significantly attenuate LF power (30). In support of a cardiac sympathetic nerve activity contribution to LF power, heart rate and sympathetic nerve activity power spectra are coherent in the LF range (31, 32), and LF power increases following baroreceptor unloading, mental stress and exercise, which can in turn be ameliorated with β-adrenoceptor antagonists (33–36). On the other hand, removal of the cardiac stellate ganglia does not always affect LF power (28, 37), and in heart failure patients, where cardiac sympathetic nerve activity is documented to be overactive, LF power is either paradoxically reduced or unchanged (38–40). Overall, this suggests that LF power reflects the balance between sympathetic and vagal inputs to the heart. A reduction in LF power should therefore be interpreted as an inability for the autonomic nervous system in general to control heart rate, whereas an increase in LF could be interpreted as a predominance of cardiac sympathetic inputs over vagal inputs, particularly if additional pharmacological evidence can be provided to indicate that in response to administration of a beta-blocker, the decrease in heart rate was greater and there was a reduction in LF power.

The HF component of HRV is well accepted to represent vagal control of heart rate. Administration of atropine abolishes HF power, whereas pyridostigmine, which increases acetylcholine levels, augments HF power (28, 33, 41, 42). Additionally, respiration also has a strong influence over HF oscillations in heart rate, and

HF power is considered a measure of respiratory sinus arrhythmia (43). Accordingly, increasing the depth of breathing or decreasing the frequency of breathing results in an increase in HF power, and vice versa (43, 44), and the amplitude of respiratory sinus arrhythmia is linearly related to the amount of HF power (43). Therefore, a reduction in HF power can be interpreted as a reduction in vagal tone and/or a reduction in respiratory sinus arrhythmia, while an increase in HF power may be interpreted as an increase in vagal tone and/or an increase in respiratory sinus arrhythmia.

4.2.2. Baroreflex Sensitivity

BRS estimates are easier to interpret and reflect the speed with which heart rate changes in response to activation or inhibition of the baroreflex at resting blood pressure. The greater the BRS estimate, the greater the ability for the baroreflex to change heart rate; whereas, the smaller the baroreflex estimate, the less able the baroreflex is to change heart rate, an indicator of a potential disease state. Two methods for estimating BRS were presented here—the sequence method and the α-coefficient. The sequence method algorithm we used looks for sequences of three successive changes in blood pressure that are opposed by a change in heart rate. Sequences that are three beats in length reflect baroreflex control of vagal outflow to the heart. Longer sequences can reflect the ability of the baroreflex to control both vagal and sympathetic outflow to the heart (9, 45). Therefore, a reduction in BRS estimated using the sequence method provided in this chapter indicates a reduced ability for the parasympathetic nervous system to evoke baroreflex changes in heart rate. The α-coefficient method correlates LF oscillations in heart rate and blood pressure. Given that both sympathetic and vagal inputs to the heart can be detected at LF oscillations (8, 16, 30), a reduction in BRS estimated using this method reflects a reduced ability for the baroreflex to evoke changes in heart rate through altering vagal and/or sympathetic inputs to the heart.

5. Conclusion

Frequency domain estimates of HRV, and sequence method and α-coefficient estimates of BRS provide an accessible method for investigating the ability of the autonomic nervous system to control heart rate. Acquisition of data and its analysis for HRV is a straightforward procedure. While interpretation must be considered in light of the caveats discussed, it provides a method for assessing the level of vagal control of heart rate and the balance between sympathetic and vagal inputs to the heart. The two approaches we presented here, for calculating BRS, allow determination of baroreflex control of vagal (sequence method) and/or sympathetic (α-coefficient) outflow to the heart. The methods for estimating

HRV and BRS provided are readily applicable to both anaesthetised and conscious animal preparations where a stable recording of blood pressure can be obtained and, with the appropriate software, have the ability to identify if autonomic regulation of the heart differs amongst various animal models of health and disease.

References

1. Mancia G, Ferrari A, Gregorini L, Parati G, Pomidossi G, Bertinieri G, Grassi G, di Rienzo M, Pedotti A, Zanchetti A (1983) Blood pressure and heart rate variabilities in normotensive and hypertensive human beings. Circ Res 53:96–104
2. Parati G, Di Rienzo M, Bertinieri G, Pomidossi G, Casadei R, Groppelli A, Pedotti A, Zanchetti A, Mancia G (1988) Evaluation of the baroreceptor-heart rate reflex by 24-hour intra-arterial blood pressure monitoring in humans. Hypertension 12:214–222
3. Johansson M, Gao SA, Friberg P, Annerstedt M, Carlstrom J, Ivarsson T, Jensen G, Ljungman S, Mathillas O, Nielsen FD, Strombom U (2007) Baroreflex effectiveness index and baroreflex sensitivity predict all-cause mortality and sudden death in hypertensive patients with chronic renal failure. J Hypertens 25:163–168
4. Oikawa K, Ishihara R, Maeda T, Yamaguchi K, Koike A, Kawaguchi H, Tabata Y, Murotani N, Itoh H (2009) Prognostic value of heart rate variability in patients with renal failure on hemodialysis. Int J Cardiol 131:370–377
5. Broadley AJ, Frenneaux MP, Moskvina V, Jones CJ, Korszun A (2005) Baroreflex sensitivity is reduced in depression. Psychosom Med 67:648–651
6. La Rovere MT, Bigger JT Jr, Marcus FI, Mortara A, Schwartz PJ (1998) Baroreflex sensitivity and heart-rate variability in prediction of total cardiac mortality after myocardial infarction. ATRAMI (Autonomic Tone and Reflexes After Myocardial Infarction) Investigators. Lancet 351:478–484
7. Stauss HM (2003) Heart rate variability. Am J Physiol Regul Integr Comp Physiol 285: R927–R931
8. Task Force of the ESC and NASPE (1996) Heart rate variability: standards of measurement, physiological interpretation and clinical use. Circulation 93:1043–1065
9. Parati G, Di Rienzo M, Mancia G (2000) How to measure baroreflex sensitivity: from the cardiovascular laboratory to daily life. J Hypertens 18:7–19
10. Zaza A, Lombardi F (2001) Autonomic indexes based on the analysis of heart rate variability: a view from the sinus node. Cardiovasc Res 50:434–442
11. Guzik P, Piskorski J, Krauze T, Schneider R, Wesseling KH, Wykretowicz A, Wysocki H (2007) Correlations between the Poincare plot and conventional heart rate variability parameters assessed during paced breathing. J Physiol Sci 57:63–71
12. Elghozi JL, Girard A, Laude D (2001) Effects of drugs on the autonomic control of short-term heart rate variability. Auton Neurosci 90:116–121
13. Padley JR, Overstreet DH, Pilowsky PM, Goodchild AK (2005) Impaired cardiac and sympathetic autonomic control in rats differing in acetylcholine receptor sensitivity. Am J Physiol Heart Circ Physiol 289:1985–1992
14. Baudrie V, Laude D, Elghozi J-L (2007) Optimal frequency ranges for extracting information on cardiovascular autonomic control from the blood pressure and pulse interval spectrograms in mice. Am J Physiol Regul Integr Comp Physiol 292:R904–R912
15. Laude D, Elghozi JL, Girard A, Bellard E, Bouhaddi M, Castiglioni P, Cerutti C, Cividjian A, Di Rienzo M, Fortrat JO, Janssen B, Karemaker JM, Leftheriotis G, Parati G, Persson PB, Porta A, Quintin L, Regnard J, Rudiger H, Stauss HM (2004) Comparison of various techniques used to estimate spontaneous baroreflex sensitivity (the EuroBaVar study). Am J Physiol Regul Integr Comp Physiol 286:R226–R231
16. Stauss HM, Moffitt JA, Chapleau MW, Abboud FM, Johnson AK (2006) Baroreceptor reflex sensitivity estimated by the sequence technique is reliable in rats. Am J Physiol Heart Circ Physiol 291:H482–H483
17. Watkins LL, Grossman P, Sherwood A (1996) Noninvasive assessment of baroreflex control in borderline hypertension. Comparison with the phenylephrine method. Hypertension 28: 238–243
18. Mancia G, Parati G, Castiglioni P, di Rienzo M (1999) Effect of sinoaortic denervation on frequency-domain estimates of baroreflex sensitivity in conscious cats. Am J Physiol 276: H1987–H1993

19. Di Rienzo M, Castiglioni P, Parati G, Mancia G, Pedotti A (1996) Effects of sino-aortic denervation on spectral characteristics of blood pressure and pulse interval variability: a wideband approach. Med Biol Eng Comput 34:133–141
20. Di Rienzo M, Castiglioni P, Mancia G, Parati G, Pedotti A (1997) Critical appraisal of indices for the assessment of baroreflex sensitivity. Methods Inf Med 36:246–249
21. Pitzalis MV, Mastropasqua F, Passantino A, Massari F, Ligurgo L, Forleo C, Balducci C, Lombardi F, Rizzon P (1998) Comparison between noninvasive indices of baroreceptor sensitivity and the phenylephrine method in post–myocardial infarction patients. Circulation 97:1362–1367
22. Hildreth CM, Padley JR, Pilowsky PM, Goodchild AK (2008) Impaired serotonergic regulation of heart rate may underlie reduced baroreflex sensitivity in an animal model of depression. Am J Physiol Heart Circ Physiol 294:H474–H480
23. Leon LR, Walker LD, DuBose DA, Stephenson LA (2004) Biotelemetry transmitter implantation in rodents: impact on growth and circadian rhythms. Am J Physiol Regul Integr Comp Physiol 286:R967–R974
24. Guild SJ, Barrett CJ, McBryde FD, Van Vliet BN, Malpas SC (2008) Sampling of cardiovascular data: how often and how much? Am J Physiol Regul Integr Comp Physiol 295:R510–R515
25. Okamoto K (1969) Spontaneous hypertension in rats. Int Rev Exp Pathol 7:227–270
26. Phillips JK, Hopwood D, Loxley RA, Ghatora K, Coombes JD, Tan YS, Harrison JL, McKitrick DJ, Holobotvskyy V, Arnolda LF, Rangan GK (2007) Temporal relationship between renal cyst development, hypertension and cardiac hypertrophy in a new rat model of autosomal recessive polycystic kidney disease. Kidney Blood Press Res 30:129–144
27. Overstreet DH, Friedman E, Mathe AA, Yadid G (2005) The Flinders Sensitive Line rat: a selectively bred putative animal model of depression. Neurosci Biobehav Rev 29:739–759
28. Akselrod S, Gordon D, Ubel FA, Shannon DC, Berger AC, Cohen RJ (1981) Power spectrum analysis of heart rate fluctuation: a quantitative probe of beat-to-beat cardiovascular control. Science 213:220–222
29. Taylor JA, Carr DL, Myers CW, Eckberg DL (1998) Mechanisms underlying very-low-frequency RR-interval oscillations in humans. Circulation 98:547–555
30. Japundzic N, Grichois ML, Zitoun P, Laude D, Elghozi JL (1990) Spectral analysis of blood pressure and heart rate in conscious rats: effects of autonomic blockers. J Auton Nerv Syst 30:91–100
31. Montano N, Lombardi F, Gnecchi Ruscone T, Contini M, Finocchiaro ML, Baselli G, Porta A, Cerutti S, Malliani A (1992) Spectral analysis of sympathetic discharge. R-R interval and systolic arterial pressure in decerebrate cats. J Auton Nerv Syst 40:21–31
32. Pagani M, Montano N, Porta A, Malliani A, Abboud FM, Birkett C, Somers VK (1997) Relationship between spectral components of cardiovascular variabilities and direct measures of muscle sympathetic nerve activity in humans. Circulation 95:1441–1448
33. Pomeranz B, Macaulay RJ, Caudill MA, Kutz I, Adam D, Gordon D, Kilborn KM, Barger AC, Shannon DC, Cohen RJ et al (1985) Assessment of autonomic function in humans by heartrate spectral analysis. Am J Physiol 248:H151–H153
34. Girard A, Meilhac B, Mounier-Vehier C, Elghozi JL (1995) Effects of beta-adrenergic blockade on short-term variability of blood pressure and heart rate in essential hypertension. Clin Exp Hypertens 17:15–27
35. Grillot M, Fauvel JP, Cottet-Emard JM, Laville M, Peyrin L, Pozet N, Zech P (1995) Spectral analysis of stress-induced change in blood pressure and heart rate in normotensive subjects. J Cardiovasc Pharmacol 25:448–452
36. Jaffe RS, Fung DL, Behrman KH (1994) Optimal frequency ranges for extracting information on autonomic activity from the heart rate spectrogram. J Auton Nerv Syst 46:37–46
37. Rimoldi O, Pierini S, Ferrari A, Cerutti S, Pagani M, Malliani A (1990) Analysis of short-term oscillations of R-R and arterial pressure in conscious dogs. Am J Physiol 258:H967–H976
38. Hasking GJ, Esler MD, Jennings GL, Burton D, Johns JA, Korner PI (1986) Norepinephrine spillover to plasma in patients with congestive heart failure: evidence of increased overall and cardiorenal sympathetic nervous activity. Circulation 73:615–621
39. Kingwell BA, Thompson JM, Kaye DM, McPherson GA, Jennings GL, Esler MD (1994) Heart rate spectral analysis, cardiac norepinephrine spillover, and muscle sympathetic nerve activity during human sympathetic nervous activation and failure. Circulation 90:234–240
40. Notarius CF, Butler GC, Ando S, Pollard MJ, Senn BL, Floras JS (1999) Dissociation between microneurographic and heart rate variability estimates of sympathetic tone in normal subjects and patients with heart failure. Clin Sci (Lond) 96:557–565

41. Ramaekers D, Beckers F, Demeulemeester H, Aubert AE (2002) Cardiovascular autonomic function in conscious rats: a novel approach to facilitate stationary conditions. Ann Noninvasive Electrocardiol 7:307–318
42. Soares PP, da Nobrega AC, Ushizima MR, Irigoyen MC (2004) Cholinergic stimulation with pyridostigmine increases heart rate variability and baroreflex sensitivity in rats. Auton Neurosci 113:24–31
43. Laude D, Weise F, Girard A, Elghozi JL (1995) Spectral analysis of systolic blood pressure and heart rate oscillations related to respiration. Clin Exp Pharmacol Physiol 22:352–357
44. Schachinger H, Oelke M, Curio I, Langewitz W, Ruddel H, Schulte W (1991) Impact of respiratory frequency on short-term blood pressure and heart rate variability. J Hypertens Suppl 9:S330–S331
45. Reyes del Paso GA, Hernandez JA, Gonzalez MI (2004) Differential analysis in the time domain of the baroreceptor cardiac reflex sensitivity as a function of sequence length. Psychophysiology 41:483–488

Chapter 13

Neurophysiological Recording of the Compound Muscle Action Potential for Motor Unit Number Estimation in Mice

Shyuan T. Ngo and Mark C. Bellingham

Abstract

Motor unit number estimation (MUNE) techniques estimate the number of motor units innervating a single muscle or a group of muscles. As such, MUNE techniques are commonly used to quantify the loss of motor units in disease. MUNE requires obtaining and analysing neurophysiological recordings of compound muscle action potentials (CMAPs) evoked by stimulation of a peripheral nerve. Here, we describe a neurophysiological method that allows for the generation of stimulus–response curves that are achieved through increasing incremental electrical stimulation of the right sciatic nerve of mice and recording of the CMAP in the right gastrocnemius muscle. Performing this neurophysiological technique for subsequent MUNE analysis has vast implications for determining motor unit numbers where motor unit loss is thought to occur.

Key words: Neurophysiology, Compound muscle action potential, Peripheral nerve stimulation, Motor unit number estimation, MUNE

1. Introduction

There are many neurophysiological techniques for quantitatively, and qualitatively, measuring peripheral nerve and muscle function. These methods include: nerve conduction velocity studies, needle electromyography, motor unit number estimation (MUNE) and repetitive nerve stimulation (1). Such electrodiagnostic techniques are commonly used to determine peripheral nervous system function in amyotrophic lateral sclerosis (motor neuron disease), traumatic nerve injury, polyneuropathy and neuromuscular junction diseases (e.g. myasthenia gravis) (1).

Neurophysiology techniques require fine control of the intensity of the stimulus in order to obtain reproducible all-or-nothing responses (2–4). MUNE is used to analyse muscle function, by determining the extent of motor unit loss in diseases like amyotrophic

Paul M. Pilowsky et al. (eds.), *Stimulation and Inhibition of Neurons*, Neuromethods, vol. 78,
DOI 10.1007/978-1-62703-233-9_13,

lateral sclerosis (2, 3, 5–9). Because MUNE is achieved by analysing compound muscle action potentials (CMAPs) in response to peripheral nerve stimulation, different MUNE methods are dependent upon different neurophysiological recording and stimulation protocols (9–14).

Current MUNE methods include incremental MUNE, which requires incremental increases in the intensity of the stimuli delivered to the nerve in order to generate a series of evoked responses. However, in this method, the excitation of different axon combinations (alternation) may lead to MUNE overestimations. Multiple point stimulation (MPS) was introduced to overcome alternation, but requires the relocation of the stimulating electrode throughout the procedure. MPS neurophysiology techniques are technically demanding and call for an appropriate number of stimuli to be delivered to guarantee that the recorded response is representative of a single axon (2, 3, 15). Statistical MUNE (Poisson MUNE) is an approach that overcomes the problem of alternation, because it is able to sample a large range of single motor unit potentials, and does not require movement of the stimulating electrode. However, as it is dependent on the assumption that recorded responses follow a Poisson distribution, variations in response amplitudes may lead to MUNE overestimates, and the stimulus current delivered can greatly affect the final MUNE (2, 3, 16).

By delivering graded electrical stimulation to a motor nerve from minimum to supramaximal threshold levels, the neurophysiological method described here permits the stable recording of CMAPs generated in response to the gradual recruitment of motor units into the maximal muscle CMAP response in mice. This allows for MUNE analysis of a CMAP stimulus–response curve that incorporates the contribution of all motor units innervating the recorded muscle (12).

2. Materials

For this procedure, gaseous anaesthetic in combination with medical oxygen is required for the induction and maintenance of anaesthesia (see Sect. 3.1). Induction of mouse anaesthesia requires a vessel in which this can be quickly and easily accomplished. A semi-transparent sealable box that is large enough to house an adult mouse comfortably (approximately 9 cm H, 21 cm W, 28 cm L) whilst allowing the flow of air is suitable. This induction chamber must contain taps that can be opened and closed so as to allow for the cessation of airflow in order to maintain the gaseous anaesthetic within the box. Animal anaesthesia should be maintained through a facemask/nose cone. Gaseous anaesthesia requires a suitable extraction system (e.g.

Nederman Extraction kit original 2000, Helsingborg, SWE; http://www.nederman.com/Home.aspx) (see Sect. 3.1).

Animals must be maintained at 36.5°C on a heated pad (e.g. ATC1000, World Precision Instruments (WPI), FL, USA) (see Sect. 3.2). Monitoring of animal temperature can be achieved via a rectal probe, which should be coated with an appropriate lubricant (e.g. liquid paraffin light; available from most chemists) just prior to use. Small animal clippers (Wahl, Stylique Designer/Trimmer, IL, USA) and Nair hair removal cream are required to remove animal fur (see Sect. 3.2). Monopolar disposable stainless steel EEG subdermal needle electrodes (019–409700, Nicolet™ Biomedical, CareFusion, CA, USA) are used as both stimulating and recording electrodes. A digital ring electrode (6032-TP, Nicolet™ Biomedical, CareFusion) can be used as a ground electrode (see Sect. 3.2). Needle electrodes should be sterile, and the ring electrode should be cleaned thoroughly. Electrode cream (Signa Crème, Parker Laboratories Inc., NJ, USA) is required for the ground electrode only. A stimulator (WPI, A360 stimulus isolator) that allows small incremental increases in stimulus intensity and an AC-coupled amplifier (e.g. WPI, DAM50 differential amplifier) are required for delivery of the stimulus and amplification of the muscle CMAP response, respectively (see Sect. 3.3).

For signal recording, a digitizer capable of acquiring two channels of data at a sampling rate of >20 kHz per channel, a computer and software capable of recording episodic (sweeps) of data are needed. Our laboratory uses a Digidata 1200B digitizer (Molecular Devices, CA, USA) in a PC running Windows XP Professional and PClamp7 data acquisition software (Molecular Devices); this digitizer requires a computer motherboard with a full-length ISA slot, which is no longer manufactured. Alternative and more economical data acquisition boards and software suitable for current computer motherboards could include National Instruments (E series or M series boards) and Strathclyde Electrophysiological Software (Windows 32 bit freeware from Dr. John Dempster, http://spider.science.strath.ac.uk/sipbs/software_ses.htm, supports the Digidata 1200B and 1320/22, several Cambridge Electronic Design digitizers, Instrutech ITC-14 and ITC-16 and National Instruments Lab-PC, E or M series boards) or Axograph X (Mac OS X or Windows versions, http://axographx.com/index.html, supports Instrutech ITC-14, ITC-16 and ITC-1600, Molecular Devices 1320 and 1440A and some National Instruments digitizers). For analysis of recorded signals, we use the Clampfit suite (Clampfit 10, Molecular Devices); Strathclyde Electrophysiological Software, Axograph X and Signal (Cambridge Electronic Design) all include appropriate signal analysis. Any digitizer/software combination meeting the basic criteria at the beginning of this paragraph can be utilised.

3. Methods

The following procedure will allow neurophysiology to be conducted on the right gastrocnemius muscle of C57Bl6/J mice ranging from 30 to 180 days of age. All experiments must be conducted in accordance with National and Institutional Guidelines for the Care of Use of Laboratory Animals and approved by the relevant Animal Ethics Committee.

3.1. Anaesthesia

Connect an isoflurane anaesthetic system (see Note 1) to the anaesthesia induction box with silicon tubing (see Note 2). The tap to which this tube is connected should be in the open position. Open the second tap and attach a second piece of silicon tubing to this tap. Place the other end of this tube under the arm of extraction system. This will allow free flow of anaesthetic gas mixture through the box whilst allowing extraction of the gaseous anaesthetic (see Note 3; see Fig. 1a). Start the isoflurane system and open the oxygen tank to allow the flow of air (0.6–0.8 L/min). Deliver isoflurane in oxygen (5%) to the anaesthesia induction box. Remove the animal from its home cage and quickly place it in the box (see Note 4). Leave the animal in the box for a minimum of 5 min. Turn the vessel taps to the closed position, close off the oxygen tank and cease the flow of isoflurane.

Connect the isoflurane delivery system to the facemask/nose cone. Remove the animal from the vessel, place it onto the heating pad in a prone position and place its head into the facemask/nose cone. A Faraday cage can be useful to reduce noise (see Note 5). Adjust the flow of isoflurane in oxygen to 1.5–2.5% (see Note 6). Place the outlet tube from the isoflurane system under the extraction arm (see Note 7; see Fig. 1a). Ensure appropriate anaesthesia in the animal is achieved before proceeding (see Note 8).

3.2. Animal Preparation

Clip the hair from the right hindlimb with small animal clippers and remove any remaining fur with Nair hair removal cream (see Note 9). Coat the rectal probe with liquid paraffin light and insert it into the animal, taping the probe lead to the animal's tail for security. Turn on the heat pad and allow the animal's temperature to reach 36.5°C (see Note 10). Tape both hindlimbs to the heat pad at an approximate 45° angle from the midline with adhesive tape. Attach two sterile monopolar needle electrodes to the stimulator. Insert the electrodes perpendicularly into the animal near the right sciatic notch (see Note 11). Securely tape the electrode leads onto the heating pad to ensure that electrodes are not displaced during delivery of the stimulus. Attach two sterile monopolar needle electrodes to the amplifier. Insert these electrodes perpendicularly into the right gastrocnemius muscle and Achilles tendon

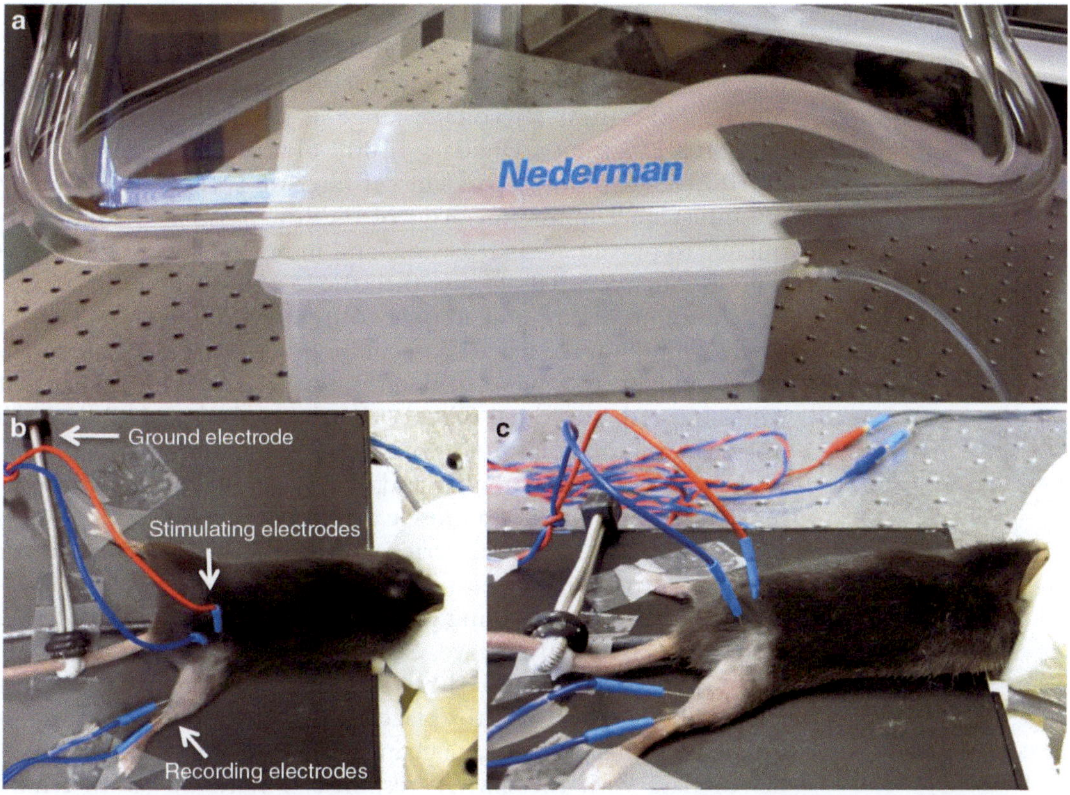

Fig. 1. Neurophysiology set-up. (**a**) Placement of the anaesthesia induction chamber and anaesthesia outlet tube under a Nederman extraction arm will allow sufficient extraction of gaseous anaesthetic. (**b**, **c**) Stimulating electrodes (*upper blue* and *red lead*) should be inserted at the sciatic notch and recording electrodes (*lower blue leads*) through the gastrocnemius muscle and Achilles tendon. The ground ring electrode should be coated with electrode cream and secured to the tail of the mouse. All electrodes can be secured to the heat pad with tape.

(see Note 12). Securely tape the recording electrode leads to the heating pad to ensure that electrodes are not displaced during recording of the response. To assure good contact between the digital ring ground electrode and the animal, coat the digital ring ground electrode with electrode cream before securing it around the tail of the mouse (see Note 13). Tape the lead of the digital ring electrode to the heat pad to ensure that it is kept in place (see Fig. 1b, c).

3.3. Neurophysiology Recordings

CMAPs are amplified (usually ×1,000; 1–10 mA range) by the amplifier and then recorded in AC mode on a PC using PClamp7 software (see Note 14). The settings for initial signal amplification are a high-pass cut-off filter of 0.1 Hz, a low-pass cut-off filter of 1 kHz, 100–1,000×gain and recording in AC mode. We commonly further amplify our CMAP response digitally (2–8× using software settings for the Digidata 1200B digitizer; this feature may not be available for other digitizers). These amplification factors are set so that the final recorded maximal signal utilises as much of the digitizer voltage range as possible. We record our data on a PC using a Digidata 1200B digitizer at a sampling rate of 10 kHz with a low-pass cut-off filter of 1 kHz; the sampling rate must be a minimum of ½ of the low-pass cut-off filter frequency to prevent signal aliasing, and in practice, we set the sampling frequency 5–10× higher than the low-pass cut-off frequency to achieve high temporal accuracy of the recorded signal. Deliver single stimuli (at 0.3 Hz, 0.5 ms pulse duration, 0–5-mA stimulation range) to the nerve to determine the level of stimulus that will achieve the maximal CMAP, viz. the evoked CMAP that will not increase in amplitude with increasing stimulus strengths. Reduce the stimulus intensity and determine the minimum stimulus that does not induce a muscle action potential response, but is just subthreshold for the recruitment of the first motor unit for muscle contraction (see Note 15). The stimulus–response curve is now generated by delivering graded electrical stimuli of increasing strength to the sciatic nerve. Stimulus intensity delivered to the tissue is indirectly measured by recording the voltage drop across a low value (~16 Ω) resistor placed in parallel across the stimulus isolator outputs; this voltage change is recorded as a second channel of data with the CMAP response. Firstly, 50 subthreshold responses at the same stimulus intensity are recorded to determine the baseline and baseline noise of the recording (see Note 16). Following this, increase the stimulus strength by an increment of <1% of the difference between minimal and maximal levels (as determined above), recording a minimum of ten responses at each new stimulus level. Continue incremental stimulation until a supramaximal response is reached and the maximal CMAP is obtained (see Note 17) (Fig. 2).

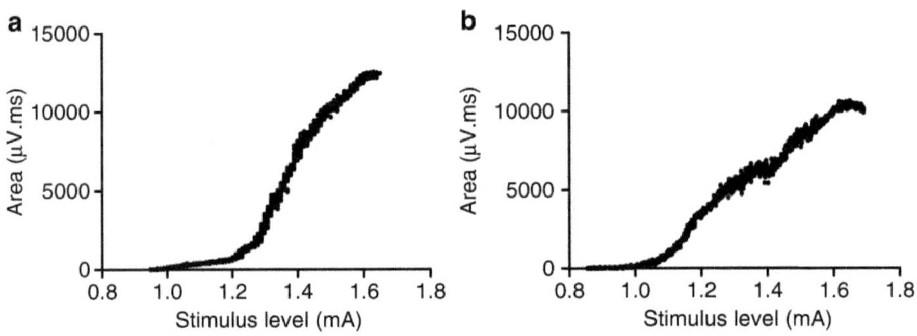

Fig. 2. Stimulus–response curves. Sample stimulus–response curves obtained from wild-type C56Bl6/J animals. Delivery of graded electrical stimuli of increasing strength to the sciatic nerve led to increases in the area under the curve of the recorded compound muscle action potential in (**a**) a 175-day-old female mouse and (**b**) a 67-day-old female mouse.

3.4. Data Analysis

To begin analysis of the recorded data, open the CMAP traces in your chosen software program. Set the baseline of all visible CMAP traces to zero (e.g. Fig. 3). Mark an area before the stimulus artefact (e.g. Fig. 2, cursors 1 and 2). Identify the beginning and end of the CMAP curve (e.g. Fig. 2, cursors 3 and 4). Average the first 50 traces to provide a baseline subthreshold recording and subtract this averaged trace from all other CMAP traces. Determine the peak of, and area under, the CMAP curve. Transfer the data to an excel spreadsheet. Similar to measuring the peak of the CMAP curve, measure the peak of each stimulus trace. Include this data on the excel spreadsheet containing the CMAP data. Convert data into a suitable form for MUNE analysis.

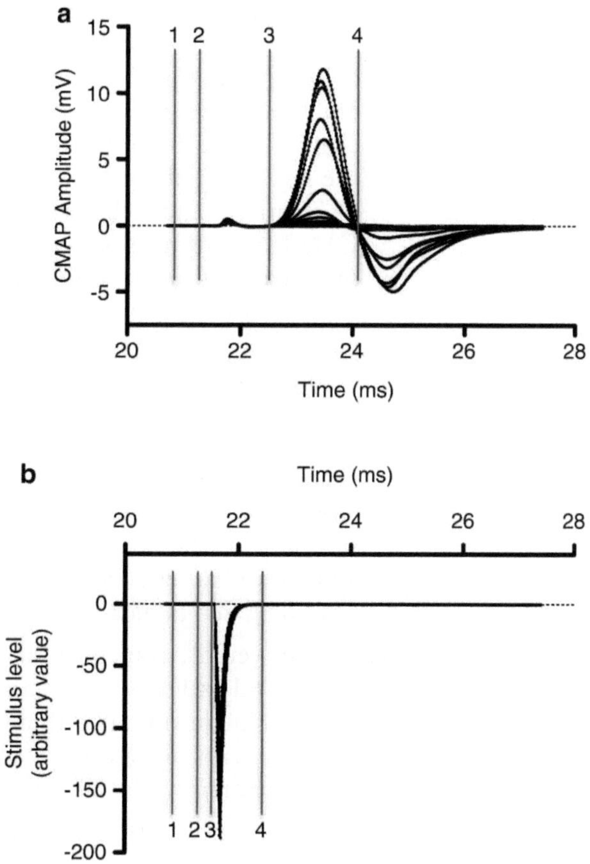

Fig. 3. Analysis of compound muscle action potential (*CMAP*) traces. CMAP traces can be analysed in Clampfit 10. (**a**) Sample CMAP traces detailing a baseline, which has been set to zero, and placement of cursors 1–4 for analysis of the CMAPs. Cursors 1 and 2 should be placed before the stimulus artefact and cursors 3 and 4 placed at the beginning and end of the CMAP curve, respectively. (**b**) Sample stimulus traces that correspond to CMAP traces in (**a**). Cursors 3 and 4 should be moved to the beginning and end of the stimulus traces, respectively, prior to measuring the peak of each stimulus trace. The stimulus traces should be converted into positive values (inverted by multiplying by a factor of −1 if using Clampfit 10).

4. Notes

1. We use a Sigma Baxter isoflurane vaporiser system with a 2.5-cm diameter facemask (Baxter, IL, USA) in combination with a Nederman arm for extraction of gaseous anaesthetic. Due to mouse size, a latex glove can be pulled tightly over the face mask and secured in place with an elastic band. Make a small incision in the glove that is sufficient to fit the head of the mouse. The induction and subsequent maintenance of a constant level of anaesthesia for up to 3 h is paramount. In each case, the level of anaesthesia must be monitored. Observe animal respiration rate and depth, and ensure the absence of any reflexes throughout the entire protocol. Anaesthetic agents that permit: (1) minimal disturbance to the animal's well-being, (2) minimal movement of the animal throughout the entirety of the procedure and (3) no disturbance to the positioning of both stimulating and recording electrodes during the recording phase of the procedure are preferred. A mixture of ketamine and xylazine may be used successfully for short CMAP recording sessions, but this method is unsuitable for our protocol with >1,000 stimuli at 0.3 Hz, lasting >1 h. Delivery of ketamine/xylazine as a single bolus dose via intraperitoneal injection may lead to erratic breathing, due to animals being too deeply anaesthetised or too lightly anaesthetised. This may in turn lead to unexpected death, or premature awakening of the animal. Further 'top-up' injections of ketamine/xylazine for maintenance of anaesthesia may be needed; this would require a restarting of the recording protocol and repositioning of the animal. Thus, it is highly recommended that animal anaesthesia be induced and maintained at a constant level by gaseous means.
2. A lightweight, transportable and transparent vessel with an airtight lid and taps that control airflow is sufficient. The use of flexible silicon tubing with an internal diameter that is suitable for these taps allows for positioning of the vessel in a convenient location whilst allowing easy control of airflow.
3. In accordance with Occupational Health and Safety requirements, excess gaseous anaesthetic must be extracted by an appropriate system according to the rules of your institution. The user should monitor the efficiency of the filter, and hours of use should be strictly recorded.
4. When dealing with animals, it is important to reduce unnecessary stress. Housing of animals in a cage with environmental enrichment (such as plastic tubing or toilet paper rolls) is preferred. Inappropriate handling of animals increases stress (17, 18). It is therefore advisable that animals be gently coaxed into the

toilet paper roll, which can then be picked up and placed into the box for the induction of anaesthesia.

5. A Faraday cage made from conductive material connected to a suitable ground will minimise extraneous electrical noise. It is important to note that if the user must reach into the Faraday cage, they should not wear any clothing or accessories that produce static charge, as this will introduce noise in the recording and cause a shift in the baseline even in AC-coupled recordings.
6. Animal anaesthesia should be monitored throughout the procedure. We use 1.5–2.5% isoflurane in oxygen to maintain anaesthesia. Anaesthesia is dependent on the age and size of the animal. Some animals may require a lower or higher percentage of isoflurane. Anaesthesia may be adjusted during the course of the experimental period to prevent anaesthetic overdose.
7. The Nederman arm should not be placed directly above the animal or within the vicinity of the electrodes. This has been found to introduce noise to the physiology recording.
8. We determine appropriate anaesthesia by ensuring complete suppression of the pedal-withdrawal reflex. The criteria for ensuring anaesthesia may vary between research groups and will be dependent upon the ethical guidelines agreed upon by the relevant Animal Ethics Committee.
9. Take care to ensure that the use of animal clippers does not lead to breaking of the skin. Broken skin will result in dehydration of the underlying muscle. Nair hair removal cream should not be left on the animals for an excessive period of time as this will lead to burning of the skin.
10. The maintenance of animal temperature is of upmost importance. Under-heating of an animal will affect the conduction velocity along the nerve and confound neurophysiological recordings. Overheating of the animal will lead to excessive dehydration and death. We place a Styrofoam slab under the heating pad; this serves to reduce heat loss from the pad to the steel bench surface underneath and as an electrical insulator between the pad and the bench surface, so that the animal and pad can be grounded to a common point and cannot form ground loops.
11. The placement of electrodes near the sciatic notch requires practise and care. The sciatic notch can be found by gently pushing around the hip joint of mice. To reduce tissue and nerve damage, stimulating electrodes should not be moved once inserted into the animal.
12. The placement of electrodes into the gastrocnemius requires care. Do not to damage the saphenous vein as this will cause

bleeding and clotting around the electrode. It is also important not to pierce the soleus muscle, as this will result in the recording of CMAPs originating from the soleus.

13. An animal ground electrode will minimise extraneous electrical noise. We use a common ground busbar system, consisting of a single brass bar which has multiple contact points receiving ground leads from the animal, instruments and Faraday cage. The bar is connected in turn by a single lead to a suitable electrical ground point; we use the ground connection of an oscilloscope. This grounding system prevents the spurious formation of ground loops if all the elements grounded to the busbar are electrically isolated from each other. Careful attention to grounding can achieve very low noise recordings. An alternative is the use of active digital subtraction of electrical noise by devices such as the Hum Bug (Quest Scientific); this technique is most useful for periodic noise sources, such as AC line frequency (50 or 60 Hz depending on the country of supply).

14. The A360 stimulator allows for small incremental increases in stimulus intensity, and the DAM50 amplifier allows for a high-pass cut-off filter of 0.1 Hz, a low-pass cut-off filter of 1 kHz, 100–1,000 × gain and recording in AC mode. These specifications are necessary for the neurophysiology recordings described here.

15. Whilst the minimum and maximum threshold will vary between animals, it is important to note that the minimum and maximum level may also vary within a given animal. Consequently, the minimum and maximum levels may need to be redetermined if a muscle response is not observed within the first 60 experimental stimuli. The user must continue the train of stimulations to a supramaximal threshold in order to guarantee that the maximum CMAP has been acquired.

16. A minimum of 50 baseline stimuli is required. If a motor unit response is detected within the first 50 baseline stimuli, the recording should be stopped. A new minimum level should be determined prior to the start of the next recording.

17. The recording of the maximal CMAP must produce a plateau at the top end of the stimulus–response curve. It is not uncommon to occasionally observe CMAP 'decrement', whereby the area under the curve of the CMPA might decrease with increasing stimulation. Similarly, an absence in the increase of the area under the curve of the CMAP may be evident with increasing stimulation. It is imperative that the user continues the recording to provide a stimulus that is over and above the predetermined maximal threshold.

References

1. Gilchrist JM, Sachs GM (2004) Electrodiagnostic studies in the management and prognosis of neuromuscular disorders. Muscle Nerve 29:165–190
2. Bromberg MB, Brownell AA (2008) Motor unit number estimation in the assessment of performance and function in motor neuron disease. Phys Med Rehabil Clin N Am 19:509–532, ix
3. Daube JR (2006) Motor unit number estimates—from A to Z. J Neurol Sci 242:23–35
4. McComas AJ, Fawcett PR, Campbell MJ, Sica RE (1971) Electrophysiological estimation of the number of motor units within a human muscle. J Neurol Neurosurg Psychiatry 34:121–131
5. Arasaki K, Tamaki M, Hosoya Y, Kudo N (1997) Validity of electromyograms and tension as a means of motor unit number estimation. Muscle Nerve 20:552–560
6. Ridall PG, Pettitt AN, Henderson RD, McCombe PA (2006) Motor unit number estimation—a Bayesian approach. Biometrics 62:1235–1250
7. Rowland LP, Schneider LA (2001) Amyotrophic lateral sclerosis. N Engl J Med 344:1688–1700
8. Mitchell JD, Borasio GD (2007) Amyotrophic lateral sclerosis. Lancet 369:2031–2041
9. Shefner JM, Cudkowicz ME, Brown RH Jr (2002) Comparison of incremental with multipoint MUNE methods in transgenic ALS mice. Muscle Nerve 25:39–42
10. Hegedus J, Jones KE, Gordon T (2009) Development and use of the incremental twitch subtraction MUNE method in mice. Suppl Clin Neurophysiol 60:209–217
11. Hegedus J, Putman CT, Gordon T (2007) Time course of preferential motor unit loss in the SOD1 G93A mouse model of amyotrophic lateral sclerosis. Neurobiol Dis 28:154–164
12. Henderson RD, Ridall GR, Pettitt AN, McCombe PA, Daube JR (2006) The stimulus–response curve and motor unit variability in normal subjects and subjects with amyotrophic lateral sclerosis. Muscle Nerve 34:34–43
13. Nandedkar SD, Barkhaus PE, Stalberg EV (2010) Motor unit number index (MUNIX): principle, method, and findings in healthy subjects and in patients with motor neuron disease. Muscle Nerve 42:798–807
14. van Dijk JP, Schelhaas HJ, Van Schaik IN, Janssen HM, Stegeman DF, Zwarts MJ (2010) Monitoring disease progression using high-density motor unit number estimation in amyotrophic lateral sclerosis. Muscle Nerve 42:239–244
15. Kadrie HA, Yates SK, Milner-Brown HS, Brown WF (1976) Multiple point electrical stimulation of ulnar and median nerves. J Neurol Neurosurg Psychiatry 39:973–985
16. Lomen-Hoerth C, Olney RK (2001) Effect of recording window and stimulation variables on the statistical technique of motor unit number estimation. Muscle Nerve 24:1659–1664
17. Steyn FJ, Huang L, Ngo ST, Leong JW, Tan HY, Xie TY, Parlow AF, Veldhuis JD, Waters MJ, Chen C (2011) Development of a method for the determination of pulsatile growth hormone secretion in mice. Endocrinol 152:3165–3171
18. Xu J, Bekaert AJ, Dupont J, Rouve S, Annesi-Maesano I, De Magalhaes Filho CD, Kappeler L, Holzenberger M (2011) Exploring endocrine GH pattern in mice using rank plot analysis and random blood samples. J Endocrinol 208:119–129

Part V

Clinical Focus and Application

Chapter 14

Animal Models of Neuropathic Pain Due to Nerve Injury

Paul J. Austin and Gila Moalem-Taylor

Abstract

Chronic neuropathic pain is a frequent outcome of nervous system injury affecting the somatosensory system, and its pathobiology is dependent on activation and disinhibition of nociceptive neurons. It is characterised by spontaneous pain, dysaesthesia/paraesthesia as well as hypersensitivity to normally non-painful (allodynia) and painful (hyperalgesia) mechanical and thermal stimuli. Behavioural disabilities, such as depression, insomnia and alterations in social behaviours, are also co-morbid with changes in sensation. To investigate the mechanisms which result in neuropathic pain, animal models of neuropathy were developed by performing standardised, reproducible nerve injuries via surgical manipulation of a peripheral nerve. Here, we present four commonly used models in laboratory rodents for the study of neuropathic pain: (1) chronic constriction injury of the sciatic nerve, (2) partial sciatic nerve ligation, (3) L5 and/or L6 spinal nerve ligation and (4) the spared nerve injury, where two of the three terminal sciatic branches are cut. Rodents which have undergone any one of these procedures routinely display increased pain responses, such as allodynia and hyperalgesia of the hindpaw, lasting up to several months. Investigators should be aware however that such animal models suffer from several limitations including inconsistency in predicting the clinical success of novel therapeutics, poor correlation with clinical neuropathic pain in terms of prevalent symptoms and timescale and the need to assess operant/affective behavioural responses in addition to reflexive withdrawals from mechanical and thermal stimuli. Despite these limitations, animal models of peripheral nerve injury combined with testing of pain hypersensitivity remain necessary to investigate the pathophysiological mechanisms and identify novel therapeutic agents to treat chronic neuropathic pain.

Key words: Chronic neuropathic pain, Peripheral nerve injury, Chronic constriction injury, Partial sciatic nerve ligation, Spinal nerve ligation, Spared nerve injury

1. Introduction

Stimulation of nociceptive neurons usually causes the perception of pain. Pain is defined as an unpleasant sensory and emotional experience associated with potential or actual tissue damage, or described in terms of such damage (International Association for the Study of Pain), for example, a sensation that occurs as a result of a noxious stimulus such as stubbing a toe. That said, it is a necessary signal, carried by the nervous system, serving as a warning mechanism to

Paul M. Pilowsky et al. (eds.), *Stimulation and Inhibition of Neurons*, Neuromethods, vol. 78,
DOI 10.1007/978-1-62703-233-9_14, © Springer Science+Business Media, LLC 2013

prevent further injury and exacerbation of tissue damage during healing. Furthermore, following tissue damage, the immediate response characterised by inflammation, pain and disruption to normal behaviours is completely functional and, in most cases, is followed by a period of recovery, with diminishing inflammation and pain, and a return to normal behaviour. Unfortunately, however, in about one in five adults, pain persists despite the injury healing, resulting in a state of chronic pain (1). Chronic pain that arises as a direct consequence of a lesion or disease affecting the somatosensory system and involving nervous system injury is termed neuropathic pain (2) and is rather prevalent in the general population (3, 4). Common causes of neuropathic pain are traumatic injury such as nerve compression, spinal cord injury and nerve damage following surgery (i.e. iatrogenic injury), or disease states such as diabetes, postherpetic neuralgia and cancer. Retrospective analyses of postsurgical outcomes show a high frequency of neuropathic pain with 10–50% of patients reporting persistent pain after common operations such as amputation, lumpectomy and mastectomy, thoracotomy, inguinal hernia repair, coronary bypass surgery and Caesarean section (5).

Symptoms of neuropathic pain are often severely debilitating including, spontaneous pain (burning, tingling, electric shock-like and diffuse), dysaesthesia/paraesthesia (unpleasant sensations), allodynia (pain resulting from normally non-painful stimuli), hyperalgesia (an increased response to painful stimuli) as well as the co-morbidity of behavioural disabilities (e.g. insomnia, depression, sexual dysfunction and disturbances in social and familial interactions). It is essential that any experimental model of neuropathic pain mimics at least some of these principle features. Wall and Gutnick (6) were the first to attempt the development of an animal model of neuropathic pain; by cutting the sciatic nerve, they observed the formation of a neuroma, which resulted in a steady barrage of nerve impulses. This resulted in further attempts to create more standardised, reproducible models, and in 1979, Wall et al. (7) developed two surgical procedures: firstly, full transection of 5 mm of the sciatic nerve at mid-thigh level and, secondly, complete sciatic nerve ligation. The result in both cases was severe self-mutilation of the hindlimb (autotomy) and lack of reflexive paw withdrawal responses, resulting from complete deafferentation. Since allodynia and hyperalgesia could not be demonstrated and autotomy is not a feature of clinical neuropathic pain, other investigators attempted to create models of peripheral nerve injury which preserved some axons and therefore maintained partial hindlimb sensation. This led, in subsequent years, to the development of partial nerve injury models, which more closely mimicked the causes of clinical neuropathic pain. Some of the most common models of neuropathic pain induced by peripheral nerve injury, shown in

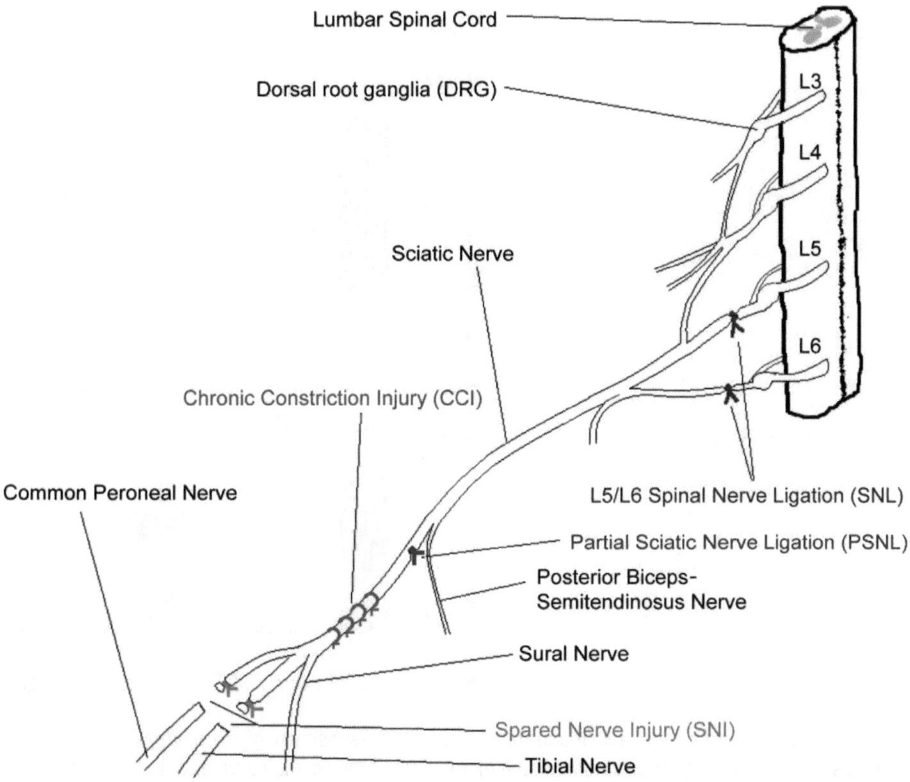

Fig. 1. A schematic illustrating the positioning of four common peripheral nerve injuries in rodent models of neuropathic pain. Chronic constriction injury (*CCI*) involves placement of four loose chromic gut ligatures on the sciatic nerve, at mid-thigh level. Partial sciatic nerve ligation (*PSNL*) is performed by tightly ligating 30–50% of the sciatic nerve axons, at high-thigh level. Spinal nerve ligation (*SNL*) involves tight ligation of the L5 and L6 spinal nerves, just distal to the dorsal root ganglia (*DRG*). The spared nerve injury (*SNI*) requires transection of two branches of the sciatic nerve, the tibial and common peroneal nerves, whilst the sural nerve is left intact. All these nerve injuries result in damage to only a portion of the afferents from the foot, sparing some axons, therefore allowing pain hypersensitivity to be measured by testing reflexive withdrawal responses of the hindpaw.

Fig. 1 and listed here, will be discussed in some considerable detail in this chapter: (1) the chronic constriction injury (CCI), most commonly, but not exclusively, of the sciatic nerve (8); (2) partial ligation of the sciatic nerve (PSNL) (9); (3) spinal nerve ligation (SNL) involving ligation of one or more of the spinal nerves which contribute to the sciatic (10); and (4) the spared nerve injury (SNI), where the tibial and peroneal sciatic nerve branches are ligated, but the sural is left intact (11). These four models result in visible signs of spontaneous pain and, by maintaining reflexive withdrawal responses, demonstrate reproducible development of mechanical and thermal pain hypersensitivity (allodynia and hyperalgesia) of the ipsilateral hindpaw, which persists for several months (12, 13). An example of pain hypersensitivity to mechanical and thermal stimuli after CCI is demonstrated in Fig. 2.

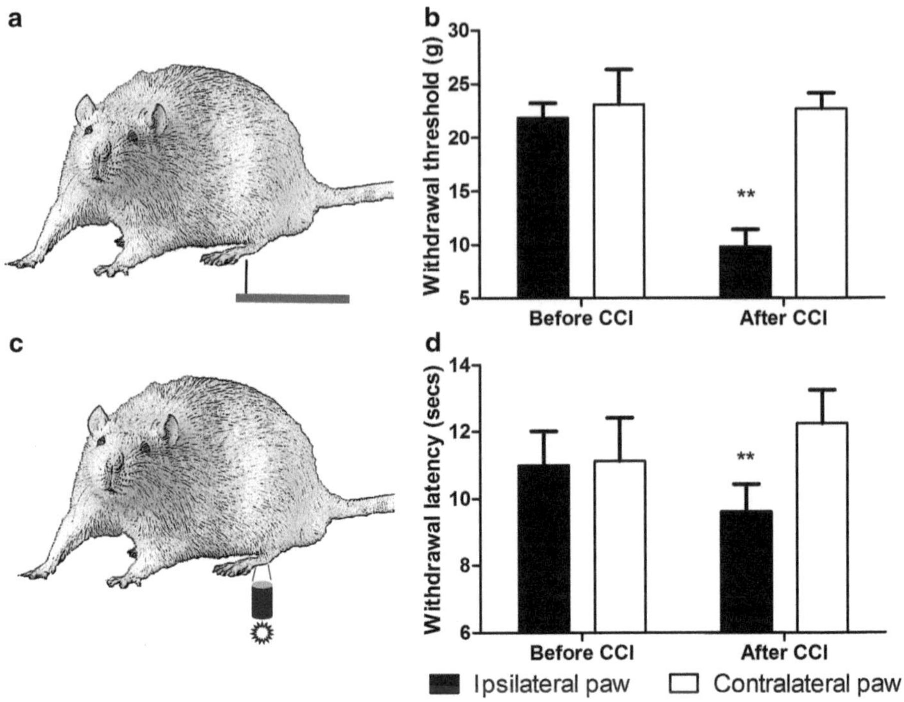

Fig. 2. Cartoon showing testing for mechanical and thermal pain hypersensitivity of the hindpaw, as well as evidence of increased pain responses following chronic constriction injury (*CCI*) of the sciatic nerve in Wistar rats. Pain testing is performed by (**a**) the application of a von Frey filament or (**c**) the application of a radiant heat source to the plantar surface of the hindpaw. Nine days after CCI, there is a significant reduction in paw withdrawal threshold to mechanical stimuli (**b**) and paw withdrawal latency to thermal stimuli (**d**) in the injured (ipsilateral) hindpaw compared to the uninjured (contralateral) hindpaw ($P<0.01$, unpaired Student's *t*-test). No such difference in mechanical thresholds and thermal latencies is observed before CCI (**b**, **d**). Data show mean ± s.e.m, $n=7$. (secs=seconds; g=grams).

As a consequence of the development of animal models of peripheral nerve injury, investigators have not only been able to test pain behaviours but also assess expression of receptors, ion channels, pain and inflammatory mediators and the infiltration/activation of immune and glial cells in post-mortem tissue, and make electrophysiological recordings in tissues and/or animals with established neuropathic pain, thus allowing the underlying mechanisms of neuropathic pain to be investigated. In particular, considerable effort was made to characterise the properties of neurons following peripheral nerve injury, ultimately leading to the proposal that peripheral and central sensitisation are important disease mechanisms. Central to this hypothesis is the existence of aberrant signals from both injured and intact nociceptors which amplify the responses to noxious and innocuous stimuli, underpinning the development of hyperalgesia and allodynia. Peripheral and central amplification is mediated by several injury-induced mechanisms, including:

1. Altered expression of receptors (e.g. glutamate receptors) and ion channels (e.g. sodium channels) by neurons.

2. Uncontrolled release of nociceptive mediators from neurons (e.g. neurotransmitters including 5-HT, substance P and glutamate), as well as from infiltrating immune and activated glial cells (e.g. prostaglandins, histamine, reactive oxygen species, bradykinin and cytokines).
3. Increased neuronal membrane excitability and ectopic generation of action potentials.
4. Sympathetically maintained pain, due to sympathetic-sensory neuronal coupling.
5. Neuronal cell death.
6. Facilitation and disinhibition of synaptic transmission in the dorsal horn of the spinal cord.
7. Synaptic plasticity.
8. Reorganisation of spinal and supraspinal nociceptive circuitry (e.g. alterations in the descending inhibitory pathways from the rostral ventromedial medulla).

For detailed mechanisms of neuropathic pain, which are outside the remit of this chapter, we encourage the reader to look at some of the excellent reviews available (14–16). In addition to these findings, animal models have provided an important advancement in the management of neuropathic pain, allowing preclinical evaluation of novel treatment strategies and their mechanisms of action. This chapter provides a detailed overview of the animal models themselves, in particular the common methods of performing experimental peripheral nerve injuries in rodents, which result in measurable neuropathic pain symptoms.

2. General Surgical Practices and Sciatic Nerve Exposure

Aseptic techniques should be used for all surgical procedures described herein. This includes disinfecting the surgical work surface with 70% ethanol and preparation of sterile instruments, gauzes, staples and swabs by autoclaving. Surgical instruments should be cleaned and re-sterilised between animals with 70% ethanol or a dry bead steriliser at 220°C. We recommend the use of commercially available sterile sutures. The investigator is advised to wear a surgical mask, hair bonnet and sterile gloves. If multiple surgeries are performed, the surgeon should change gloves between animals.

Anaesthesia is necessary for surgery, the depth of which should be assessed to ensure it is not too light or too deep. If the anaesthetic plane is too light, the animal may start moving or struggling, and if the anaesthetic plane is too deep, the animal may die. The anaesthetic plane can be firstly assessed by pinching the animal's toe, tail or ear. For example, limb withdrawal in response to pinching the toes indicates anaesthesia is too light, and additional anaesthetic should be given. Anaesthetised animals must be monitored during the procedure by visual inspection of respiratory pattern and frequency, which should be regular and even, and by the colour of mucous membranes and exposed tissues, which should be bright pink to red, not grey or blue indicative of anaesthesia that is too deep. During anaesthesia, hypothermia is the biggest cause of animal mortality, and the use of a thermoregulated heating mat maintaining temperature at 37°C is recommended in time-consuming surgeries.

All peripheral nerve injury models are carried out unilaterally on either the left or right side with equivalent outcomes in pain behaviours. Since the CCI, PSNL and SNI models of neuropathic pain involve exposure of the sciatic nerve or its terminal branches, the method in Sect. 2.2 can be applied to all three, with slight modifications, whereas the method for SNL will be covered in Sect. 6.2.

2.1. Materials and Supplies for Surgery

1. Anaesthetic, inhaled isoflurane/halothane (2–3%) delivered in 100% O_2 (1 L/min) or intraperitoneal sodium pentobarbital (50 mg/kg). A cocktail of ketamine (60 mg/kg)/xylazine (5–10 mg/kg) is an alternative injectable anaesthetic; however, due to its mechanism of action, we suggest avoiding its use in these models (see Sect. 7.2). For inhaled anaesthetics, an anaesthetic regulator with induction chamber and small animal face mask are also required.
2. Sterilised surgical consumables (drape, swabs, gauzes, sutures, staples, masks, gloves and hair bonnets).
3. Lubricating ophthalmic gel (PolyGel, Lubricating Eye Gel, Alcon).
4. Shaver.
5. Surgical scrub solutions: 70% isopropyl alcohol and povidone-iodine solution (Riodine, 1%w/v Iodine, Orion).
6. Disinfected surgery table, with sterile waterproofed pad.
7. Sterilised surgical instruments (i.e. blunt-ended scissors, micro-scissors, scalpel, blades, forceps (#5 watchmaker and curved blunt-ended), haemostat and retractor).
8. Thermoregulated heating mat.
9. Surgical microscope and multi-directional light source.
10. Masking tape.
11. Bead steriliser (optional).

2.2. Exposure of the Sciatic Nerve (or Branches) in the Rat

1. Prepare the rat for surgery using the selected inhalant gas anaesthetic or injectable anaesthetic. Ensure that the rat is deeply anaesthetised by confirming lack of the paw pinch reflex (retraction of the hindpaw in response to toe pinching). Apply ophthalmic gel to prevent drying of the cornea during surgery.
2. On the side to be operated, shave the rat's hind leg and lower back. Sterilise the shaved area by applying surgical scrub solution; scrub in circular pattern from the interior of the shaved area to the exterior with alternate applications of iodine solution and 70% isopropyl alcohol.
3. Position the rat on the surgical area, with the rat lying on its thorax and the surgical site facing up. Elevate the hind leg (with a rod made of taped gauze) and hold in position with the femur at 90° to the spine, using masking tape to hold the foot down. Then, drape the animal with a waterproof disposable paper drape.
4. Using a scalpel, make a 2–3 cm incision in the skin parallel, but 3–4 mm posterior to the femur, and carefully free the skin from the muscle surrounding the incision by cutting through the connective tissue with blunt scissors. For CCI, this incision will be at mid-thigh level, for PSNL at high-thigh level, whilst for SNI, it will be more lateral, exposing the popliteal fossa/sciatic trifurcation.
5. Carefully cut through the connective tissue between the gluteus superficialis and the biceps femoris muscles using microscissors. This should be possible with minimal damage to either muscle. A further blunt dissection with blunt-ended scissors can be made through the muscle, according to the level of the desired injury.
6. Widen the gap in the muscle layer with a retractor, allowing clear visualisation of the sciatic nerve or its branches, as appropriate to the desired injury. However, carefully avoid stretching the sciatic nerve, posterior biceps-semitendinosus nerve (PBSN) or the femoral artery, which supplies the nerves circulation and is usually present near the top of the incision.
7. Gently free approximately 7 mm of the sciatic nerve (or its branches) from the two sheets of fascia which sandwich it. For CCI and PSNL, this should be proximal to the sciatic trifurcation, whilst for SNI this should be distal to the trifurcation, thus freeing the sciatic nerve branches. This can be achieved, under a dissecting microscope, by picking up the fascia on one side with watchmaker forceps and cutting alongside the nerve with microscissors and repeating on the other side. Then carefully introduce blunt-ended forceps underneath the nerve to ensure the nerve is completely freed, removing additional connective tissue if necessary.

3. Chronic Constriction Injury of the Sciatic Nerve

First described in 1988 by Bennett and Xie (8), Chronic Constriction Injury (CCI) of the sciatic nerve in rats involves placement of four loose chromic catgut ligatures around the sciatic nerve, to just occlude but not arrest the epineural blood flow. Here, we describe the method of unilateral sciatic nerve CCI.

3.1. Material for CCI

1. Chromic gut (4.0, Ethicon, USA) for nerve ligatures.
2. Sutures (5.0, Mersilk, Ethicon) for closing muscles.
3. Sterile saline.
4. Staple gun and staples (Autoclip, 9-mm stainless steel, Becton Dickinson).
5. Surgical scrub solution, such as povidone-iodine solution (Riodine, 1%w/v Iodine, Orion).
6. Surgical microscope and multi-directional light source.
7. Surgical equipment, as described in Sect. 2.1.

3.2. Step-by-Step Method for CCI

1. Using a surgical microscope and a good light source (steps 1–7), expose the sciatic nerve, as described above (see Sect. 2.2). It should be noted that in a small number of animals, one of the sciatic nerve branches may leave the common sciatic nerve much more proximal than the usual position at the popliteal fossa, and these animals should not be used for CCI.
2. Prepare 3 cm lengths of chromic gut suture, quickly immersing them in sterile saline, to prevent them drying out and losing pliability.
3. Insert one of the 3 cm lengths of 4.0 chromic gut suture under the nerve, approximately 3 mm proximal to the sciatic nerve trifurcation. This can be achieved by introducing curved blunt-ended forceps underneath the nerve (without lifting and stretching the nerve), and then with the other hand holding a length of chromic gut suture in a second pair of blunt-ended forceps, advance them under the nerve into the jaws of the first pair of forceps and then grasp and pull the suture under.
4. Tie the suture in a flat single knot, and pull the two ends using forceps so the suture is just barely in contact with the nerve. The correct level of tightness will slow but not arrest epineural blood flow. This can be achieved experimentally as the tightness at which the suture will no longer freely move up and down the length of the nerve as it is gently pulled. Constriction of the nerve should be stopped immediately if a brief twitch is observed.
5. A second knot is tied directly over the first forming a square knot. The purpose of the second knot is to hold the first in

place, not to loosen or tighten first knot. Care should also be taken to ensure the correct orientation of the second knot, so it lies flat over the first and does not cause excessive additional constriction of the nerve due to twisting of the suture.

6. Cut excess chromic gut suture from the ends of the knot to about 1 mm.
7. Repeat steps 3–6 until a total of four loose chromic gut sutures have been placed, each 1 mm more proximal to the previous.
8. Make sure the ligated nerve is sitting flat, before closing the muscle layers with sutures.
9. Close the skin layer with staples, taking care to avoid trapping hair in the staples.
10. Sterilise the wound with iodine solution.
11. Each rat should be closely observed during the anaesthesia recovery period and allowed to recover in a separate cage with flat paper bedding such as sterile paper towels (instead of the standard animal husbandry bedding), in order to prevent the unconscious animal choking.

3.3. CCI General Considerations and Modifications

The actual cause of the nerve damage in CCI is not the ligature itself but from the self-strangulation beneath the ligature as the nerve swells due to oedema. It is believed to damage myelinated afferent axons much more severely than unmyelinated fibres (17). Overtightening the ligatures is the most common cause of variation in the response and can lead to complete axotomy and autotomy, both undesirable outcomes which prevent successful reflexive withdrawal testing. That said, if ligatures are too loose, little or no oedema will result, and minimal/short-lived sensory abnormality is the likely outcome. However, sciatic nerve CCI undertaken by an experienced researcher aiming for minimal but sufficient constriction leads within 24 h to a reproducible response characterised by:

1. Decreased threshold in paw withdrawal following thermal (hot and cold), mechanical and chemical stimulation (i.e. the presence of allodynia and hyperalgesia).
2. Injury-specific spontaneous pain behaviours (i.e. observations of repeated shaking and licking of the injured hindpaw).
3. Altered motor patterns specific to the nerve damaged (i.e. sciatic nerve: limping, altered gait and reduced weight bearing).
4. Abnormal 'guarding' posture of the injured hindpaw (i.e. toes held together and plantar flexed and paw everted).

Although motor difficulties subside within a few days, pain hypersensitivity to thermal and mechanical stimulation, as well as behavioural disabilities, can persist for 8–12 weeks (8, 12, 17–22). Following CCI, some investigators have demonstrated the presence of complex behavioural disabilities, such as changes in social interactions, disruption of sleep-wake cycles, as well as anxiety-like and depressive-like behaviours (23–27). However, this is in contrast to others who have failed to find any such disabilities (28, 29). Whilst no changes in mechanical and thermal withdrawal thresholds are generally reported in the contralateral side, at least one study has suggested subtle pain hypersensitivity and behavioural signs of pain in the contralateral hindlimb of some animals 1–2 weeks after CCI (18). Pain due to CCI appears to be sympathetically independent, at least in the first postsurgery week; however, over the following weeks, there appears to be some degree of sympathetic neuronal coupling that contributes to pain hypersensitivity (12). The immune-mediated response to the chromic gut itself has been implicated in the behavioural and neurochemical changes following CCI (30, 31); hence, other suture material is likely to be less effective.

The CCI model has also been modified for use in mice, where the procedure is identical, except for the use of 2 or 3 chromic gut/Prolene/nylon sutures, 6.0–8.0 (29, 32, 33). CCI of the infraorbital nerve, pioneered by Vos et al. (34), is a common alternative to sciatic nerve CCI and is a model of severely debilitating trigeminal neuropathy in humans. Infraorbital nerve CCI in rats results in thermal hyperalgesia of the face and snout (with stimulation causing brisk head withdrawal and avoidance behaviour), as well as spontaneous pain behaviour (increased face-grooming activity with face wash strokes directed to the injured nerve territory) (34, 35). In addition, a recent modification of the CCI model was developed by ligating the sciatic nerve with 0, 1, 2 or 4 chromic gut sutures producing a graded allodynia in rats (36).

4. Partial Ligation of the Sciatic Nerve

Partial Ligation of the Sciatic Nerve (PSNL) in the rat was developed by Seltzer et al. (9), and involves tight ligation of $^{1}/_{3}$ to $^{1}/_{2}$ of the sciatic nerve at high-thigh level just distal to the PBSN, resulting in axotomy of the ligated fibres. Below, we describe the method for performing PSNL in rats.

4.1. Materials for PSNL

1. Fine silk sutures (7.0 or 8.0) with 3/8 curved, reverse cutting needle (Ethicon, USA) for nerve ligature.
2. Sutures (5.0, Mersilk, Ethicon) for closing muscles.
3. Staple gun and staples (Autoclip, 9-mm stainless steel, Becton Dickinson).
4. Surgical scrub solution such as povidone-iodine solution (Riodine, 1% w/v Iodine, Orion).
5. Surgical microscope and multi-directional light source.

6. Surgical equipment, as described in Sect. 2.1.

4.2. Step-by-Step Method for PSNL

1. Expose the sciatic nerve, as described above (see Sect. 2.2).
2. Under the surgical microscope (steps 2–4), carefully free the dorsal surface of the sciatic nerve of connective tissue near the trochanter, just distal to the branch point of the PBSN.
3. Using watchmaker forceps, grasp the epineurium to stabilise the nerve (without stretching the nerve itself), and carefully pass the needle (with attached fine silk suture) through the sciatic nerve (just distal to the branch point of the PBSN) at a position which splits the nerve with approximately 60% of fibres in lateral ventral portion and 40% ($^1/_3$ to $^1/_2$) of fibres in the medial dorsal portion.
4. Once the needle and fine suture thread has been passed through the centre of the nerve, tightly ligate the 40% ($^1/_3$ to $^1/_2$) of medial dorsal nerve fibres caught in the suture by tying a double knot, and cut the excess suture.
5. Make sure the nerve is sitting flat, before closing the muscle layers with sutures and the skin wounds with skin staples, taking care to avoid trapping hair in the staples.
6. Sterilise the wound with iodine solution.
7. Place the animal in a separate cage with flat paper bedding (not the standard husbandry bedding), and observe closely during the anaesthetic recovery period.

4.3. PSNL General Considerations and Modifications

PSNL, like CCI, produces a partial denervation of the sciatic nerve; however, unlike CCI, it is believed to cause equal damage to all types of axons (myelinated and unmyelinated). Within a few hours of surgery, and lasting several months, PSNL results in spontaneous pain behaviour, manifested by licking and guarding of the ipsilateral hindpaw (9). Mechanical, thermal and chemical withdrawal thresholds also decrease soon after surgery (i.e. presence of allodynia and hyperalgesia), with these effects lasting for 6–8 weeks (9, 12, 13). To a lesser extent, reductions in mechanical and thermal thresholds have been observed in the contralateral hindpaw (9, 12), reminiscent of clinical 'mirror image pain'. Although complex behavioural disabilities may occur following PSNL, several studies have failed to find changes in anxiety and depression, in both rats and mice (26, 37). Due to the fact at least 50% of the axons are preserved in PSNL, autotomy has not been reported (9). It was also established, through sympathectomy, that during the first post-surgical week, pain is sympathetically independent: however, from week 2 onwards, it becomes sympathetically dependent (12).

A modified version of PSNL, using 9.0 silk for the ligation, was developed in mice (38), with the investigators reporting that whilst mechanical pain hypersensitivity persisted for 10 weeks, thermal pain hypersensitivity resolved by 7 weeks. The main drawback with

PSNL is the fact that it is impossible to ensure that exactly the same number of axons, in exactly the same part of the nerve, is ligated in each animal. Hence, each animal has a slightly different level of nerve damage, which may have greater bearing on the degree of pain hypersensitivity than any other experimental variable (e.g. pharmacological treatment).

5. Spared Nerve Injury

The SNI model of neuropathic pain was developed in 2000 by Decosterd and Woolf (11) and involves transection of two terminal branches of the sciatic nerve in rats. After exposure of the sciatic nerve at the level of its trifurcation, the tibial and the common peroneal branches are tightly ligated with 5.0 silk sutures, before 2 mm of the distal nerve stump is removed, thus axotomising these branches but 'sparing' the sural nerve. Below, we describe the common method used by Decosterd and Woolf and many others since.

5.1. Materials for SNI

1. Sutures (5.0, Mersilk, Ethicon) for nerve ligature and closing muscles.
2. Staple gun and staples (Autoclip, 9-mm stainless steel, Becton Dickinson).
3. Surgical scrub solution such as povidone-iodine solution (Riodine, 1%w/v Iodine, Orion).
4. Surgical microscope and multi-directional light source.
5. Surgical equipment, as described in Sect. 2.1.

5.2. Method for SNI

1. Expose the sciatic nerve and its three terminal branches (the sural, the common peroneal and the tibial nerves), as above (see Sect. 2.2).
2. Under a surgical microscope, carefully free the tibial and the common peroneal nerves from connective tissue, with watchmaker forceps, taking great care to avoid any contact with, or stretching of, the intact sural nerve.
3. Tightly ligate the tibial and common peroneal nerves with fine 5.0 silk suture, 2 mm distal to the sciatic nerve trifurcation.
4. Transect a 2 mm portion of each ligated branch, distal to the ligature.
5. Make sure the nerve branches are sitting flat, before closing the muscle layers with sutures and the skin wounds with skin staples, taking care not to trap any fur under the staples.
6. Sterilise the wound with iodine solution.

7. Place the animal in a separate cage with flat paper bedding (not the standard husbandry bedding), and observe closely during the anaesthetic recovery period.

5.3. SNI Comments and Modifications

SNI is considered a robust model of neuropathic pain, mimicking several features of the clinical presentation including evidence of spontaneous pain behaviour (e.g. a sudden sustained spontaneous withdrawal of the hindpaw in the absence of an external stimulus) and a decrease in mechanical withdrawal threshold, occurring within 2 days, and maintained for over 7 months (11). Although thermal withdrawal latency is not affected by SNI, the duration of withdrawal, once evoked, is increased, an effect maintained for 3 months. Following SNI, there are no signs of changes in social interactions or sleep-wake cycles in rats (11); however, two recent studies have reported increased depressive-like and anxiety-like behaviour (39, 40).

SNI is considered to be more reproducible than PSNL and CCI, because the surgical procedure is simple and standardised, resulting in an identical injury between different animals and investigators. Another major advantage over other partial sciatic nerve axotomy models (such as, CCI, PSNL and SNL) is that intact axons (i.e. sural and saphenous) and degenerating axons (i.e. tibial and common peroneal) innervate different skin territories. Sensory thresholds measured from intact sural areas, adjacent to denervated areas, showed increased responsiveness to noxious heat, as well as non-noxious mechanical and thermal stimulation (11). Similar but less prominent observations were seen in the region innervated by the saphenous nerve, which originates from the L3 spinal nerve, running separate to the common sciatic nerve. These findings highlight the fact that uninjured neurons must be present and contribute to the abnormal activity, such as aberrant/ectopic firing, seen in neuropathic pain. Pain produced by SNI is independent of the sympathetic system (41).

Two common variations of SNI, which are used predominately in mice (42, 43), are based on ligation of different combinations of sciatic nerve branches: (1) the common peroneal and the sural nerves are sectioned, leaving the tibial nerve (t) intact (SNIv (t)), and (2) the tibial nerve is injured leaving the sural (s) and common peroneal (cp) nerves intact (SNIv (s,cp)). Both variants are reported to show only mechanical pain hypersensitivity (42, 43). Further, other deviations from SNI include alternative manipulations of the sciatic nerve branches. For example, the tibial and sural nerve transection (TST), where both these branches are tightly ligated and a 2 mm segment of nerve distal to each ligature is transected, whilst the common peroneal nerve is maintained (44). TST produces robust pain hypersensitivity to mechanical, thermal and chemical stimulation, as well as spontaneous pain behaviours (44), and like SNI, it is considered a model of sympathetically independent pain (45). Finally, ligation of the common peroneal nerve

(CPN) was demonstrated to produce mechanical and thermal pain hypersensitivity lasting up to 28 days in mice (46). Furthermore, changes in pain sensitivity following CPN ligation are resistant to morphine, similar to some clinical cases of neuropathic pain.

6. Spinal Nerve Ligation

In 1992, Kim and Chung developed the Spinal Nerve Ligation (SNL) model of neuropathic pain, with the stated aim to standardise the number of axons damaged (10), which is widely variable in CCI and PSNL. They ligated both L5 and L6 spinal nerves, sparing L4, which has an abundance of motor fibres, with its ligation resulting in severe motor deficits (10). Below, we describe Kim and Chung's original method for undertaking SNL in rats.

6.1. Material for SNL

1. Fine silk suture (6.0, Ethicon, USA) for nerve ligature.
2. Sutures (5.0, Mersilk, Ethicon) for closing muscles.
3. Staple gun and staples (Autoclip, 9-mm stainless steel, Becton Dickinson).
4. Surgical scrub solutions: 70% isopropyl alcohol and povidone-iodine solution (Riodine, 1%w/v Iodine, Orion).
5. Surgical microscope and multi-directional light source.
6. Surgical equipment, as described in Sect. 2.1
7. Heating mat with rectal probe.

6.2. Method for SNL

1. Under anaesthesia, shave the back of the rat and disinfect the area using multiple applications of iodine solution and isopropyl alcohol. Place the rat in the prone position on a heating mat, regulated at 37°C with a rectal probe.
2. Make a 2–3 cm longitudinal incision 5 mm lateral to the midline, at the level of the iliac crest, exposing the left paraspinal muscles. For the rest of the procedure (3–6), use a surgical microscope and a good light source.
3. Using watchmaker forceps, carefully remove the left paraspinal muscles from the L4 spinous processes to the rostral part of the dorsal sacrum (i.e. the fused bone of S1–S2), fully exposing the dorsal surface of the L6 transverse process.
4. With a small rongeur, completely remove the L6 transverse process close to the vertebrae, to visualise the L4 and L5 spinal nerves. Special care should be taken since the L4 and L5 spinal nerves run directly beneath the transverse process, and any minor damage (touch, stretch or entrapment) to the L4 nerve can disrupt the development of pain hypersensitivity by damag-

ing motor neurons. In some animals, there may be a thin layer of muscle under the transverse process which must be removed to visualise the L4 and L5 spinal nerves.

There is great variability in the position where the L4 and L5 spinal nerves join together, such that in some animals, they will have merged together quite proximally, at the level exposed by removal of the L6 transverse process. Since only the L5 is to be ligated, the nerves must be carefully separated before ligation, and if this requires excessive manipulation of the L4 spinal nerve, it is advisable to remove these animals from the ligation group.

5. Again using a rongeur, carefully remove the rostral edge of the sacrum to visualise the L6 spinal nerve.
6. At a point distal to the dorsal root ganglia, using a piece of fine 6.0 silk suture, tightly ligate first the L5 and then the L6 spinal nerve, thereby lesioning all the axons within them.
7. Suture the muscle layers, and use staples to close the skin layers and iodine solution to sterilise the wound.
8. Place the animal in a separate cage with flat paper bedding (not the standard husbandry bedding), and observe closely during the anaesthetic recovery period. Given the more invasive nature of the SNL model, extra care should be given to the animals during the anaesthesia recovery period. Short-term analgesia may be considered, and we advise consultation with your veterinarian. However, investigators should be aware that analgesics given post-operatively usually blunt reflexive withdrawal responses (47) and may decrease the development of neuropathic pain (see Sect. 7.2).

6.3. SNL General Considerations and Modifications

In the SNL model, all sciatic nerve fibres carried by the L5 and L6 spinal nerves are denervated, which includes axons of all sizes, whilst the sciatic nerve axons carried by L4 and L3 spinal nerves are preserved. SNL is characterised by spontaneous pain behaviour, as well as mechanical and thermal pain hypersensitivity, that develop in the ipsilateral hindlimb within 24–48 h and persist for 10–16 weeks, longer than CCI and PSNL (10, 12, 48). There is also a decrease in mechanical threshold in the contralateral limb; however, this is less severe and has a slower onset compared to the ipsilateral hindlimb (10). Following SNL, the presence of complex behavioural disabilities is unclear; rats tested in the elevated plus maze showed no signs of anxiety or depression (28), whilst mice showed depressive-like and anxiety-related behaviour in the open field, light–dark exploration, elevated plus maze and forced swim tests (49).

SNL offers two main advantages over the CCI and PSNL; they are (1) increased reproducibility and a reduction of inter-investigator variation and (2) the specific spinal segments that are injured versus intact are known, allowing clearer investigation of the central (spinal cord) mechanisms of neuropathic pain. However, one disadvantage of SNL is the fact that the surgery is considerably more invasive than CCI, PSNL and SNI, which rely solely on sciatic nerve manipulations. L5/L6 SNL is considered a model of sympathetically dependent pain, since pain behaviours in the SNL model are relieved by chemical sympathetic block and by surgical sympathectomy (12, 50).

A common alternative to ligation of L5/L6 SNL is L5 SNL, a second animal group described by Kim and Chung (10), and used by several subsequent investigators (51–53). The L5 SNL produces a similar response to L5/L6 SNL; however, the degree of mechanical pain hypersensitivity is somewhat reduced, lasting only 7 weeks (10). Another alternative is L5 spinal nerve transection, where 1–2 mm of the nerve is removed to prevent reconnection (54, 55). This model has been primarily used in mice to look at spinal cord mechanisms of neuropathic pain at the L5 level and results in robust mechanical and thermal pain hypersensitivity, lasting at least 30 days (55).

7. General Considerations

7.1. Gender, Age and Strain Differences

It has been demonstrated experimentally that gender is an important factor in pain sensitivity (56, 57), being affected by the oestrous cycle and gender differences in the opioid system. In order to minimise variation, it is common to use only male rodents in these models. However, this approach has been criticised since clinical neuropathic pain is more common in females (58). Another important factor is age, as neuropathic pain behaviour does not develop in infant rats until they are over 3 weeks old (59), and young, old and aged rats develop differential pain hypersensitivity following peripheral nerve injury (60–63). In addition, the influence of neuropathic pain on affective and cognitive behaviours is age dependent, with mid-aged (10 months old) rats being more susceptible to depression and cognitive deterioration induced by experimental neuropathy (SNI) than young (3 months old) and older (21 months old) rats (40). Due to the effect genetics plays on the development of neuropathic pain, significant strain differences are reported (64–66). It is recommended investigators thoroughly check the literature to confirm the strain of choice for any given model, as well as being consistent on the source of animals throughout any series of experiments, since even substrains from different suppliers respond differently to the same injury, displaying different degrees of pain behaviours (66).

7.2. Anaesthetic and Analgesia

Most inhaled or intraperitoneal anaesthetics are suitable for performing animal models of neuropathic pain; however, ketamine, an *N*-methyl-D-aspartate receptor antagonist, should be avoided

(17), since these glutamate receptors likely play a role in the development of neuropathic pain (67, 68).

Another important consideration is the use of post-operative analgesia which is likely to interfere with the development of neuropathic pain and measurements of reflexive withdrawal responses. For example, following SNL in rats, 4 days post-operative treatment with buprenorphine, oxymorphone or carprofen resulted in insensitivity to mechanical stimuli during treatment, which was not fully reversed until day 14 (47). In another study, following SNI, 3 days of post-operative treatment with the opioid, fentanyl, prevented development of dynamic mechanical allodynia, although rats treated with the nonsteroidal anti-inflammatory, flunixin, developed pain hypersensitivity similar to untreated rats (69). Even locally applied anaesthetics, such as lidocaine, infused at the site of skin incision during surgery can modulate development of neuropathic pain, reducing pain hypersensitivity following CCI (70). Therefore, due to reduced sensitivity in common pain tests, and the fact allowing an analgesic washout period may not reverse the effects during the critical establishment phase of neuropathic pain, post-operative analgesia is rarely used in laboratories utilising animal models of neuropathic pain (17).

7.3. Mechanical and Thermal Sensitivity Testing

Several reliable quantitative measures of pain hypersensitivity testing are established to test the reflexive withdrawal response to different pain stimuli:

1. Heat hyperalgesia: using a commercially available plantar analgesia metre based on the method of Hargreaves (71) to measure the withdrawal latency to heat stimulation (see Fig. 2c, d) or the tail-flick test.
2. Mechanical allodynia: using a series of manual von Frey filaments with the quantitative up-down method (72), or a commercially available electronic von Frey aesthesiometer to measure paw withdrawal threshold to mechanical stimulation (see Fig. 2a, b).
3. Mechanical hyperalgesia: using the pinprick test (73) or a commercially available Randall-Selitto instrument.
4. Cold allodynia: using an ice plate maintained at 4°C (8), a cold plate (74) or an acetone drop on the paw (48).
5. Hyperalgesia to other noxious stimuli, such as chemical (e.g. mustard oil or capsaicin) or electrical stimulation.

Instruments for many of these pain tests are available commercially (e.g. Ugo Basile, IITC, Stoelting or Somedic). Their uses, in combination with animal models of neuropathic pain, allow for quantitative comparisons of paw sensitivity to different pain modalities and following different treatment options. The area of the hindpaw used for such pain behavioural testing depends on the pain model. Since co-mingling of distal intact and degenerating sciatic nerve axons exists in the CCI, PSNL and SNL models, the mid-plantar surface of the hindpaw, just posterior to the footpads, is usually used. This area is mostly innervated by the tibial nerve. However, for the SNI model, the lateral surface of the hindpaw (non-injured, sural nerve skin area) should be used. In all models, the heel and footpads should be avoided due to differential skin sensitivity.

Whilst the tests of reflexive withdrawal from mechanical and thermal stimuli are the most widely used measures of neuropathic pain in animal models, there are several criticisms of their use, including (1) a poor correlation with human symptoms (58, 75); (2) to some extent, testing the response of motor neurons, not pain *per se* (76); (3) the absence of supraspinal processing (76); and (4) the involvement of considerable experimenter bias (58). To avoid bias and to include supraspinal processing, operant reward-conflict assays, producing mechanical and thermal stimuli that are not experimenter initiated and result in behaviour that is indicative of pain intensity, were developed (77, 78). Additional paradigms which evaluate the affective-motivational aspects of pain such as elevated plus maze, forced swim test, resident–intruder social interaction testing and open field exploration (21, 23–27), as well as monitoring of spontaneously emitted behaviours (e.g. vocalisations and changes in facial expression) (21, 79) and assessing conditioned place preference for a context associated with pain relief (80) are also recommended.

8. Conclusion

Animal models of partial peripheral nerve injury (e.g. CCI of the sciatic nerve, PSNL, SNI and SNL) involving stimulation of nociceptive neurons, as well as peripheral and central sensitisation, result in pain hypersensitivity predominantly in the ipsilateral hindpaw. All these types of nerve injury produce a simple and useful animal model of neuropathic pain, but each leads to somewhat different balance of pain symptoms (12). Despite their widespread use in conjunction with testing of cutaneous evoked withdrawal responses, there are several drawbacks in the outcome and implementation of these models, which question their relevance to clinical neuropathic pain (75). These concerns are: (1) measures for highly prevalent symptoms of clinical pain, such as sensory loss, paraesthesia, dysaesthesia and spontaneous pain, are difficult to quantify or have not yet been developed in rodents; (2) behavioural disabilities often associated with chronic pain syndromes occur in rodents with some degree of variability, seen by some (21, 23–27, 39, 40, 49) and not by others (28, 29, 37), but they are rarely considered by most investigators; (3) whilst the majority of animals develop thermal and mechanical pain hypersensitivity following nerve injury, not all patients develop pain hypersensitivity, for example, thermal hyperalgesia does not occur in postherpetic neuralgia (14); (4) the timescale of clinical neuropathic pain is years, whereas animal models exist on a scale of weeks or at most months; and (5) problems with translation of in vivo efficacy in preclinical studies to the clinic. Indeed, whilst animal models have confirmed the efficacy of pharmacological agents commonly used to treat clinical neuropathic pain, for example, gabapentin, amitriptyline and fluoxetine (52), and in the case of ziconotide, predicted efficacy (81), there are several instances of failed translation, including neurokinin 1 antagonists, glycine-site antagonists and sodium channel blockers (58). In spite of all these concerns, animal models remain an essential tool, not only in the investigation of potential novel therapies but also to uncover the mechanisms of neuropathic pain, such as the activation and inhibition of central and peripheral neurons, which underlie the pathological amplification of noxious and non-noxious stimuli.

References

1. Blyth FM, March LM, Brnabic AJ, Jorm LR, Williamson M, Cousins MJ (2001) Chronic pain in Australia: a prevalence study. Pain 89(2–3):127–134
2. Treede RD, Jensen TS, Campbell JN, Cruccu G, Dostrovsky JO, Griffin JW, Hansson P, Hughes R, Nurmikko T, Serra J (2008) Neuropathic pain: redefinition and a grading system for clinical and research purposes. Neurology 70(18):1630–1635
3. Bouhassira D, Lanteri-Minet M, Attal N, Laurent B, Touboul C (2008) Prevalence of chronic pain with neuropathic characteristics in the general population. Pain 136(3):380–387
4. Toth C, Lander J, Wiebe S (2009) The prevalence and impact of chronic pain with neuropathic pain symptoms in the general population. Pain Med 10(5):918–929
5. Kehlet H, Jensen TS, Woolf CJ (2006) Persistent postsurgical pain: risk factors and prevention. Lancet 367(9522):1618–1625
6. Wall PD, Gutnick M (1974) Properties of afferent nerve impulses originating from a neuroma. Nature 248(5451):740–743
7. Wall PD, Devor M, Inbal R, Scadding JW, Schonfeld D, Seltzer Z, Tomkiewicz MM (1979) Autotomy following peripheral nerve lesions: experimental anaesthesia dolorosa. Pain 7(2):103–111
8. Bennett GJ, Xie YK (1988) A peripheral mononeuropathy in rat that produces disorders of pain sensation like those seen in man. Pain 33(1):87–107
9. Ze S, Dubner R, Shir Y (1990) A novel behavioral model of neuropathic pain disorders produced in rats by partial sciatic nerve injury. Pain 43(2):205–218
10. Kim SH, Chung JM (1992) An experimental model for peripheral neuropathy produced by segmental spinal nerve ligation in the rat. Pain 50(3):355–363
11. Decosterd I, Woolf CJ (2000) Spared nerve injury: an animal model of persistent peripheral neuropathic pain. Pain 87(2):149–158
12. Kim KJ, Yoon YW, Chung JM (1997) Comparison of three rodent neuropathic pain models. Exp Brain Res 113(2):200–206
13. Dowdall T, Robinson I, Meert TF (2005) Comparison of five different rat models of peripheral nerve injury. Pharmacol Biochem Behav 80(1):93–108
14. Campbell JN, Meyer RA (2006) Mechanisms of neuropathic pain. Neuron 52(1):77–92
15. Costigan M, Scholz J, Woolf CJ (2009) Neuropathic pain: a maladaptive response of the nervous system to damage. Annu Rev Neurosci 32:1–32
16. Latremoliere A, Woolf CJ (2009) Central sensitization: a generator of pain hypersensitivity by central neural plasticity. J Pain 10(9):895–926
17. Bennett GJ, Chung JM, Honore M, Seltzer Z (2003) Models of neuropathic pain in the rat.

Current Protocols in Neuroscience 9.14.1–9.14.16 Copyright © 2003 by John Wiley & Sons, Inc. Wiley, New York

18. Attal N, Jazat F, Kayser V, Guilbaud G (1990) Further evidence for 'pain-related' behaviours in a model of unilateral peripheral mononeuropathy. Pain 41(2):235–251
19. Bennett GJ (1993) An animal model of neuropathic pain: a review. Muscle Nerve 16(10): 1040–1048
20. Kupers RC, Nuytten D, De Castro-Costa M, Gybels JM (1992) A time course analysis of the changes in spontaneous and evoked behaviour in a rat model of neuropathic pain. Pain 50(1): 101–111
21. Monassi CR, Bandler R, Keay KA (2003) A subpopulation of rats show social and sleep-waking changes typical of chronic neuropathic pain following peripheral nerve injury. Eur J Neurosci 17(9):1907–1920
22. Paulson PE, Casey KL, Morrow TJ (2002) Long-term changes in behavior and regional cerebral blood flow associated with painful peripheral mononeuropathy in the rat. Pain 95(1–2):31–40
23. Austin PJ, Beyer K, Bembrick AL, Keay KA (2010) Peripheral nerve injury differentially regulates dopaminergic pathways in the nucleus accumbens of rats with either 'pain alone' or 'pain and disability'. Neuroscience 171(1): 329–343
24. Keay KA, Monassi CR, Levison DB, Bandler R (2004) Peripheral nerve injury evokes disabilities and sensory dysfunction in a subpopulation of rats: a closer model to human chronic neuropathic pain? Neurosci Lett 361(1–3):188–191
25. Jesse C, Wilhelm E, Nogueira C (2010) Depression-like behavior and mechanical allodynia are reduced by bis selenide treatment in mice with chronic constriction injury: a comparison with fluoxetine, amitriptyline, and bupropion. Psychopharmacology 212(4):513–522
26. Roeska K, Doods H, Arndt K, Treede RD, Ceci A (2008) Anxiety-like behaviour in rats with mononeuropathy is reduced by the analgesic drugs morphine and gabapentin. Pain 139(2):349–357
27. Hu B, Doods H, Treede R-D, Ceci A (2009) Depression-like behaviour in rats with mononeuropathy is reduced by the CB2-selective agonist GW405833. Pain 143(3):206–212
28. Kontinen VK, Kauppila T, Paananen S, Pertovaara A, Kalso E (1999) Behavioural measures of depression and anxiety in rats with spinal nerve ligation-induced neuropathy. Pain 80(1–2):341–346
29. Urban R, Scherrer G, Goulding EH, Tecott LH, Basbaum AI (2011) Behavioral indices of ongoing pain are largely unchanged in male mice with tissue or nerve injury-induced mechanical hypersensitivity. Pain 152(5):990–1000
30. Maves TJ, Pechman PS, Gebhart GF, Meller ST (1993) Possible chemical contribution from chromic gut sutures produces disorders of pain sensation like those seen in man. Pain 54(1):57–69
31. Xu J, Pollock CH, Kajander KC (1996) Chromic gut suture reduces calcitonin-gene-related peptide and substance P levels in the spinal cord following chronic constriction injury in the rat. Pain 64(3):503–509
32. Sommer C, Lindenlaub T, Teuteberg P, Schäfers M, Hartung T, Toyka KV (2001) Anti-TNF-neutralizing antibodies reduce pain-related behavior in two different mouse models of painful mononeuropathy. Brain Res 913(1):86–89
33. Walczak J-S, Beaulieu P (2006) Comparison of three models of neuropathic pain in mice using a new method to assess cold allodynia: the double plate technique. Neurosci Lett 399(3):240–244
34. Vos BP, Strassman AM, Maciewicz RJ (1994) Behavioral evidence of trigeminal neuropathic pain following chronic constriction injury to the rat's infraorbital nerve. J Neurosci 14(5 Pt 1):2708–2723
35. Imamura Y, Kawamoto H, Nakanishi O (1997) Characterization of heat-hyperalgesia in an experimental trigeminal neuropathy in rats. Exp Brain Res 116(1):97–103
36. Grace PM, Hutchinson MR, Manavis J, Somogyi AA, Rolan PE (2010) A novel animal model of graded neuropathic pain: utility to investigate mechanisms of population heterogeneity. J Neurosci Methods 193(1):47–53
37. Hasnie FS, Wallace VC, Hefner K, Holmes A, Rice AS (2007) Mechanical and cold hypersensitivity in nerve-injured C57BL/6J mice is not associated with fear-avoidance- and depression-related behaviour. Br J Anaesth 98(6): 816–822
38. Malmberg AB, Basbaum AI (1998) Partial sciatic nerve injury in the mouse as a model of neuropathic pain: behavioral and neuroanatomical correlates. Pain 76(1–2):215–222
39. Norman GJ, Karelina K, Zhang N, Walton JC, Morris JS, Devries AC (2010) Stress and IL-1beta contribute to the development of depressive-like behavior following peripheral nerve injury. Mol Psychiatry 15(4):404–414
40. Leite-Almeida H, Almeida-Torres L, Mesquita AR, Pertovaara A, Sousa N, Cerqueira JJ, Almeida A (2009) The impact of age on emotional and cognitive behaviours triggered by experimental neuropathy in rats. Pain 144(1–2):57–65

41. Pertin M, Allchorne AJ, Beggah AT, Woolf CJ, Decosterd I (2007) Delayed sympathetic dependence in the spared nerve injury (SNI) model of neuropathic pain. Mol Pain 3:21
42. Bourquin AF, Suveges M, Pertin M, Gilliard N, Sardy S, Davison AC, Spahn DR, Decosterd I (2006) Assessment and analysis of mechanical allodynia-like behavior induced by spared nerve injury (SNI) in the mouse. Pain 122(1–2):14.el–14
43. Shields SD, Eckert WA, Basbaum AI (2003) Spared nerve injury model of neuropathic pain in the mouse: a behavioral and anatomic analysis. J Pain 4(8):465–470
44. Lee BH, Won R, Baik EJ, Lee SH, Moon CH (2000) An animal model of neuropathic pain employing injury to the sciatic nerve branches. Neuroreport 11(4):657–661
45. Han DW, Kweon TD, Kim KJ, Lee JS, Chang CH, Lee YW (2006) Does the tibial and sural nerve transection model represent sympathetically independent pain? Yonsei Med J 47(6):847–851
46. Vadakkan KI, Jia YH, Zhuo M (2005) A behavioral model of neuropathic pain induced by ligation of the common peroneal nerve in mice. J Pain 6(11):747–756
47. Simkins MD, Shadiack AM, Burns CA, Molino LJ, Amaratunga D, Hall J, Rogers KE, Clark LP (1998) Evaluation of post-operative analgesics in a model of neuropathic pain. J Am Assoc Lab Anim Sci 37(6):61–63
48. Choi Y, Yoon YW, Na HS, Kim SH, Chung JM (1994) Behavioral signs of ongoing pain and cold allodynia in a rat model of neuropathic pain. Pain 59(3):369–376
49. Suzuki T, Amata M, Sakaue G, Nishimura S, Inoue T, Shibata M, Mashimo T (2007) Experimental neuropathy in mice is associated with delayed behavioral changes related to anxiety and depression. Anesth Analg 104(6):1570–1577
50. Kim SH, Na HS, Sheen K, Chung JM (1993) Effects of sympathectomy on a rat model of peripheral neuropathy. Pain 55(1):85–92
51. LaBuda CJ, Fuchs PN (2000) A behavioral test paradigm to measure the aversive quality of inflammatory and neuropathic pain in rats. Exp Neurol 163(2):490–494
52. LaBuda CJ, Little PJ (2005) Pharmacological evaluation of the selective spinal nerve ligation model of neuropathic pain in the rat. J Neurosci Methods 144(2):175–181
53. LaGraize SC, Borzan J, Fuchs PN (2003) Decreased L5 spinal nerve ligation nociceptive behavior following L4 spinal nerve transection. Brain Res 990(1–2):227–230
54. Cao L, DeLeo JA (2008) CNS-infiltrating CD4+ T lymphocytes contribute to murine spinal nerve transection-induced neuropathic pain. Eur J Immunol 38(2):448–458
55. Tanga FY, Nutile-McMenemy N, DeLeo JA (2005) The CNS role of Toll-like receptor 4 in innate neuroimmunity and painful neuropathy. Proc Natl Acad Sci USA 102(16):5856–5861
56. DeLeo JA, Rutkowski MD (2000) Gender differences in rat neuropathic pain sensitivity is dependent on strain. Neurosci Lett 282(3):197–199
57. Tall JM, Stuesse SL, Cruce WL, Crisp T (2001) Gender and the behavioral manifestations of neuropathic pain. Pharmacol Biochem Behav 68(1):99–104
58. Mogil JS, Davis KD, Derbyshire SW (2010) The necessity of animal models in pain research. Pain 151(1):12–17
59. Howard RF, Walker SM, Mota PM, Fitzgerald M (2005) The ontogeny of neuropathic pain: postnatal onset of mechanical allodynia in rat spared nerve injury (SNI) and chronic constriction injury (CCI) models. Pain 115(3):382–389
60. Chung JM, Choi Y, Yoon YW, Na HS (1995) Effects of age on behavioral signs of neuropathic pain in an experimental rat model. Neurosci Lett 183(1–2):54–57
61. Crisp T, Giles JR, Cruce WL, McBurney DL, Stuesse SL (2003) The effects of aging on thermal hyperalgesia and tactile-evoked allodynia using two models of peripheral mononeuropathy in the rat. Neurosci Lett 339(2):103–106
62. Kim YI, Na HS, Yoon YW, Nahm SH, Ko KH, Hong SK (1995) Mechanical allodynia is more strongly manifested in older rats in an experimental model of peripheral neuropathy. Neurosci Lett 199(2):158–160
63. Tanck EN, Kroin JS, McCarthy RJ, Penn RD, Ivankovich AD (1992) Effects of age and size on development of allodynia in a chronic pain model produced by sciatic nerve ligation in rats. Pain 51(3):313–316
64. Lovell JA, Stuesse SL, Cruce WL, Crisp T (2000) Strain differences in neuropathic hyperalgesia. Pharmacol Biochem Behav 65(1):141–144
65. Rode F, Thomsen M, Brolos T, Jensen DG, Blackburn-Munro G, Bjerrum OJ (2007) The importance of genetic background on pain behaviours and pharmacological sensitivity in the rat spared serve injury model of peripheral neuropathic pain. Eur J Pharmacol 564(1–3):103–111
66. Yoon YW, Lee DH, Lee BH, Chung K, Chung JM (1999) Different strains and substrains of

rats show different levels of neuropathic pain behaviors. Exp Brain Res 129(2):167–171

67. Sotgiu ML, Biella G (2000) Differential effects of MK-801, a N-methyl-D-aspartate non-competitive antagonist, on the dorsal horn neuron hyperactivity and hyperexcitability in neuropathic rats. Neurosci Lett 283(2): 153–156
68. Woolf CJ (2007) Central sensitization: uncovering the relation between pain and plasticity. Anesthesiology 106(4):864–867
69. Stewart LSA, Martin WJ (2003) Influence of postoperative analgesics on the development of neuropathic pain in rats. Comp Med 53(1):29–36
70. Dougherty PM, Garrison CJ, Carlton SM (1992) Differential influence of local anesthetic upon two models of experimentally induced peripheral mononeuropathy in the rat. Brain Res 570(1–2):109–115
71. Hargreaves K, Dubner R, Brown F, Flores C, Joris J (1988) A new and sensitive method for measuring thermal nociception in cutaneous hyperalgesia. Pain 32(1):77–88
72. Chaplan SR, Bach FW, Pogrel JW, Chung JM, Yaksh TL (1994) Quantitative assessment of tactile allodynia in the rat paw. J Neurosci Methods 53(1):55–63
73. Tal M, Bennett GJ (1994) Extra-territorial pain in rats with a peripheral mononeuropathy: mechano-hyperalgesia and mechano-allodynia in the territory of an uninjured nerve. Pain 57(3):375–382
74. Allchorne AJ, Broom DC, Woolf CJ (2005) Detection of cold pain, cold allodynia and cold hyperalgesia in freely behaving rats. Mol Pain 1:36
75. Mogil JS (2009) Animal models of pain: progress and challenges. Nat Rev Neurosci 10(4):283–294
76. Vierck JCJ (2006) Animal models of pain: lessons for drug development. In: Campbell JN, Basbaum AI, Dray R, Dubner RH, Dworkin RH, Sang CN (eds) Emerging strategies for the treatment of neuropathic pain. IASP Press, Seattle, WA, pp 475–496
77. Neubert JK, Widmer CG, Malphurs W, Rossi HL, Vierck JCJ, Caudle RM (2005) Use of a novel thermal operant behavioral assay for characterization of orofacial pain sensitivity. Pain 116(3):386–395
78. Neubert JK, King C, Malphurs W, Wong F, Weaver JP, Jenkins AC, Rossi HL, Caudle RM (2008) Characterization of mouse orofacial pain and the effects of lesioning TRPV1-expressing neurons on operant behavior. Mol Pain 4:43
79. Langford DJ, Bailey AL, Chanda ML, Clarke SE, Drummond TE, Echols S, Glick S, Ingrao J, Klassen-Ross T, Lacroix-Fralish ML, Matsumiya L, Sorge RE, Sotocinal SG, Tabaka JM, Wong D, van den Maagdenberg AM, Ferrari MD, Craig KD, Mogil JS (2010) Coding of facial expressions of pain in the laboratory mouse. Nat Methods 7(6):447–449
80. King T, Vera-Portocarrero L, Gutierrez T, Vanderah TW, Dussor G, Lai J, Fields HL, Porreca F (2009) Unmasking the tonic-aversive state in neuropathic pain. Nat Neurosci 12(11):1364–1366
81. Schmidtko A, Lötsch J, Freynhagen R, Geisslinger G (2010) Ziconotide for treatment of severe chronic pain. Lancet 375(9725): 1569–1577

Chapter 15

Detection of Sensitized Nerve Responses: Dorsal Root Reflexes, Live Cell Calcium, and ROS Imaging

Karin N. Westlund and Liping Zhang

Abstract

Significant changes in afferent nerve activity and/or neuronal sensitization occur in response to peripheral injury and inflammation. One approach to study of sensitized nerve responses is to record increased peripheral nerve activity in vivo in experimental animal models with electrophysiological recording techniques. Recording of responses of nerves that have been sensitized for at least 30 min reveals development of dorsal root reflex activity, that is, nerve responses that propagate back out to the periphery. This method can also be used to examine the ability of pharmacological agents to reduce sensitization. An alternative to the use of live animals is primary culture models of inflammation. Live cell imaging of fluorescent dye conversion can be utilized to study activation events in dorsal root ganglia harvested from injury models for comparison to cells from controls. Activation responses, including evidence of reactive oxidative/nitroxidative species generation and intracellular calcium mobilization, are increased in sensitized cells and can continue for a prolonged time. These experimental methodologies will be described here and their research utility emphasized.

Key words: Pain, Arthritis model, Calcium mobilization, Reactive oxygen species, Inflammation

1. Introduction

Arthritis is a disorder characterized by inflammation of one or more joints causing swelling, stiffness, and pain. Over 80% of the population experience some form of arthritis pain, including an estimated 1.3 million suffering from rheumatoid arthritis (RA), 0.3 million afflicted with juvenile arthritis, and 27 million with osteoarthritis (Arthritis Foundation, Atlanta, GA, May 2010). Conservative predictions made by the US Center for Disease Control estimate that 67 million adults will have arthritis in 2030. Of these, a predicted 25 million, or 9.3%, of the adult US population will be limited in their activities (http://www.cdc.gov/arthritis/data_statistics). Current estimate is a societal cost totaling $128

Paul M. Pilowsky et al. (eds.), *Stimulation and Inhibition of Neurons*, Neuromethods, vol. 78,
DOI 10.1007/978-1-62703-233-9_15,

billion per year due to lost income, medical and surgical costs, as well as increased morbidity rates. While women are two to three times more likely to get RA than men, men tend to be more severely affected by this autoimmune condition. Clinical studies find long-term therapy reducing inflammatory mediator tumor necrosis factor alpha (TNFα) is effective for treating RA, but the therapy predisposes some patients to lethal septic infections.

Understanding the role of tissue microenvironmental and neuronal mechanisms that cause acute inflammation to become a chronic inflammatory pain condition will allow the development of more comprehensive treatments to halt disease progression and relieve pain for patients suffering from arthritis. One approach is to study sensitization of nerve responses induced by experimental inflammation models in anesthetized animals and to examine the ability of pharmacological agents to reduce sensitization. An alternative to the use of live animals, primary culture models of inflammation, can be utilized to study activation events in dorsal root ganglia including evidence of reactive oxidative/nitroxidative species (ROS) generation and intracellular calcium mobilization using live cell imaging of fluorescent dye conversion. These experimental methodologies will be described here and their research utility emphasized.

2. In Vivo Nerve Recordings

We have recorded responses of medial articular nerves (MANs) to von Frey fiber stimulation on the skin overlying the knee in anesthetized animals after induction of acute knee joint inflammation (1–3) (Fig. 1; (2)). Dynamic changes compared to responses in nerves of normal animals are observed. An acute inflammation of the knee joint can be induced by a single injection of the irritants 3% kaolin silt and 3% carrageenan (k/c) (100 or 50 μl in saline for rats or mice, respectively) directly into the knee joint cavity, while the animals are anesthetized with a short-acting anesthetic. In this knee joint arthritis model, the animal's own mobilization about the home cage produces a localized inflammation of the knee joint that reaches its maximum by 4 h and that remains for 24 h. During that time, behavioral studies can be carried out or ongoing afferent nerve activity can be recorded.

2.1. Dorsal Root Reflex Recording

Significant changes in afferent nerve activity and sensitization of dorsal horn neurons after knee joint inflammation result in the generation of dorsal root reflex (DRR) activity, i.e., nerve responses that propagate back out to the periphery. The DRRs originate at the central endings of the primary afferent neurons in the dorsal horn and propagate back out the afferent nerves where they can be recorded in the target tissue (4). The DRR activity recorded from

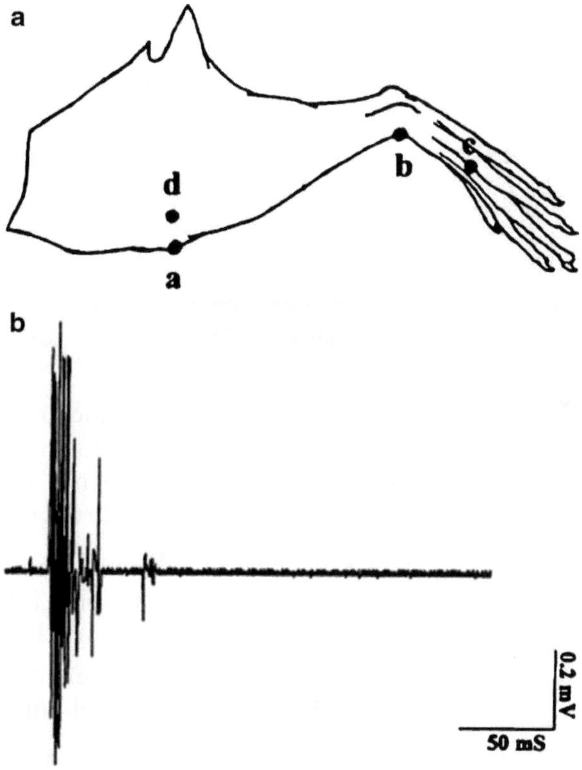

Fig. 1. (**a**) Stimulation sites on the lower limb were used to evoke DRR activity with a von Frey fiber with a 2-N bending force in animals with an acutely inflamed knee joint. Mechanical stimulation was applied in the order of anterior of knee (*a*), ankle (*b*), foot (*c*), and the lateral part of the knee (*d*). (**b**) An example of compound action potentials evoked by von Frey fiber application on the inflamed knee joint and recorded in the proximal end of the cut MAN (2) (with permission).

peripheral nerves can be used to monitor indirectly the hyperactivity of spinal dorsal horn neurons after knee joint inflammation. The DRR has been proposed to be the result of primary afferent depolarization caused by the activation of γ-aminobutyric acid ($GABA_A$) receptors on the central primary afferent nerve terminals (5). Centrally generated DRR can be distinguished from afferent terminal responses by recording from the proximal stump of the cut MAN. Unilateral dorsal rhizotomy, but not sympathectomy, results in a significant reduction of DRR activity, joint temperature, and swelling (1, 3, 6). The DRRs and spinal cord release of glutamate are also eliminated by administration of non-*N*-methyl-D-aspartate (NMDA) and $GABA_A$ receptor antagonists into the dorsal horn (3, 7, 8). All the evidence indicates that DRRs are the result of central sensitization in the dorsal horn involving ionotropic glutamate and $GABA_A$ receptor-mediated events. It has also been shown that the long-term release of glutamate and plastic changes resulting in secondary hyperalgesia in the dorsal horn potentiate responses of spinothalamic tract projection neurons in the dorsal horn of monkeys (9, 10).

2.2. Recording DRR in the Medial Articular Nerve

1. All experimental procedures were approved by the local institutional animal care and use committee. The procedures and set-up for recording DRRs in anesthetized animals are as follows:
2. Rats receive initial anesthesia using urethane (Sigma–Aldrich, St. Louis, MO; 1.3 g/kg).
3. The jugular vein is cannulated and a tracheotomy is performed.
4. The rat is paralyzed and anesthetized with a continuous intravenous infusion of a mixture of pancuronium bromide (Astra Pharmaceuticals, Wayne, PA; 4.5 mg/kg/h) and urethane (75 mg/kg/h), and artificially ventilated.
5. End-tidal CO_2 is monitored and maintained at 3.5–4.5%.
6. The depth of anesthesia is judged by monitoring the heart rate and is kept stable during the whole experiment.
7. Core body temperature is maintained at 37.5°C with a homeothermic blanket control unit linked to a rectal probe.
8. The skin on the medial side of the right thigh is incised rostrally from the inguinal fossa to a point just below the medial condyle of the tibia.
9. The MAN is dissected free and cut distally and the skin flap is retracted to form a pool filled with warmed mineral oil to prevent desiccation of the nerve.
10. The whole MAN, which is composed of <200 axons is placed on a bipolar silver electrode (11).
11. The evoked antidromic compound action potentials are generated by stimulating the inflamed knee joint with von Frey fibers and recorded.
12. Action potentials are amplified with an AC amplifier (World Precision Instruments, Sarasota, FL), displayed on an oscilloscope and digitized using a data-logging interface system (CED micro 1401, Cambridge Electronic Design Limited, Cambridge, United Kingdom).
13. The nerve spike data is then recorded in WaveMark form using the Spike2 program (Cambridge Electronic Design Limited) and stored in a Pentium (Dell) computer for offline analysis.
14. Spike rate histograms and interspike interval histograms used in this study are plotted.

After cutting the nerve adjacent to the inflamed knee joint, the proximal end of the MAN is placed on the electrode as described above. In the arthritic rat, responses to mechanical stimuli are tested at the beginning of the electrophysiological recordings, 4 h after injection of k/c. The threshold for evoking activity in the MAN is established by applying a series of von Frey fibers with increasing bending forces onto the inflamed leg in each animal. Four stimulation sites were tested as illustrated in Fig. 1a to evoke DRR activity in the proximal axons of the cut MAN (Fig. 1b).

2.3. Administration of Drugs

Delivery of chemicals directly into the dorsal horn to alter neuronal responses in control animals or to return sensitized responses of experimental animals closer to those in controls can be achieved using a microdialysis fiber. The microdialysis fiber (Hospal, Meyzieu, France) (200-μm outer diameter, 45,000-MW cut-off, Hospal AN69) is coated with epoxy glue except for a 2-mm permeable length. The fiber is inserted into the spinal dorsal horn with the permeable area in laminae III–IV of L4–L6 segments. This allows a reasonable spread of the agents throughout these spinal segments without damage of nociceptive afferents. After fiber insertion, artificial cerebrospinal fluid (aCSF) is infused for 1 h, and animals are then allowed to recover overnight. The detailed insertion of the microdialysis fiber is described elsewhere (12). During experiments, aCSF is infused through the fiber at a rate of 5μL/min. The composition of the aCSF is (in mM/L) NaCl 125, $CaCl_2$ 1.3, KCl 2.6, $MgCl_2$ 0.9, $NaHCO_3$ 21, Na_2HPO_4 1.3, and glucose 3.5 (pH 7.4).

3. In Vitro Models of Neuron Activation

Cell culture models can be used to observe, stimulate, and inhibit the following indicators of neuronal activation:

1. Intracellular calcium ($(Ca^{2+})_i$) mobilization indicative of initiation of signaling events.
2. Intracellular generation of ROS as an indicator of oxidative stress and injury response.

In vitro studies by Huang and colleagues (13) have provided strong evidence that somata of DRG neurons actively release transmitters and play a crucial role in the bidirectional communication between neurons and surrounding satellite glial cells. Their studies showed that electrical stimulation of dorsal root ganglion (DRG) neurons elicits robust vesicular ATP release from their somata. Entry of external Ca^{2+} into the cell is required for this release. Entry of calcium is mediated by reactive oxygen species. Increasing intracellular calcium can be visualized with live cell-imaging microfluorimetry that will be described below. ATP activates P2X3 and P2X7 receptors in satellite cells that enwrap each DRG neuron and triggers the communication between neuronal somata and glial cells. Blocking L-type Ca^{2+} channels eliminates intracellular Ca^{2+} increase in both neurons and satellite cells, suggesting that activation of L-type Ca^{2+} channels is essential for the communication between neuronal somata and satellite cells. Activation of P2X receptors leads to the release of TNFα from satellite cells. TNFα in turn potentiates the P2X receptor-mediated responses. TNFα also directly increases the excitability of DRG neurons through TNFR1 receptors (14). TNFα produced by activation of satellite cells also alters the excitability of neurons in chronic pain conditions but involves TNFR2. Chronically elevated levels of exogenous TNFα would sensitize DRG cells inducing increases in neuronal activity and establish a positive feedback loop between neurons and satellite cells, thus contributing to exaggerated responses under pathological conditions. We have shown that a ROS scavenger decreases TNFR1and DRG excitability in vitro after TRPV1 stimulation (see below) (15). In the following sections, methodology for DRG primary cultures, live cell calcium, and ROS microfluorimetry is provided, as well as examples of cellular responses to inflammatory conditions and oxidative stress.

3.1. General Method for DRG Primary Cultures for Live Cell Calcium and ROS Imaging

While the methods described below refer to culture of spinal DRG (16), primary neuronal cultures can be prepared from a variety of ganglia, for example, trigeminal and nodose ganglia:

1. Mice or rats are deeply anesthetized (sodium pentobarbital, 50 mg/kg, i.p.) or decapitated, the vertebral column exposed, excised, and put onto ice.
2. While on ice, the DRG is quickly exposed by dorsal laminectomy, cutting away the upper half of the spinal column with side-to-side movements with a bone cutter.
3. Lumbar DRGs (L2–L5) are carefully picked from between the vertebrae, axonal fibers cut away, and the DRG placed in ice-cold, oxygenated dissecting solution containing (in mM): NaCl 130, KCl 5, KH_2PO_4 2, $MgCl_2$ 6, $CaCl_2$ 1.5, glucose 10, and HEPES 10 (pH adjusted to 7.2 with TRIS base and osmolality adjusted with sucrose to 305 mOsm (17)).
4. The DRGs are minced into small pieces and digested in 0.1% collagenase D (Roche Diagnostics, Indianapolis, IN, USA) and 0.06% trypsin (Sigma–Aldrich, St. Louis, MO, USA) in a 37°C water bath shaker (86 rpm) for 1 h.
5. After three washes in culture media solution (DMEM with FBS, antibiotics) alternated with centrifugation (50×*g*) (about 300 rpm, 6 min), cells are reconstituted in 1-ml culture media.
6. Cells are plated (0.1 ml) onto polylysine-coated (MW 70–150 K, SIGMA) cover slips (Ø 12 mm) inside 35-mm culture dishes for 30–60 min.

7. Alternatively, cover slips can be coated with 0.1-mg/ml poly-L-ornithine and 5 μg/ml mouse laminin.
8. Culture medium (2 ml) is added to the wells, and the cells incubated over night in an incubator (5% CO_2, 37°C).
9. Two large DRGs or nodose ganglia are sufficient to plate 6 cover slips at the appropriate plating density for live cell imaging. Four to six thoracic level ganglia would be required.

3.2. Live Cell Calcium Mobilization Microfluorimetry

Cellular activation through downstream signaling pathways is initiated by entry of calcium across the cell membrane and/or release from intracellular stores. Visualization of $(Ca^{2+})_i$ mobilization, indicative of cellular activation, can be performed in primary cultures of DRG neurons. Cells are pre-loaded with Fura-2, a fluorescent dye indicator, for ratiometric microfluorimetry analysis that allows an accurate estimation of dynamic change from resting calcium levels.

3.2.1. Cytosolic Calcium Imaging

1. To assess the intracellular calcium levels, cells are loaded with a high affinity, ratiometric calcium indicator Fura-2 acetoxymethoxy ester (FURA-2 AM, 2–5 μM, Molecular Probes, Eugene, OR, USA), which diffuses across the cell membrane from the culture medium solution (without phenol red which will increase background activity) for 1 h in the same incubator at 37°C temperature in the dark prior to imaging.
2. Cells are then washed three times with external (experimental bath) recording solution. External recording solution for DRG neurons has the following composition: (in mM) NaCl 130, KCl 5, KH_2PO_4 2, $CaCl_2$ 2.5, $MgCl_2$ 1, HEPES 10, and glucose 10. The solution is adjusted to pH 7.3 with 10 N NaOH (~1 ml 10 N NaOH into 2,000-ml recording solution). Osmolarity of the external recording solution is adjusted to 295–305 mOsm by adding sucrose (~2.5 g/L).
3. After washing, the indicator is allowed to de-esterify for 30 min. In this case, the charged Ca^{2+} indicator is trapped inside the cell body. Any changes in the 340/380 ratio is an indication of $(Ca^{2+})_i$ increase.
4. The cover slip with the DRGs is mounted in a recording chamber (P3, Warner Instruments) attached to the stage of an inverted fluorescence microscope containing 1-ml external recording solution at room temperature (22–23°C), with a flow rate of 2 ml/min, to which either agonist or antagonist can be added.
5. Studies are performed on a population of cells (10–20) visible in the microscope field for each plate using a 20 or 40×0.9 quartz objective of a Nikon TE2000 inverted epi-fluorescence microscope equipped with a 340- and 380-nm excitation filter

changer (Sutter, Novato, CA, USA) and a 510-nm long pass emission filter (Molecular Devices, Sunnyvale, CA, USA).

6. Fura-2 AM-loaded DRGs are excited alternately at wavelengths of 340 nm and 380 nm. Fluorescent images and ratios are acquired every 2–10 s with a 12-bit digital monochrome-cooled CCD camera (Roper Scientific CoolSNAP HQ) connected to a Dell computer controlled by MetaFluor software v6.5 (Universal Imaging) or Nikon Elements.
7. The ratio of emitted light from the dye by excitation with 340/380 nm is detected at 510 nm and used as an indirect ratiometric measurement of intracellular Ca^{2+} activation.
8. Cells were considered responsive to activators or agonists if their F340/F380 ratio increased by more than 20%. Ratioing considerably reduces the effects of uneven dye loading, leakage of dye, and photobleaching, as well as problems associated with measuring Ca^{2+} in cells of unequal thickness (http://probes.invitrogen.com/media/pis/mp01200.pdf, Molecular Probes, Burlingame, CA, USA).
9. Experiments are replicated at least four times to obtain a total number of at least 50 cells for comparisons between groups.
10. Cell surface areas can be measured in the light transmission images acquired and analyzed offline using MetaMorph software (Molecular Devices, Sunnyvale, CA, USA).

Afferent nerve sensitivity to specific stimuli can be tested in vitro using calcium imaging of DRG cultures. The chemical agent icilin activates TRPA1 and TRPM8 ion channels. Examination of $(Ca^{2+})_i$ mobilization responses to icilin has been utilized to demonstrate sensitized responses of DRG sensory neurons from animals with sciatic nerve ligation (SNL) (Fig. 2; (16)). The magnitude of the response is calculated as a mean change in ratio during a 2-min icilin application. The ratio is significantly increased in DRGs from animals with SNL (2.09 ± 0.03) compared with DRG from naïve rats (1.27 ± 0.05) (Fig. 2, inset). In DRGs from rats with SNL, 65% of the cells responded, while only 15% of the cells respond to icilin in DRG cultures from naïve animals (16).

3.3. Imaging Reactive Oxygen/Nitroxidative Species Conversion

ROS, for example, superoxide anion, hydrogen peroxide, hydroxyl, nitric oxide, and peroxynitrite radicals, are highly reactive molecules which in low concentrations are required for many cellular regulatory and signaling mechanisms (18). However, tissue injury generates and releases excessive ROS. While unstable ROS cannot be directly measured, the conversion of ROS to less reactive metabolites can be visualized and monitored with sensitive fluorescent dyes in histological samples and in live cell imaging of DRG primary cultures as an indication of neuronal activation.

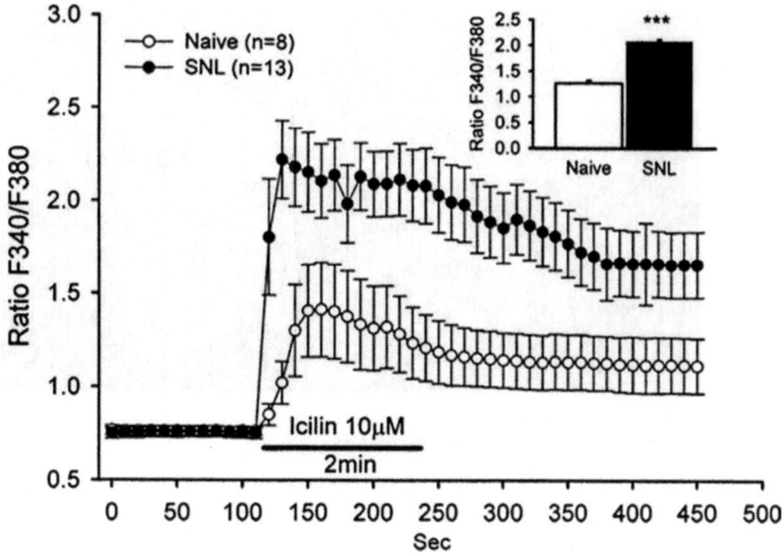

Fig. 2. TRPA1/TRPM8 agonist icilin-induced cytosolic $(Ca^{2+})_i$ mobilization responses in L4 DRG cells. Responses to icilin are reflected as changes in FURA-2 F340/F380 nm ratio. During the 2-min application of icilin (1 μM), the cytosolic $(Ca^{2+})_i$ mobilization was significantly increased in DRGs from animals with spinal nerve ligation (*SNL*) compared with that of naive animals (inset, ***$P \leq 0.001$, Mann–Whitney *U* test) (16) (with permission).

For live cell-imaging studies, primary cultures of DRGs are prepared from normal animals or from animals with chronic arthritis (19). This method allows study of functional sensitization mechanisms in DRG. Activation of cultured DRGs with TRPV1 agonists increases the generation of ROS (Fig. 3; (15)). These increases in ROS are not evident in DRG from mice deficient in TRPV1. ROS generation was also observed with similar methods in acute macrophage and synoviocyte ion channel activation studies in vitro by us and others (19–21). Conditions of induced nitroxidative stress can be mimicked in vitro by exposure to reactive oxygen species donor tertiary-butyl hydroperoxide (tBOOH), hyposmolarity, or TNFα (19, 22). We have shown in these studies that TRPV1-evoked ROS can be blocked with the ROS scavenger phenyl-*N*-tert-butylnitrone (PBN) (Fig. 3).

3.3.1. Determination of ROS Generation in Primary DRG cells with DCFDA Fluorescent Probe

1. On the day of the experiment, the ROS-sensitive fluorescent dye, 5-(and-6)-chloromethyl-2′,7′-dichlorodihydrofluorescein diacetate, acetyl ester (CM-H2-DCFDA, Invitrogen; Carlsbad, CA), is freshly reconstituted.
2. Cultured primary DRG cells are taken out of the incubator and kept at room temperature for 30 min before replacing culture medium with phosphate-buffered saline (PBS).
3. All experiments are conducted at room temperature.

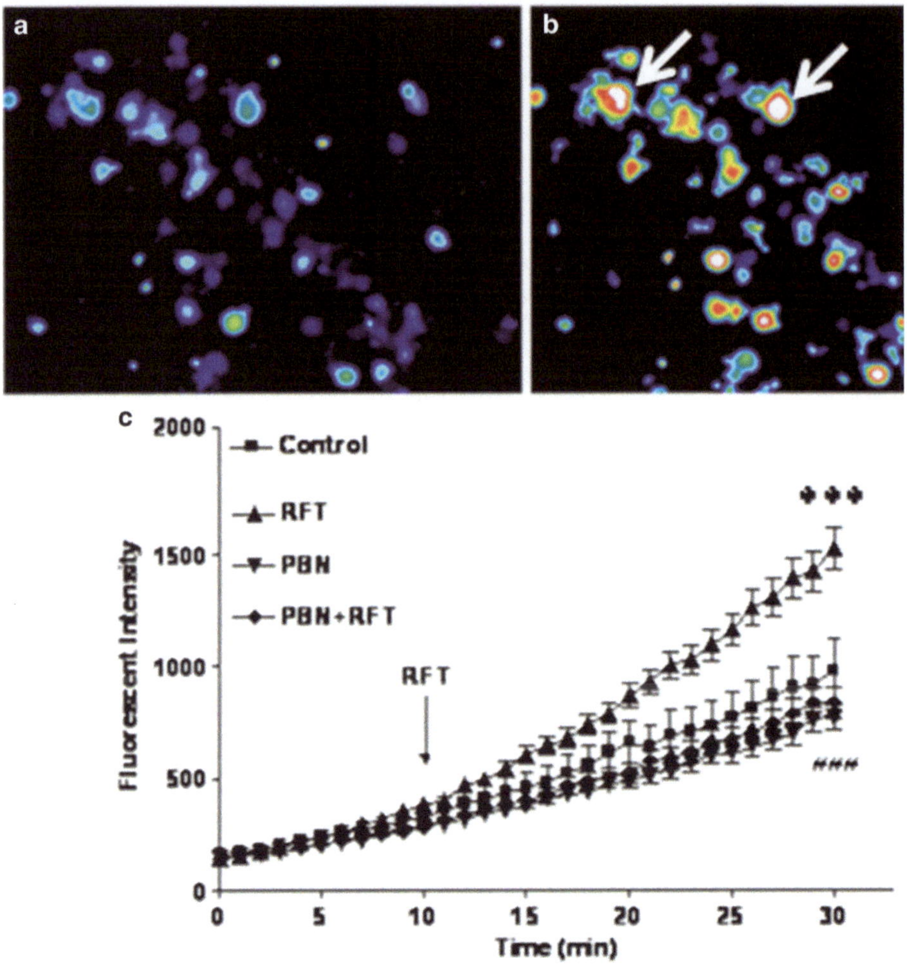

Fig. 3. ROS production is increased in primary cultures of DRG neurons in responses to TRPV1 agonist resiniferatoxin and is blocked by ROS scavenger PBN (15). (**a**) Baseline ROS imaging. (**b**) Increased ROS (white arrows) imaged 30 min after bath application of resiniferatoxin (RFT). (**c**) Fluorescent intensity for ROS generation increases over time after RFT (black arrow) but can be blocked by ROS scavenger PBN. ***p<0.001, ###p<0.001.

4. Primary DRG cells are pre-loaded with DCFDA (in PBS) at a final concentration of 4 μM for 30 min in the dark (RT).
5. The loading buffer is then removed and replaced with Dulbecco's PBS with calcium and magnesium, without phenol red (pH 7.2, Invitrogen).
6. Allow a 30-min recovery time to allow esterase activity to remove the carrier that enabled the fluorescent dye to enter the cell, trapping the dye intracellularly.
7. Thus, any fluorescent activation recorded indicates intracellular conversion of the dye.
8. Cell fluorescent imaging is performed using an Eclipse 2000-S microscope (Nikon) equipped with a high-sensitivity digital

monochrome CoolSNAP ES camera (Photometrics, Tucson, AZ, USA) controlled by NIS element (Nikon) or MetaMorph (Molecular Devices, Sunnyvale, CA, USA) software

9. The DCF dye-loaded cells are excited at 480 nm, and emitted light is collected using a 520-nm long pass filter.
10. The images are captured at 1-min intervals for 30 min.
11. The baseline fluorescence intensity of the loaded cells is determined prior to exposing the cells to experimental inducements. Nikon Elements or MetaMorph offline software (Molecular Devices, Sunnyvale, CA) is used for digital image data analysis.

3.4. Detection of ROS in Tissue Sections with a ROS-Sensitive Detector Dye

We have detected functional increases of ROS in spinal cord visualizing a ROS-sensitive detector dye, dihydroethidium (DHE) (Fig. 4), with HPLC measurement of its metabolites (19). Likewise, we have visualized nitric oxide synthase (NOS) increases in tissue sections from arthritic animals using immunohistochemistry (23).

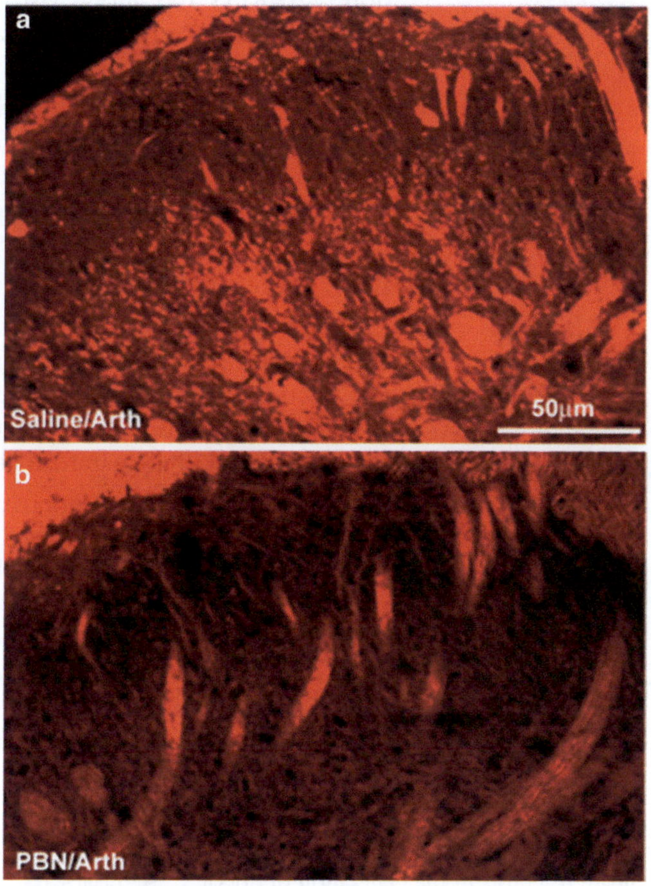

Fig. 4. (**a**) Cumulative generation of ROS in the L4 spinal cord dorsal horn after knee joint inflammation is detected by conversion of the fluorescent dye DHE. (**b**) ROS generation is blocked by pretreating the animals with the ROS scavenger PBN (i.p.) (19).

Increased nitroxidation after induction of arthritis in animal models can be reduced by intraperitoneal injection with the ROS scavenger PBN (100 mg/kg; Sigma, St. Louis, MO, USA).

Release of ROS in the spinal cord can be visualized histologically in animals subjected to an inflammation model if a ROS-sensitive dye is injected prior to injury (Fig. 4) (19). For these studies:

1. Rats are anesthetized with isoflurane (2–4%).
2. The ROS sensitive detector dye dihydroethidium (DHE, Sigma; 50 μl; 100 mM in saline:DMSO, 1:1; Molecular Probes, Eugene, OR, USA) is injected intrathecally through PE32 polyethylene tubing onto the thoracolumbar spinal cord.
3. The tubing is connected externally to a blunt 30 G needle attached to a 250 μl Hamilton syringe.
4. Animals are allowed to recover and knee joint injury is induced 24 h later.
5. The dye is converted to a stable fluorescent product by reaction with ROS and allows visualization of ROS generated in the spinal cord as a result of spinal neuron activation after knee joint inflammation.
6. This reaction product can be visualized histologically in the spinal cord sections after the animal is aldehyde fixed at the conclusion of the behavioral study.
7. In spinal cord sections cut (15 μm) and mounted onto Superfrost Plus slides (VWR, West Chester, PA, USA), the DHE fluorescence is visualized using rhodamine filter settings with a fluorescence photomicroscope.
8. ROS generation can also be visualized in fresh, snap frozen tissue sections that are incubated post-mortem with DHE (5 mM/l in PBS) in a dark humidified chamber (37°, 30 min), then post-fixed in 4% paraformaldehyde (10 min).
9. The DHE is oxidized by superoxides to ethidium bromide that binds to DNA in the cell nucleus and emits red fluorescence. Higher superoxide content can be revealed in comparisons to controls.

References

1. Rees H, Sluka KA, Westlund KN, Willis WD (1994) Do dorsal root reflexes augment peripheral inflammation? Neuroreport 5:821–824
2. Zhang LP, Chen Y, Clark BP, Sher E, Westlund KN (2000) The role of type 1 metabotropic glutamate receptors in the generation of dorsal root reflexes induced by acute arthritis or the spinal infusion of 4-aminopyridine in the anesthetized rat. J Pain 1:151–161
3. Rees H, Sluka KA, Westlund KN, Willis WD (1995) The role of glutamate and GABA receptors in the generation of dorsal root reflexes by acute arthritis in the anaesthetized rat. J Physiol 484(Pt 2):437–445
4. Eccles JC, Kozak W, Magni F (1961) Dorsal root reflexes of muscle group I afferent fibres. J Physiol 159:128–146

5. Curtis DR, Duggan AW, Felix D, Johnston GA (1971) Bicuculline, an antagonist of GABA and synaptic inhibition in the spinal cord of the cat. Brain Res 32:69–96
6. Sluka KA, Lawand NB, Westlund KN (1994) Joint inflammation is reduced by dorsal rhizotomy and not by sympathectomy or spinal cord transection. Ann Rheum Dis 53:309–314
7. Sluka KA, Westlund KN (1993) Spinal cord amino acid release and content in an arthritis model: the effects of pretreatment with non-NMDA, NMDA, and NK1 receptor antagonists. Brain Res 627:89–103
8. Sluka KA, Jordan HH, Willis WD, Westlund KN (1994) Differential effects of N-methyl-D-aspartate (NMDA) and non-NMDA receptor antagonists on spinal release of amino acids after development of acute arthritis in rats. Brain Res 664:77–84
9. Dougherty PM, Willis WD (1992) Enhanced responses of spinothalamic tract neurons to excitatory amino acids accompany capsaicin-induced sensitization in the monkey. J Neurosci 12:883–894
10. Sorkin LS, Westlund KN, Sluka KA, Dougherty PM, Willis WD (1992) Neural changes in acute arthritis in monkeys. IV. Time-course of amino acid release into the lumbar dorsal horn. Brain Res Brain Res Rev 17:39–50
11. Hildebrand C, Oqvist G, Brax L, Tuisku F (1991) Anatomy of the rat knee joint and fibre composition of a major articular nerve. Anat Rec 229:545–555
12. Sluka KA, Westlund KN (1992) An experimental arthritis in rats: dorsal horn aspartate and glutamate increases. Neurosci Lett 145:141–144
13. Zhang X, Chen Y, Wang C, Huang LY (2007) Neuronal somatic ATP release triggers neuron-satellite glial cell communication in dorsal root ganglia. Proc Natl Acad Sci USA 104:9864–9869
14. Schafers M, Sommer C, Geis C, Hagenacker T, Vandenabeele P, Sorkin LS (2008) Selective stimulation of either tumor necrosis factor receptor differentially induces pain behavior in vivo and ectopic activity in sensory neurons in vitro. Neuroscience 157:414–423
15. Ma F, Zhang L, Westlund KN (2009) Reactive oxygen species mediate TNFR1 increase after TRPV1 activation in mouse DRG neurons. Mol Pain 5:31
16. Ji G, Zhou S, Kochukov MY, Westlund KN, Carlton SM (2007) Plasticity in intact A delta- and C-fibers contributes to cold hypersensitivity in neuropathic rats. Neuroscience 150:182–193
17. Xu GY, Huang LY (2002) Peripheral inflammation sensitizes P2X receptor-mediated responses in rat dorsal root ganglion neurons. J Neurosci 22:93–102
18. Maher P, Schubert D (2000) Signaling by reactive oxygen species in the nervous system. Cell Mol Life Sci 57:1287–1305
19. Westlund KN, Kochukov MY, Lu Y, McNearney TA (2010) Impact of central and peripheral TRPV1 and ROS levels on proinflammatory mediators and nociceptive behavior. Mol Pain 6:46
20. Pfeiffer ZA, Guerra AN, Hill LM, Gavala ML, Prabhu U, Aga M, Hall DJ, Bertics PJ (2007) Nucleotide receptor signaling in murine macrophages is linked to reactive oxygen species generation. Free Radic Biol Med 42:1506–1516
21. Caporali F, Capecchi PL, Gamberucci A, Lazzerini PE, Pompella G, Natale M, Lorenzini S, Selvi E, Galeazzi M, Laghi Pasini F (2008) Human rheumatoid synoviocytes express functional P2X7 receptors. J Mol Med 86:937–949
22. Kochukov MY, McNearney TA, Yin H, Zhang L, Ma F, Ponomareva L, Abshire S, Westlund KN (2009) Tumor necrosis factor-alpha (TNF-alpha) enhances functional thermal and chemical responses of TRP cation channels in human synoviocytes. Mol Pain 5:49
23. Wu J, Lin Q, McAdoo DJ, Willis WD (1998) Nitric oxide contributes to central sensitization following intradermal injection of capsaicin. Neuroreport 9:589–592

Chapter 16

Visual Evoked Potential Recording in Rodents

Yuyi You, Alexander Klistorner, and Stuart L. Graham

Abstract

The visual evoked potential (VEP) is an electronic potential recorded from the visual cortex in response to a visual stimulus. It provides a means to examine the function of the visual pathway from the retina to the occipital cortex. The most common animals employed in VEP laboratory studies are rats. In this chapter, we describe the basic 'flash VEP' recording protocol in rodents and discuss the practical aspects of the preparation including anaesthesia methods, electrode configuration, stimulus design, dark adaptation, filter settings and signal sampling with the authors' personal experience. We also review the recent use of the VEP in laboratory researches in several optic nerve disease models, including glaucoma, ischemic optic neuropathy and optic neuritis.

Key words: Visual evoked potential, Electrophysiology, Visual cortex, Optic nerve, Optic neuritis, Glaucoma, Rodent

1. Introduction

The visual evoked potential (VEP) is an electronic potential recorded from the visual cortex in response to a visual stimulus (e.g. flash stimulus, pattern reversal stimulus, pattern onset/offset stimulus). It reflects the post-retinal function from the retina to the primary visual cortex (1). Optic nerve diseases such as optic neuritis (2) and glaucoma (3) will affect the VEP signal. The VEP provides an objective means for measuring the function of the visual pathway and is widely used to assess the visual performance in laboratory animals, including rats, mice, guinea pigs, rabbits, cats, dogs and primates (4). Rats and mice have become key animal models for ophthalmic research. They are the most common species employed in VEP studies because of their accessibility with a short growth cycle, which is important for genetic studies in mutant strains. Additionally, the visual pathway from the retina to the visual cortex of both species shares many similarities with that of

Paul M. Pilowsky et al. (eds.), *Stimulation and Inhibition of Neurons*, Neuromethods, vol. 78,
DOI 10.1007/978-1-62703-233-9_16,

humans, such as cell types and structure features (5). Because of the low spatial resolution of the rodent visual system and the high refractive index of the rodent cornea, the flash VEPs are predominantly used.

Since Creel et al. (6, 7) described the VEP waveform in rats and designated the P1, N1, P2, N2, P3 and N3 components (Fig. 1), the rodent VEP has been widely used to study ocular diseases and the effects of drugs on the visual system. However, it is important to note that in the literature the rat (6–20), as well as mouse (21–27), VEP waveforms were variable between different labs. This can be attributed, at least in part, to different strains used in these experiments. Furthermore, length of dark adaptation, anaesthesia, electrode configuration, stimulus design and filter settings also have a significant impact on the latency and amplitude of the VEP waveform (see Notes below (Sect. 4)). For example, the rat VEP amplitude we have reported (28, 29) was smaller than that from Heiduschka et al. (8), which is very likely due to the short adaptation, the albino rat strain and the monocular flash stimulation. Therefore, a standardised recording protocol needs to be determined before the rodent VEP can be compared between different labs. It is also important to understand that the early VEP components (P1–N1–P2 complex, Fig. 1) following the flash illumination are more stable and principally affected by the excitation of the primary visual cortex via retino-geniculate fibres, whereas the later components may have significant contribution from higher visual areas and therefore are affected by behavioural state or level of consciousness (7, 16).

Here, we focus on the rodent VEP recording protocol, summarise the practical procedures and discuss the use of rodent VEP in some rodent eye disease models.

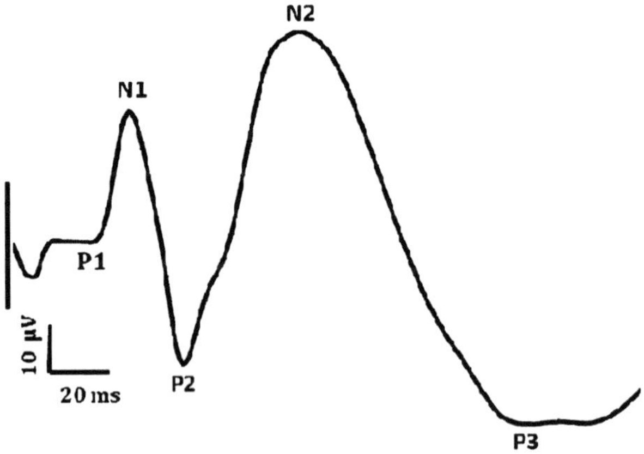

Fig. 1. An averaged flash VEP waveform from SD rats. The components are designated into positive (P1, P2, P3) and negative (N1, N2) peaks.

2. Materials

1. Animals: Rat (e.g. Sprague-Dawley, Long-Evans), mouse (e.g. C57BL/6J)
2. Anaesthetics: Ketamine (80 mg/kg) and medetomidine (0.5 mg/kg)
3. Mydriatic eye drops: 1.0% tropicamide (Alcon Laboratories)
4. Warming pad: Homoeothermic blanket system (Harvard Apparatus)
5. Impedance Meter: F-EZM5 (Grass)
6. Electrodes Screw electrodes: M1.0 × 3 mm Csk Slot M/T 304 S/S (Micro Fasteners, Australia)

   Subdermal needle electrodes: F-E3M-72 (Grass)
7. Dental cement: Rapid Repair (DeguDent GmbH)
8. Visual stimuli: Strobe photostimulator PS33-PLUS (Grass) or a Ganzfeld stimulator
9. Band-pass filter and amplifier: BMA-400 Bioamplifier (CWE, Inc.)
10. Computer-based signal analysing system: CED Power1401 and Spike 2 (Cambridge Electronic Design, Ltd.)

3. Methods

3.1. Anaesthesia

Animals are anaesthetised with an intraperitoneal injection of ketamine (80 mg/kg) and medetomidine (0.5 mg/kg) for both surgery and VEP recording (see Note (Sect. 4.1) for the choice of anaesthetics).

3.2. Screw Electrode Implantation

After anaesthesia, the skin on the head is disinfected with 75% ethanol and a longitudinal incision is made. Small burr holes are drilled manually using a micro hand drill at 7 mm behind the bregma and 3 mm lateral of the midline. Screw electrodes are implanted through the skull just into the cortex (area 17), penetrating the cortex to approximately 0.5 mm. A reference screw electrode is also implanted on the midline 3 mm rostral to the bregma. Dental cement is used to encase and fix the screws. The wound is sutured and antibiotics administrated (Fig. 2 below with photograph and diagram showing landmarks and electrode positions). At least 1 week is allowed for the rats to recover from the surgery.

3.3. Recording Procedures

The animals are anaesthetised, placed in a dark room and allowed to adapt to darkness for 5 min. Body temperature is maintained at

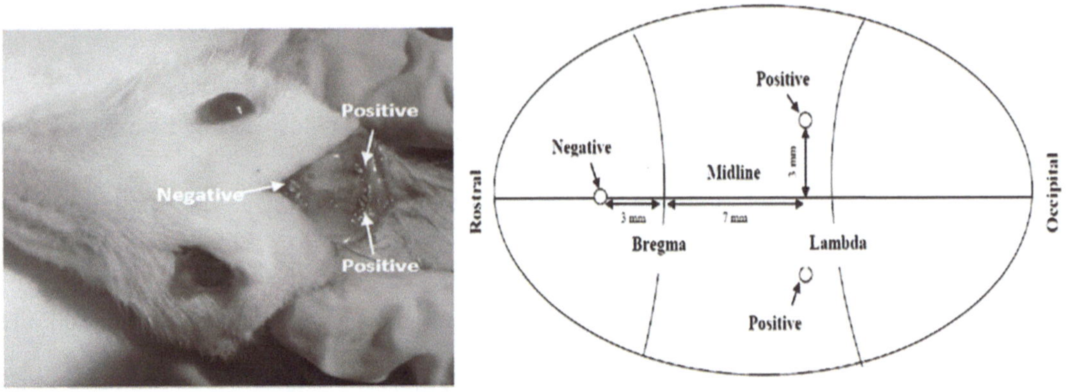

Fig. 2. Positive electrodes are implanted in area 17 of the primary visual cortex (7 mm behind the bregma and 3 mm lateral of the midline). A negative electrode is implanted on the midline 3 mm rostral to the bregma (You et al. ARVO 2011 E-Abstract 6100).

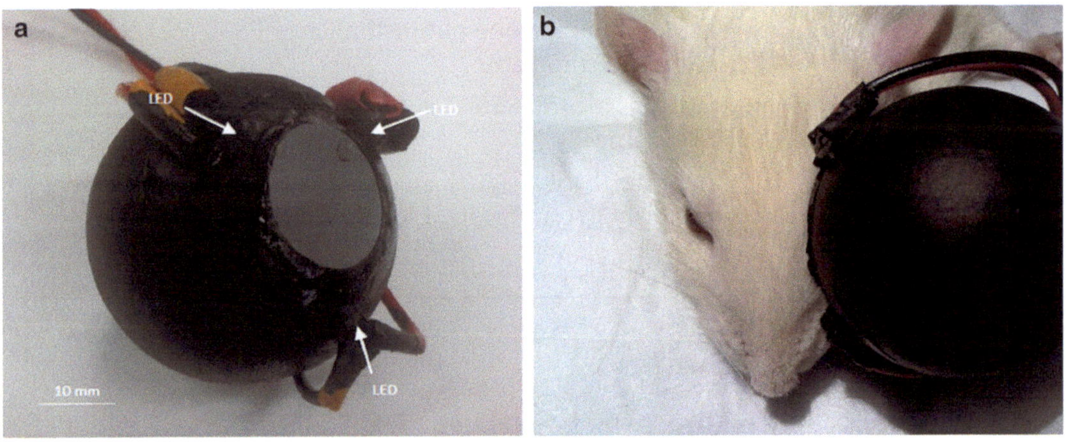

Fig. 3. (**a**) A mini-Ganzfeld stimulator; (**b**) mini-Ganzfeld with a positioned rat (You et al. (29)).

37 ± 0.5°C by the homoeothermic blanket system with a rectal thermometer probe. The pupils are dilated using 1.0% tropicamide eye drops. The skin over the skull is opened to access the preplaced in situ screws. The screw over the contralateral visual cortex of the stimulated eye and the reference screw are connected to the amplifier. A needle electrode is inserted into the tail as the ground. The electrode impedance is measured and maintained below 5 kΩ. The strobe photostimulator is fixed 15 cm from the eye. Alternatively, a mini-Ganzfeld stimulator (Fig. 3) can be used which should be placed directly on the skin around the eyelids to provide superior eye isolation. The illuminations of all stimuli need to be calibrated by a photometer. Photic stimulation is delivered 100 times at a frequency of 1 Hz. Responses are amplified 20,000 times with low and high band-pass filter settings of 1 and 100 Hz. The signal is sampled at 5 kHz.

3.4. VEP Follow-Up in an Individual Animal

After the recording, the wound is closed and antibiotics administrated. The animals are allowed to recover on a warming pad. The VEP can be repeatedly recorded on individual animals to study the time course of diseases or to assess the effects of therapeutic agents.

4. Notes

4.1. Anaesthesia and Body Temperature Maintenance

Different types of anaesthetics and the depth of anaesthesia can affect the VEP waveforms (4, 27). The VEP latency with pentobarbital anaesthesia was longer than that with ketamine in rabbits (30). Meeren et al. (16) compared the rat VEP recorded during various sleep–wake states, and they found state-dependent alterations of the N2–P3 components. In our lab, we routinely use ketamine (31) in combination with an alpha-adrenoceptor agonist such as xylazine (15 mg/kg) or medetomidine diluted in saline and injected intraperitoneally. Combination ketamine/medetomidine anaesthesia provides stable anaesthesia (60–90 min) that can quickly be reversed by administration of an alpha-receptor antagonist such as atipamezole (2.5 mg/kg). Medetomidine is a more selective agent than xylazine (32) and is therefore used in preference. Top-up doses should contain ketamine only (10% of initial dose). Pentobarbital sodium (Nembutal®) also has been widely used for VEP recording in rodents. However, it is important to note that pentobarbital is not permitted for use in rodent recovery experiments in some countries according to the local animal ethics legislations. Corneal clouding and cataracts in rodents can reduce the light available to the retina and affect the reliability of the VEP signal. The cataracts mainly result from the dehydration of the cornea in anaesthetised animals (33, 34). The authors recommend that moisture eye drops (e.g. Optive®) be used regularly during the recording procedures.

Body temperature can also have a significant impact on the speed of nerve conduction, especially in demyelinated nerves (Uhthoff phenomena) (35). Temperature-dependent changes in visual evoked potentials were observed in rats (36, 37)—both flash and pattern VEP latencies were significantly longer as body temperature was lowered (37). In our lab, the body temperature of the rodents is maintained at 37 ± 0.5°C during the experiments by a homoeothermic blanket system with a rectal thermometer probe.

4.2. Electrode Configuration

Electrode positioning requires consideration of the anatomy of the rodent visual cortex. The recorded VEPs are the net postsynaptic potentials in large pyramidal cells oriented perpendicular to the visual cortical surface (38). Unlike humans, the primary visual cortex of rodents is relatively flat and exposed on the surface of the brain. Cytoarchitectonic area 17 occupies an area of 2–3 mm^2 and is the

Table 1
Intersession reproducibility analysis of the rat VEP latency with subdermal needle and implanted screw electrodes (both with 3 cd/m² strobe flash) (29)

	Sw (ms)	CoV (%)	ICC
Needle ($n=6$)	7.95	11.14	0.02
Screw ($n=5$)	3.83	6.74	0.73
P value	0.024*	0.013	0.001*

Sw within-subject standard deviations; *CoV* coefficient of variation; *ICC* intraclass correlation coefficient
$^*P<0.05$

only cortical area that receives its principle input from the dorsal lateral geniculate nucleus (4, 39). We routinely implant the screw electrodes through the skull into area 17 (7 mm behind the bregma and 3 mm lateral of the midline) to serve as the active electrodes (Fig. 2). A variety of the negative electrode positions are described in the literature for rodent VEP recording (6–27). We have found that a screw electrode on the midline 3 mm rostral to the bregma provided superior signal with high signal-to-noise ratios. For the ground lead, it is convenient to use a stainless needle electrode and insert into the animal's tail, two-thirds from its base.

Some laboratories also use subdermal needles as the active electrodes to measure the VEP in rodents non-invasively (26). However, it has been reported that the electrode positioning significantly affected the latency and amplitude of the VEP waveform (15) and the position of the needle electrode is relatively more variable. We found that the intersession reproducibility of the VEP parameters was significantly higher with screw electrodes than with subdermal needle electrodes (Table 1) (29). Therefore, the choice of the electrodes is dependent on the objective of the research. If the project aims at quantifying the nerve conduction by the VEP or monitoring the visual system function over time, we suggest that only screw electrodes be used. Furthermore, the amplitude of the VEP with subdermal electrodes is significantly smaller than that with screws.

4.3. Dark Adaptation

Both rod and cone pathways contribute to the VEP (5). The purpose of dark adaptation is to ensure higher photoreceptor sensitivity, and the amplitude of the VEP significantly increased in dark-adapted rats. In our lab, the animals are placed in a dark room and allowed to adapt to dark for 5 min before the actual recording (28, 29). We found overnight dark adaptation is not necessary for the VEP, which is a routine procedure for electroretinogram (ERG) recording (34).

First, the amplitude of the VEP after 5-min adaptation is already quite acceptable and repeatable for analysis. Secondly, it is not convenient to perform VEP recording in fully dark-adapted animals with a standard flash (3 cd s/m^2), as VEP recording needs about 100 repeated flashes to minimise the noise from the cortical activity (electroencephalogram, EEG)—the flashes rapidly light-adapt the animals. Thirdly, the VEP is designed to investigate post-retinal function; therefore, it is not important to separate the rods and cones in most cases. However, to obtain more reliable VEP signals, we highly recommend the recording be performed under dim background lighting with constant luminance.

4.4. Visual Stimuli

The flash VEP in rodents is well described using various visual stimuli, including a strobe photostimulator (7, 11, 15) and the full-field (Ganzfeld) system (8, 9, 12, 24). These conventional visual stimuli work well for rodents. As the flash illumination can significantly affect the latency and amplitude of the VEP, it is important to always calibrate the stimulus before recording. The stimulus intensity can be measured in cd s/m^2. 3.0 cd s/m^2 is the International Society for Clinical Electrophysiology of Vision (ISCEV) standard flash (40). In many eye disease models, one eye is treated, leaving the other to serve as an internal control. Using the conventional visual stimuli, we found it difficult to entirely isolate the treated eye and the control eye even when well covered, especially in some albino species. This may be due to light penetration through the surrounding scalp and tissue. We designed a mini-Ganzfeld stimulator for the rodents, which consists of a 40-mm diameter ball, three LEDs and an intensity controller (Fig. 3), and it provided superior repeatability in both latency and amplitude.

We compared VEP signals from this full-field stimulus with a focal flash (light-emitting diode, LED) and found the full-field flash showed significantly better repeatability than the LED (Fig. 4 and Table 2) (29). Considering that rat photoreceptors distribute more peripherally than human (41), we suggest that only a full-field stimulus should be used for rodent flash VEP recording and the pupils should always be fully dilated. Rodent VEP responses may also be obtained using patterned stimuli (25). A pattern stimulus is used to assess several aspects of the visual system, such as visual acuity, cortical magnification factor and contrast threshold. However, it is important to consider the low spatial resolution and high corneal refraction of rodents when designing the pattern stimuli (26).

4.5. Filter Settings and Signal Averaging

The VEP is an EEG-based signal; however, it is small compared to the EEG. Therefore, the stimulation usually has to be repeated ~100 times to obtain an adequate signal-to-noise ratio from the EEG. It is estimated that after 100 repetitions, noise is reduced tenfold (38). The settings of low and high band-pass filter also have effects on the rat flash VEP (42). According to the literature,

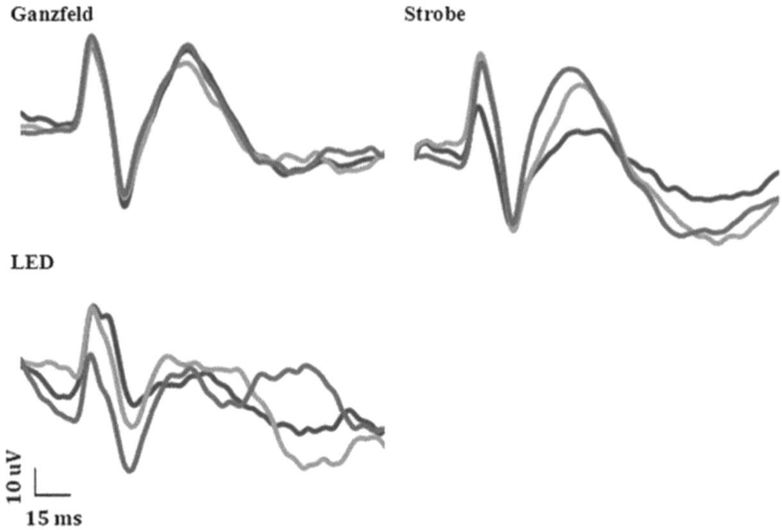

Fig. 4. Overlays of the VEP traces recorded from an individual rat using different flash stimulators.

Table 2
Intraclass correlation coefficient of the rat VEP using different visual stimuli ($n=10$) (29)

		Ganzfeld	Strobe	LED
Intrasession	Latency	0.85	0.72	0.59
	Amplitude	0.96	0.83*	0.70*
Intersession	Latency	0.70	0.62	0.42*
	Amplitude	0.81	0.70	0.44*

*$P<0.05$ compared to Ganzfeld

the low band-pass setting is usually 0.1–1 Hz and the high band-pass is 100–200 Hz. We routinely sample VEP waveforms using filter settings of 1–100 Hz, with which the signal should be sampled at least about 250–300 Hz to ensure that more than two samples are collected by the computer during each cycle of the highest frequency (f_{MAX}) (Nyquist rate $=2\times f_{MAX}=200$ Hz; $2.3\times f_{MAX}=230$ Hz is for practical use) (38). We usually set the sampling rate at 5 kHz, as we also use this system to record the ERG, the f_{MAX} of which is 1,000–2,000 Hz.

4.6. Examples of VEPs in Eye Disease Models

The VEP is used extensively in pharmacological and toxicological experiments, as it provides a non-invasive means of monitoring neural activity and sensory processing in vivo.

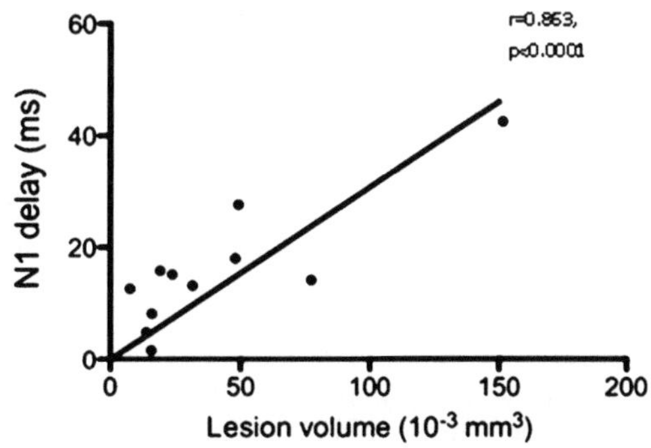

Fig. 5. The VEP is proposed to assess the integrity of the visual pathway in optic neuritis patients by Halliday et al. (2). The amplitude of the VEP is believed to reflect the number of functional optic nerve fibres (46). It was also suggested that the effect of conduction change can be qualitatively measured by the latency delay of the VEP (2, 47, 48). There was an assumption that initial latency prolongation of the VEP corresponds to the size of the demyelinated area of the optic nerve in optic neuritis, while subsequent shortening of latency was thought to represent the process of optic nerve remyelination. The VEP is also used to investigate visual pathway function in some laboratory models of multiple sclerosis (49, 50). In these studies, a correlation between VEPs and histopathologic results was observed. However, it is hard to reproduce and to quantify the demyelination in the visual pathway in a multiple sclerosis model. We recently develop a rat model of focal demyelination in the optic nerve. The figure shows linear regression analysis of the latency delay of N1 peak and the lesion volume in the optic nerve ($r = 0.863$, $p < 0.0001$) (You et al. (28)).

The VEP is also employed in many eye disease models, especially optic nerve disease models to assess the visual function. In a retinal ischemia-reperfusion model of rat, the reduction in VEPs was correlated with the duration of ischemia (12). In another paper, the VEP was used to assess the neuroprotective effect of transcorneal electrical stimulation on optic nerve injury (11). The VEP was also useful in developing a rodent model of anterior ischemic optic neuropathy (19). Recently, we used the VEP to assess the amount of demyelination in a rat model of focal demyelination in the optic nerve and observed a strong correlation between the latency delay and the lesion size (Fig. 5) (28).

ISCEV has published several standards for VEP and ERG recording in humans. A dog electroretinography recording guideline was also proposed (43). As mentioned above, in the literature rodent VEP waveforms were relatively variable. A VEP recording standard is necessary before rodent VEPs can be compared between different labs. A multifocal VEP (44) has not been reported in animals, which may also be useful in assessing the visual function of specific regions of the visual field. The blue-on-yellow VEP may be useful in investigating the koniocellular pathway in primates (45).

References

1. Creutzfeldt O, Maekawa K, Hosli L (1969) Forms of spontaneous and evoked postsynaptic potentials of cortical nerve cells. Prog Brain Res 31:265–273
2. Halliday A, McDonald W, Mushin J (1972) Delayed visual evoked response in optic neuritis. Lancet 1:982–985
3. Towle V, Moskowitz A, Sokol S, Schwartz B (1983) The visual evoked potential in glaucoma and ocular hypertension: effects of check size, field size, and stimulation rate. Invest Ophthalmol Vis Sci 24:175–183
4. Ridder W (2006) Visual evoked potentials in animals. In: Heckenlively JR, Arden GB (eds) Principles and practice of clinical electrophysiology of vision. The MIT Press, Cambridge, pp 935–947
5. Ridder W, Nusinowitz S (2006) The visual evoked potential in the mouse—origins and response characteristics. Vision Res 46:902–913
6. Creel D, Dustman R, Beck E (1973) Visually evoked responses in the rat, guinea pig, cat, monkey, and man. Exp Neurol 40:351–366
7. Creel D, Dustman RE, Beck EC (1974) Intensity of flash illumination and the visually evoked potential of rats, guinea pigs and cats. Vision Res 14:725–729
8. Heiduschka P, Schraermeyer U (2008) Comparison of visual function in pigmented and albino rats by electroretinography and visual evoked potentials. Graefes Arch Clin Exp Ophthalmol 246:1559–1573
9. Iwamura Y, Fujii Y, Kamei C (2003) The effects of certain H1-antagonists on visual evoked potential in rats. Brain Res Bull 61:393–398
10. Iwamura Y, Fujii Y, Kamei C (2004) The effects of selective serotonin-reuptake inhibitor on visual evoked potential in rats. J Pharmacol Sci 94:271–276
11. Miyake K-I, Yoshida M, Inoue Y, Hata Y (2007) Neuroprotective effect of transcorneal electrical stimulation on the acute phase of optic nerve injury. Invest Ophthalmol Vis Sci 48:2356–2361
12. Jehle T, Wingert K, Dimitriu C, Meschede W, Lasseck J, Bach M, Lagreze WA (2008) Quantification of ischemic damage in the rat retina: a comparative study using evoked potentials, electroretinography, and histology. Invest Ophthalmol Vis Sci 49:1056–1064
13. Papathanasiou ES, Peachey NS, Goto Y, Neafsey EJ, Castro AJ, Kartje GL (2006) Visual cortical plasticity following unilateral sensorimotor cortical lesions in the neonatal rat. Exp Neurol 199:122–129
14. Tomita H, Sugano E, Yawo H, Ishizuka T, Isago H, Narikawa S, Kugler S, Tamai M (2007) Restoration of visual response in aged dystrophic RCS rats using AAV-mediated channelopsin-2 gene transfer. Invest Ophthalmol Vis Sci 48:3821–3826
15. Onofrj M, Harnois C, Bodis-Wollner I (1985) The hemispheric distribution of the transient rat VEP: a comparison of flash and pattern stimulation. Exp Brain Res 59:427–433
16. Meeren H, Van Luijtelaar E, Coenen A (1998) Cortical and thalamic visual evoked potentials during sleep-wake states and spike-wave discharges in the rat. Electroencephalogr Clin Neurophysiol 108:306–319
17. Goto Y, Furuta A, Tobimatsu S (2001) Magnesium deficiency differentially affects the retina and visual cortex of intact rats. J Nutr 131:2378–2381
18. Yargicoglu P, Yaras N, Agar A, Gumuslu S, Bilmen S, Ozkaya G (2003) The effect of vitamin E on stress-induced changes in visual evoked potentials (VEPs) in rats exposed to different experimental stress models. Acta Ophthalmol Scand 81:181–187
19. Bernstein S, Guo Y, Kelman S, Flower R, Johnson M (2003) Functional and cellular responses in a noval rodent model of anterior ischemic optic neuropathy. Invest Ophthalmol Vis Sci 44:4153–4162
20. Herr D, Kim T, King D, Boyes W (1996) Possible confounding effects of strobe "clicks" on flash evoked potentials in rats. Physiol Behav 59:325–340
21. Green D, Herreros de Tejada P, Glover M (1994) Electrophysiological estimates of visual sensitivity in albino and pigmented mice. Vis Neurosci 11:919–925
22. Lopez L, Brusa A, Fadda A, Loizzo S, Martinangeli A, Sannita WG, Loizzo A (2002) Modulation of flash stimulation intensity and frequency: effects on visual evoked potentials and oscillatory potentials recorded in awake, freely moving mice. Brain Res 131:105–114
23. Mazzucchelli A, Conte S, D'Olimpio F, Ferlazzo F, Loizzo A, Palazzesi S, Renzi P (1995) Ultradian rhythms in the N1-P2 amplitude of the visual evoked potential in two inbred strains of mice: DBA/2J and C57BL/6. Behav Brain Res 67:81–84
24. Peachey NS, Roveri L, Messing A, McCall MA (1997) Functional consequences of oncogene-induced horizontal cell degeneration in the retinas of transgenic mice. Vis Neurosci 14:627–632
25. Porciatti V, Pizzourusso T, Maffei L (1999) The visual physiology of the wild type mouse determined with pattern VEPs. Vision Res 39:3071–3781

26. Peachey NS, Ball SL (2003) Electrophysiological analysis of visual function in mutant mice. Doc Ophthalmol 107:13–36
27. Henry K, Rhoades R (1978) Relation of albinism and drugs to the visual evoked potential of the mouse. J Comp Physiol Psychol 92: 271–279
28. You Y, Klistorner A, Thie J, Graham S (2011) Latency delay of visual evoked potential is a real measurement of demyelination in a rat model of optic neuritis. Invest Ophthalmol Vis Sci 52:6911–6918
29. You Y, Klistorner A, Thie J, Graham S (2011) Improving reproducibility of VEP recording in rats: electrodes, stimulus source and peak analysis. Doc Ophthalmol 123:109–119
30. Castro-Junior J, Resende L, Bertotti M, Fonseca R, Zanchetta S, Schelp A (2005) Comparision of the effects of barbiturate, benzodiazepine and ketamine on visual evoked potentials in rabbits. Electromyogr Clin Neurophysiol 45:259–262
31. Anis N, Berry S, Burton N, Lodge D (1983) The dissociative anaesthetics, ketamine and phencyclidine, selectively reduce excitation of central mammalian neurones by N-methyl-aspartate. Br J Pharmacol 79:565–575
32. Virtanen R, Savola J, Saano V, Nyman L (1988) Characterization of the selectivity, specificity and potency of medetomidine as an alpha 2-adrenoceptor agonist. Eur J Pharmacol 150:9–14
33. Ridder W, Nusinowitz S, Heckenlively J (2002) Cause of cataract development in anesthetized mice. Exp Eye Res 75:365–370
34. Weymouth AE, Vingrys AJ (2008) Rodent electroretinography: methods for extraction and interpretation of rod and cone responses. Prog Retin Eye Res 27:1–44
35. Smith KJ, Waxman SG (2005) The conduction properties of demyelinated and remyelinated axons. In: Waxman SG (ed) Multiple sclerosis as a neuronal disease. Elsevier Academic Press, Amsterdam, pp 85–100
36. Boyes W, Padilla S, Dyer R (1985) Body temperature-dependent and independent actions of chlordimeform on visual evoked potentials and axonal transport in optic system of rat. Neuropharmacology 24:743–749
37. Hetzler B, Boyes W, Creason J, Dyer R (1988) Temperature-dependent changes in visual evoked potentials of rats. Electroencephalogr Clin Neurophysiol 70:137–154
38. Fahle M, Bach M (2006) Origin of the visual evoked potentials. In: Heckenlively J, Arden G (eds) Principles and practice of clinical electrophysiology of vision. MIT Press, Cambridge, pp 207–234
39. Leamey CA, Protti DA, Dreher B (2008) Comparative survey of the mammalian visual system with reference to the mouse. In: Chalupa LM, Williams RW (eds) Eye, retina, and visual system of the mouse. MIT Press, Cambridge, pp 35–60
40. Odom JV, Bach M, Brigell M, Holder GE, McCulloch DL, Tormene AP, Vaegan (2010) ISCEV standard for clinical visual evoked potentials (2009 update). Doc Ophthalmol 120:111–119
41. McCall M, Robinson S, Dreher B (1987) Differential retinal growth appears to be the primary factor producing the ganglion cell density gradient in the rat. Neurosci Lett 79:78–84
42. Shaw N (1992) The effects of low-pass filtering on the flash visual evoked potential of the albino rat. J Neurosci Methods 44:233–240
43. Narfstrom K, Ekesten B, Rosolen S, Spiess B, Percicot C, Ofri R (2002) Guidelines for clinical electroretinography in the dog. Doc Ophthalmol 105:83–92
44. Klistorner A, Crewther D, Crewther S (1997) Separate magnocellular and parvocellular contributions from temporal analysis of the multifocal VEP. Vision Res 37:2161–2169
45. Klistorner A, Graham S, Martins A, Grigg J, Arvind H, Kumar R, James A, Billson F (2007) Multifocal blue-on-yellow visual evoked potentials in early glaucoma. Ophthalmology 114:1613–1621
46. Klistorner A, Arvind H, Nguyen T, Garrick R, Paine M, Graham S, O'Day J, Grigg J, Billson F, Yiannikas C (2008) Axonal loss and myelin in early ON loss in postacute optic neuritis. Ann Neurol 64:325–331
47. Klistorner A, Graham S, Fraser C, Garrick R, Nguyen T, Paine M, O'Day J, Arvind H, Billson F (2007) Electrophysiological evidence for heterogeneity of lesions in optic neuritis. Invest Ophthalmol Vis Sci 48:4549–4556
48. Youl B, Turano G, Miller D, Towell A, MacManus D, Moore S, Jones S, Barrett G, Kendall B, Moseley I (1991) The pathophysiology of acute optic neuritis: an association of gadolinium leakage with clinical and electrophysiological deficits. Brain 114:2437–2450
49. Meyer R, Weissert R, Diem R, Storch MK, de Graaf KL, Kramer B, Bahr M (2001) Acute neuronal apoptosis in a rat model of multiple sclerosis. J Neurosci 21:6214–6220
50. Diem R, Demmer I, Boretius S, Merkler D, Schmelting B, Williams S, Sattler M, Bahr M, Michaelis T, Frahm J, Bruck W, Fuchs E (2008) Autoimmune optic neuritis in the common marmoset monkey: comparison of visual evoked potentials with MRI and histopathology. Invest Ophthalmol Vis Sci 49:3707–3714

Chapter 17

The Visual Evoked Potential in Humans

Stuart L. Graham and Alexander Klistorner

Abstract

The visual evoked potential (VEP) is recorded in humans in response to flash or pattern stimuli with appropriately placed electrodes over the occipital scalp. In recent years the multifocal VEP (mfVEP) has been developed as a means of obtaining focal responses rather than just a summed response from the central field. The mfVEP has been used as a form of objective perimetry to detect visual field loss in glaucoma. This chapter covers conventional VEP recording and some principles of the multifocal technique.

Key words: Visual evoked potential, Electrophysiology, Visual cortex, Optic nerve, Optic neuritis, Glaucoma

1. Introduction

The visual evoked potential (VEP) or visual evoked response is a signal generated in the visual cortex in response to either flash or pattern stimuli presented to one or both eyes. The recording electrodes are placed on the scalp, with the main signal detected over the occipital region. Responses are recorded using single or multiple channels. The ISCEV International Standards (1) define the preferred recording conditions for standard single channel flash and pattern VEP recording. For investigating the visual pathway in disease states, 3-channel recording gives more information about pre-chiasmal versus post-chiasmal pathology (see below). There are also several different types of VEP stimulation that can be used, mainly for research purposes (see Table 1). For more detailed topographic mapping of VEPs, the multifocal multichannel VEP (mfVEP) has recently been developed as a means for detecting focal defects in diseases such as glaucoma or tracking focal demyelination and recovery in optic neuritis.

Paul M. Pilowsky et al. (eds.), *Stimulation and Inhibition of Neurons*, Neuromethods, vol. 78,
DOI 10.1007/978-1-62703-233-9_17, © Springer Science+Business Media, LLC 2013

Table 1
Types of visual evoked potentials (VEP)

Flash VEP
Pattern reversal VEP (conventional)
Steady state VEP
Pattern-onset VEP (includes sparse presentation)
Sweep VEP
Chromatic VEP
Hemifield VEP
Binocular or dichoptic VEP (includes VR goggles)
Multifrequency VEP
Multifocal VEP

For conventional recording, a pattern stimulus provides a less variable response between subjects than a flash stimulus. The pattern stimulus should cover at least 15° of central visual field and be presented on a screen with uniform brightness properties (some screens can vary by as much as 30% in different parts). Typically a pattern reversal stimulus is used, with equal numbers of black and white checks such that the mean luminance of the screen is not changed with each reversal. The reversal rate is usually 1–3 reversals/s producing a transient VEP. Fast rates of reversal can produce a steady state response with a waveform resembling a sine wave. A pattern-onset–offset stimulus can also be used on a background of equal luminance, with 1–2 onsets/s. ISCEV standards suggest the check size be defined in visual angle and recommend at least two sizes such as 15- and 60-min angle in any recording protocol. Michelson contrast should be >75%. Recording should be monocular with the other eye occluded, and the subject optimally refracted for the working distance—typically around 60 cm. Electrode positions for single channel should be active Oz, reference Fz according to the International 10–20 system. The ground electrode can be forehead, vertex, mastoid or ear lobe. If 3-channel recording is done, there are two additional lateral electrodes placed either side of the midline Oz electrode at O1 and O2.

A typical waveform derived from a slowly repetitive stimulus and recorded with a monopolar electrode contains a primary response at around 70 ms (N70) and a secondary response around 100 ms (P100), with a third peak around 135 ms, followed by rhythmic after waves. The components reflect the activity of the V1 striate cells and possibly extrastriate areas (2–4). Amplitude is

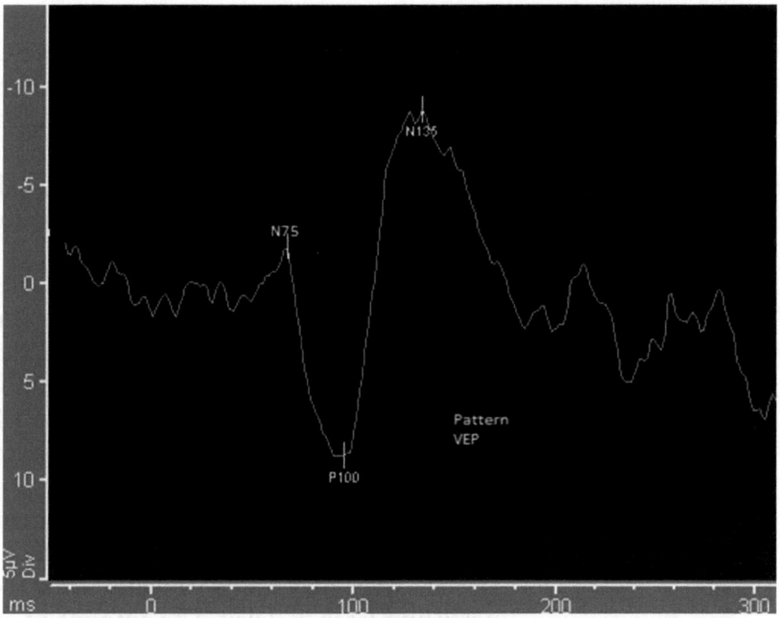

Fig. 1. Pattern VEP trace. Shows peaks at around 75, 100 and 135 ms, which are termed N75, P100 and N135. Traces may be shown inverted by different conventions; in this case, P100 is a negative wave.

measured from the major positive to the major negative deflection, and latency as the time from stimulus onset to the positive peak (usually P100). Figure 1 shows a typical VEP response.

For 3-channel recording, the responses from the two lateral electrodes provide additional information about the site of the pathology. However, interpretation of the findings is complicated and very much dependant on the position of the electrodes and the size of the stimulated area. Paradoxical lateralisation (see below) leads to an ipsilateral signal that has to be interpreted with caution. A certain degree of anatomical asymmetry can occur in normals. As a general rule, if the disease is localised to the eye or pre-chiasmal optic nerve, the responses from both lateral electrodes will be similarly reduced when testing the eye monocularly. In the case of albinism where a higher proportion of nerve fibres decussate in the chiasm than normal, there will be a 'crossed asymmetry' where the signal from the lateral electrode over the opposite hemisphere to the eye being tested is larger, and this holds when testing either eye. A different type of 'crossed asymmetry' can also occur when there is a bitemporal hemianopia (5), for example, from a pituitary lesion, where the pattern of asymmetry reverses with stimulation of each eye. If however there is post-chiasmal disease, then there may be an 'uncrossed asymmetry' where the pattern is similar for stimulation of each eye. It is important to note, however, that the signal generators in the lateral electrodes are not solely from the cortex immediately underlying each electrode. Half field stimulation studies (6) have shown a paradoxical lateralisation, such that there is both an ipsilateral complex and a contralateral activity constituting the responses obtained at each of the lateral electrodes. Thus, a significant component of the signal actually arises from generators in the opposite hemisphere, probably due to their orientation in the striate cortex.

In disease states, there may be variable effects on the amplitude or latency of the response. For example, retinal pathology such as macular lesions, macular degeneration, diabetic retinopathy or retinal dystrophies will reduce signal generation in the eye and therefore reduce the resulting amplitude of the VEP. Optic nerve pathology (e.g. compressive lesions, glaucoma, ischaemic, toxic or traumatic optic neuropathy) can all reduce the amplitude of the response with variable effects on latency. Inflammatory demyelination (optic neuritis) will in the acute stages reduce amplitude and prolong latency, with later stages showing some recovery of amplitude with persisting latency delay. The amount of axonal loss will determine the final amplitude. We have shown that in a rat model of demyelination, the amount of delay is in proportion to the size of the area of demyelination, and the amplitude reduction is in proportion to the axonal loss (7). Finally, the VEP can also be affected by cortical lesions such as strokes or tumours and can be reduced acutely by drugs suppressing cortical function such as sedatives and anaesthetic agents.

2. Multifocal Recording Techniques

Since the occipital area of the cortex predominantly subserves macular function, the conventional VEP signal is mainly derived from the macular region (8). To better evaluate local responses from the retina and visual cortex and to map them topographically, Sutter and Tran (9) introduced a technique which enables the presentation of a multifocal stimulus with calculation of the corresponding local responses. This has been adapted to record multifocal multichannel VEPs (mfVEPs) which can be used to objectively detect visual field defects (10–13).

Multifocal stimulation is now available commercially in several different electrophysiological systems. These include the VERIS Scientific system (Electro-Diagnostic Imaging, Redwood City, CA), the RETIscan system (Roland Instruments, Wiesbaden, Germany), the MetroVision (MetroVision, Perenchies, France), the SHIL Multifocal Imager (Scottish Health Innovations, Ltd.,

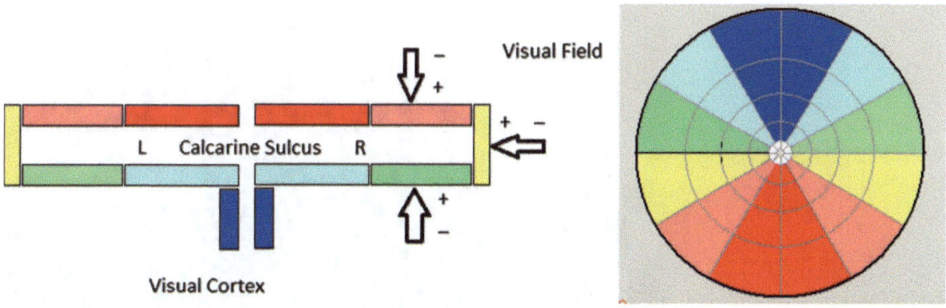

Fig. 2. Schematic posterior view of V1 visual cortex with arrows showing dipoles of VEP signals. Areas of the visual field (*right*) are represented by correspondingly *coloured* regions of cortex around the calcarine fissure. As a result, the direction of dipole varies with different field locations, generating different signal results at the surface recording electrode. Multiple recording channels are therefore needed.

Glasgow, United Kingdom) and the AccuMap (ObjectiVision, Sydney, Australia). The visual stimulus is usually generated on a CRT screen (e.g. 22-in. high-resolution display), but with faster refreshing rates, flat screens (e.g. LCD and plasma) can now be used. We have also used virtual reality goggles and twin LCD screens to present the stimulus dichoptically (14) (see below).

Due to the variable orientation of underlying dipoles in the striate cortex (see Fig. 2), we reported that it is necessary to have an array of bipolar electrodes arranged over the occipital cortex, with at least two orthogonal channels, in order to maximise detection of all signals (15). We usually use 4 channels, with a vertical, horizontal and two oblique channels. An electrode cross with headband serves to hold the electrodes in position. An ear electrode serves as ground.

The visual stimuli typically consist of preset numbers of hexagons in arrays for multifocal ERG or segments in a dartboard pattern for multifocal VEP. Flash or pattern stimuli can be presented within each hexagon or dartboard segment. The stimulus areas can be retinally or cortically scaled. This means that the area of each hexagon or dartboard segment increases with eccentricity, in proportion to cone density for retinal stimulation, or to a cortical magnification factor (M-scale) for VEP recording (Fig. 3).

For the pattern stimulus, the multifocal technique utilises a pseudorandom binary exchange of two opposite checkerboard pattern conditions at each of the test sites of the visual field. According to this sequence, there is a 50% probability for the checkerboard pattern to reverse its polarity with every frame of the stimulating display (e.g. every 15 ms). Each input (stimulation site) is modulated in time according to the same pseudorandom binary m-sequence, but shifted in time (by an integer number of stimulus frames).

Fig. 3. Cortically scaled pattern stimulus for the multifocal VEP. Central number is randomly changing fixation target.

This allows computation of cross-correlation kernels with very low-level 'contamination' by adjacent stimulus sectors, as well as characterisation of nonlinear interactions between sequential visual events (16, 17). The first-order 'kernel' is similar to a conventional impulse response. Second-order kernels, however, show the difference between the anticipated summation response in a linear system when consecutive flashes (or polarity reversals) are presented, and the response that is actually recorded. It is a measure of the temporal nonlinearity of the visual system. The 'slices' of the second-order kernel represent nonlinearities observed when the consecutive stimulus event occurs in subsequent frames (e.g. at 15, 30, 45 ms). This technique of nonlinear analysis is important since the human visual system exhibits temporal nonlinearities. Thus, this technique enables characterisation of these effects. Alternatively, different pseudorandom sequences can be used to drive the temporal sequence at each individual stimulus element using families of different binary or ternary sequences and subsequent cross-correlations performed to extract the local responses. The latter technique does not allow for examination of higher-order kernels.

A multifocal flash electroretinogram (mfERG) is very useful in delineating areas of outer retinal damage in many diseases (for review, see Hood (18, 19)). When recording the mfVEP, however, responses to pattern stimuli are superior to flash, being more uniform throughout the field. Figure 4 shows a pattern mfVEP recorded to a cortically scaled stimulus pattern of checks. With appropriate scaling to reduce intersubject variability (20), probability maps can be constructed using a normative database to graphically

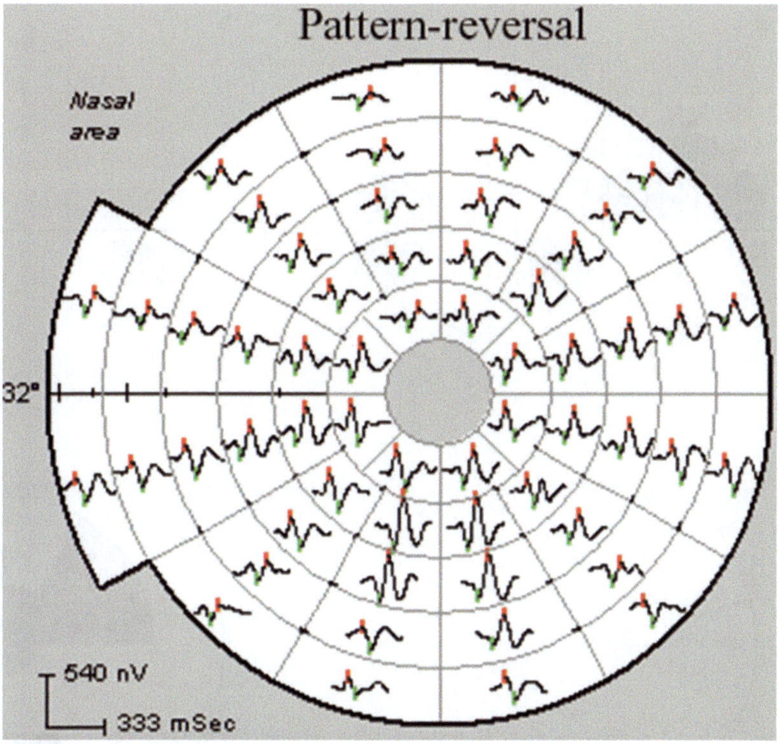

Fig. 4. Example of normal mfVEP recording from right eye.

represent areas of reduced signal and demonstrate clinical scotomas. Comparison of signals between eyes is useful to delineate early defects (12). Figure 5 shows an example of a glaucoma patient with an arcuate scotoma identified on visual field testing and on mfVEP.

3. Conventional VEP and mfVEP Recording in Disease States

While several anatomical and perimetric studies have demonstrated a diffuse component of damage in glaucoma, the most characteristic visual field defects are spatially localised within the midperiphery. This gives the mfVEP a clinical advantage for objectively detecting visual field loss. The conventional flash or pattern VEP is generally agreed to be abnormal in glaucoma (21–23), but many cases are not successfully detected by VEP and its sensitivity is usually reported as only moderate. Due to the large interindividual variability in VEP amplitude, most studies have concentrated on VEP latency. In an early study with an age-corrected cohort of

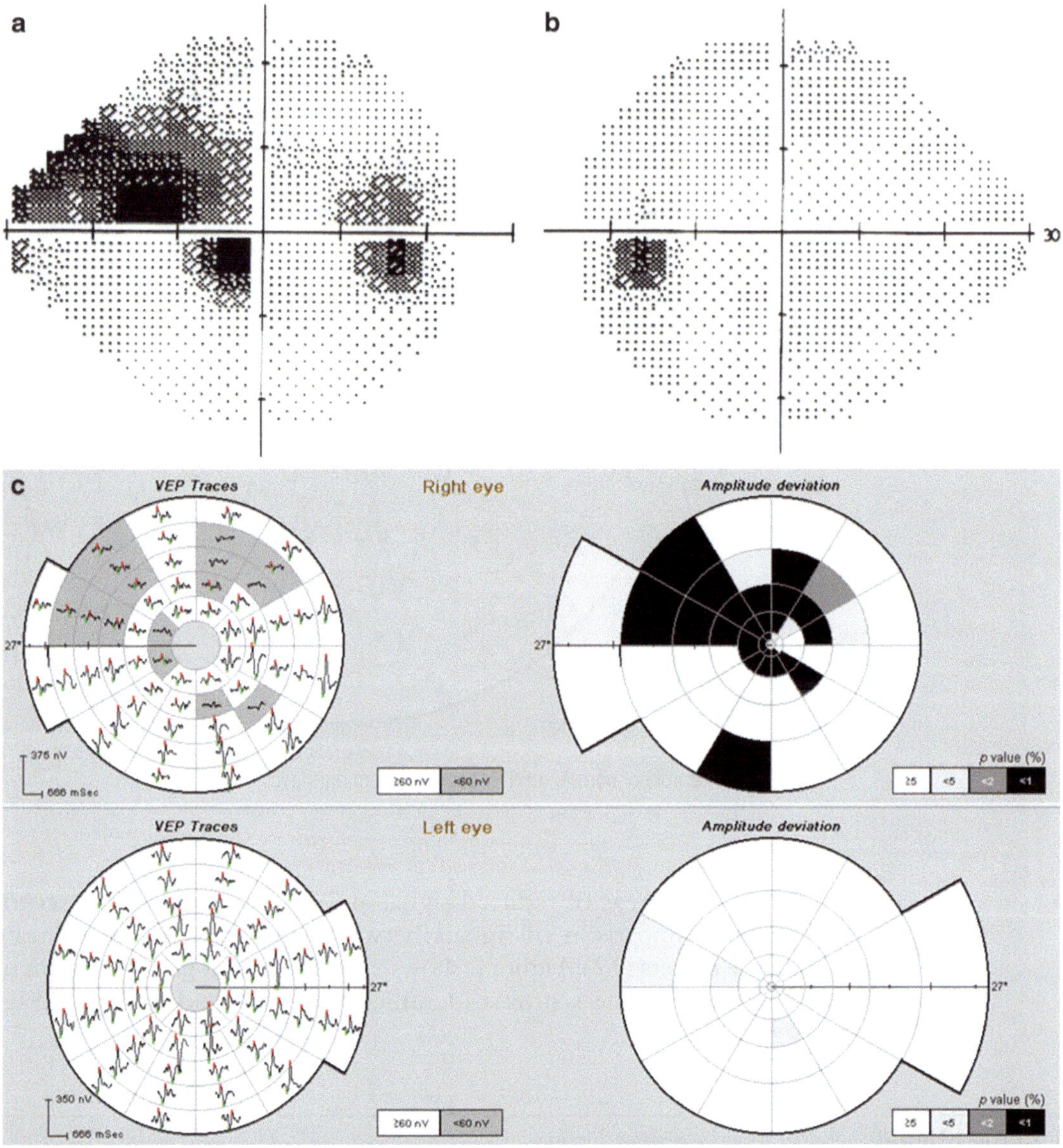

Fig. 5. Example of glaucoma patient with right superior arcuate scotoma and inferior paracentral defect and normal left eye. (**a**) and (**b**) show right and left eye Humphrey perimeter visual fields. (**c**) shows mfVEP with upper row right eye with good correspondence, lower row normal left eye responses.

patients with OAG and OHT, the central field pattern VEP showed about 50 and 25% of patients, respectively, to have a delay in latency compared to normals (24). However, Parisi et al. have reported 100% sensitivity using VEP latency for glaucoma detection (25). In monkeys with monocular laser-induced glaucoma, there was a reduction in amplitude, but no phase shift seen in the pattern VEP (26).

The likely reason for variable results in glaucoma is that the pattern VEP is predominantly generated in the visual cortex by areas receiving projections from the central retina (27–29), and the signal is dominated by the lower visual field (13). A small, unified check size, which is commonly used for stimulation, is another factor that tends to bias the central response (30). This limits application of the standard full-field VEP in detection of peripheral field changes in glaucoma. In fact, it is possible for a subject with a dense superior scotoma to have a normal VEP response (31). The same subject recorded with a mfVEP, however, will show a clear reduction of signals in the corresponding area of field loss. We therefore recommend mfVEPs be used for investigation of glaucoma in all cases, unless the subject cannot fixate. In subjects with reduced acuity from other causes such as cataract, the central and paracentral responses will be increasingly reduced once the acuity drops below 6/12, but the peripheral responses will remain unaffected.

Another advantage of mfVEPs arises in optic neuritis, where it is possible for a subject to have lost the response from a conventional VEP due to a central scotoma. In some of these cases, the mfVEP can demonstrate preservation of function of the more peripheral field and provide evidence of latency delay (see Fig. 6). In serial monitoring of optic neuritis, we first see recovery of amplitude, then amplitude asymmetry, then latency and finally latency asymmetry, but this may all occur to a varying extent and may reflect the underlying prognosis for the patient with regard to multiple sclerosis (32). Figure 7 shows an example of mfVEP recovery over a 12-month period. The mfVEP also limits the cancellation effects of variously oriented underlying dipoles which may lead to waveform distortion and consequently abnormal latency detection, which remains a problem with conventional full-field VEPs.

4. New Stimuli for mfVEP

The possibility of simultaneous binocular (dichoptic) mfVEPs was recently realised using virtual reality goggles (14) or with twin LCD screens split by 45° mirrors (33). The advantage of a simultaneous binocular technique is that inter-eye comparisons are more valuable due to the identical recording and noise conditions for each eye. Limitations will remain, however, with patients who have underlying strabismus or another disparity in the fixation angle between their eyes. The use of a spatially sparse stimulation (34) pattern has the potential to provide better signal-to-noise ratios, at least in the central field, and possibly greater sensitivity to early glaucoma.

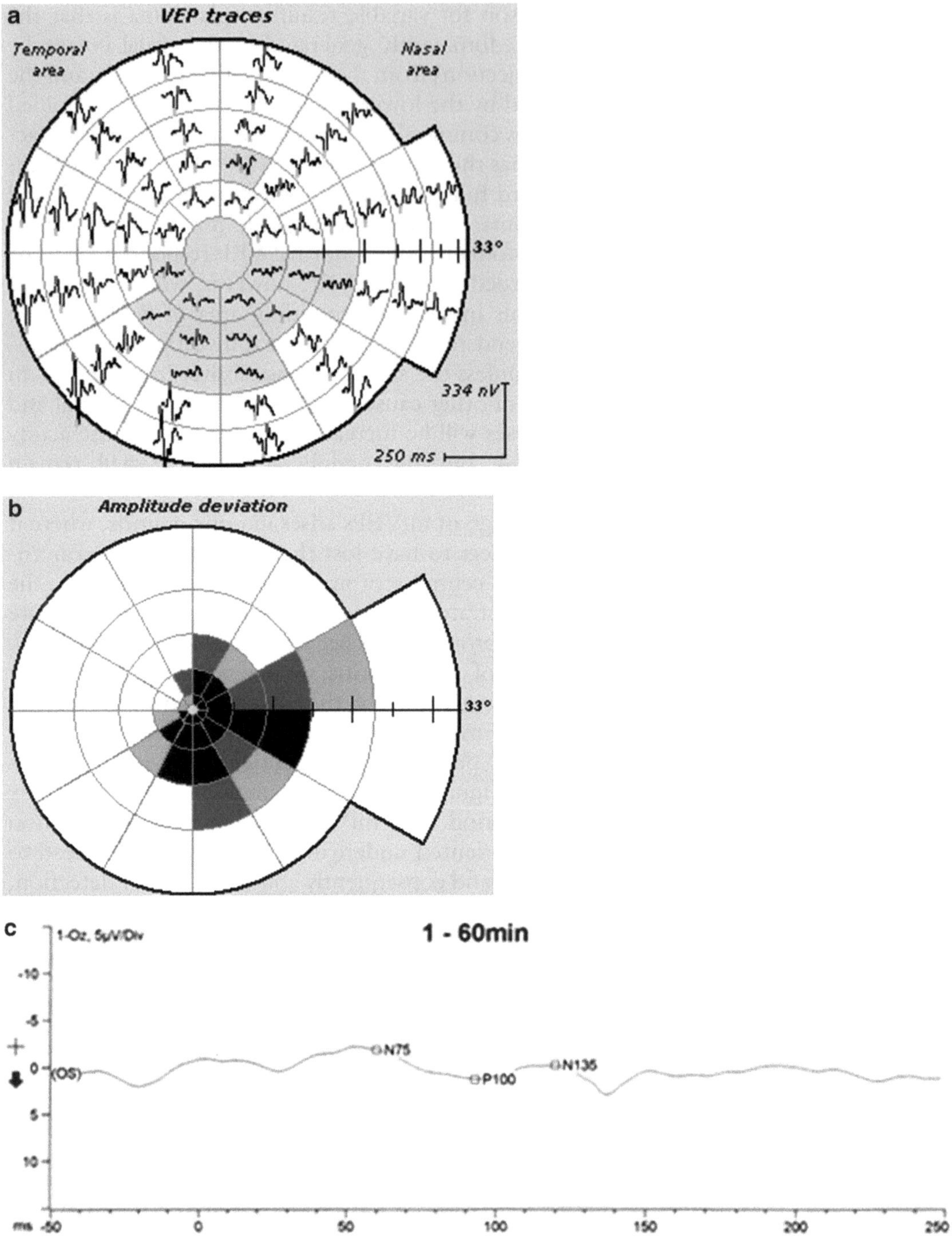

Fig. 6. Optic neuritis case with central scotoma. The mfVEP shows loss of central signals but preservation of more peripheral responses which can be analysed for latency delay. Conventional VEP is flat.

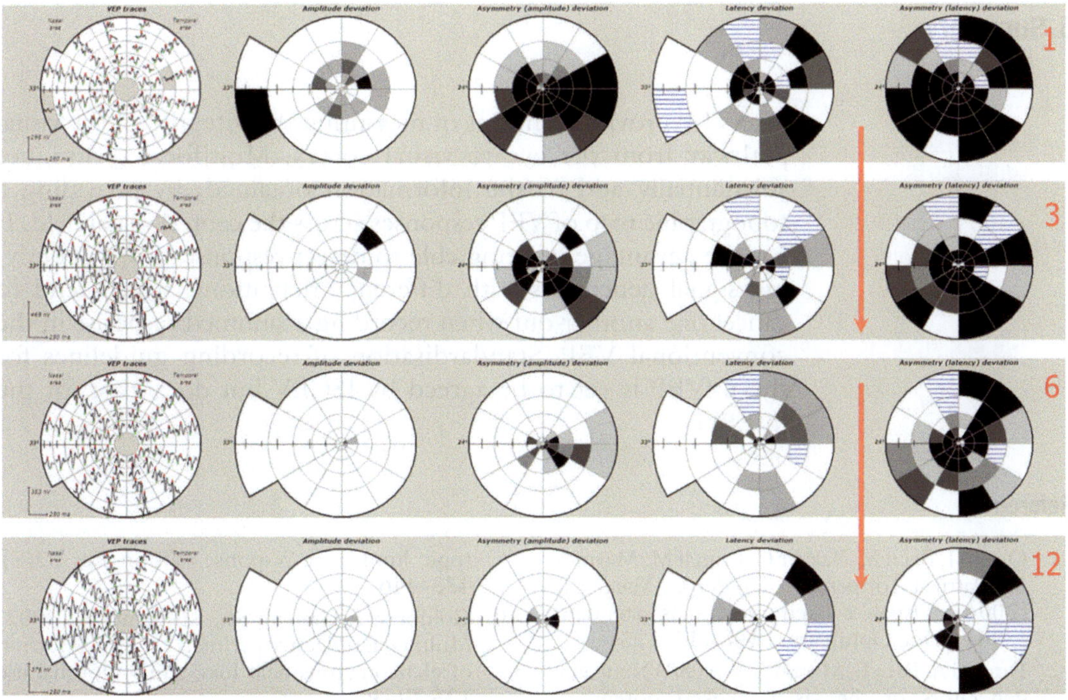

Fig. 7. mfVEPs recorded at 3-month follow-up interval in an acute optic neuritis patient tracking recovery. *Rows* represent 1, 3, 6 and 12 months postattack. First column mfVEP traces, second column amplitude deviation plots, third column amplitude asymmetry, fourth column latency deviation plot and fifth is latency asymmetry. Shows initial recovery of amplitudes at around 3 months, with slower recover of latency and then latency asymmetry by 12 months.

We recently reported on a high-resolution stimulus recording with 120 test zones of 3 × 3 checks arranged in eight rings (vs. 4 × 4 checks in five rings for conventional mfVEP) (35). The recordings showed slightly reduced variability across the field and improved scotoma definition (but not initial detection). There was no loss of sensitivity despite the lower signal-to-noise ratio with small testing zones. The testing time was longer, however, so this technique may be reserved for eyes with small focal or paracentral defects due to its much better topographic definition.

A blue-on-yellow mfVEP has also been described using a sparse blue pattern-onset stimulus (instead of pattern reversal) on a yellow-adapting background. The goal was to target the koniocellular pathway. This approach showed good sensitivity and displayed more extensive scotomas than the conventional black-white mfVEP (92.2% sensitivity), and it correlated well with SAP (36, 37). The advantage of the blue-on-yellow stimulus over standard mfVEP, however, was probably more likely related to the fact that the pattern-onset stimulus was spatially sparse (38) (likely due to less lateral inhibition) and was of low luminance, both of which we have demonstrated may increase sensitivity (39).

5. Summary

The VEP provides a means of measuring the integrity of the visual pathway from the eye to visual cortex. Multifocal techniques substantially add to the information obtained, by providing a topographic map of VEP responses across the central visual field. It has the advantage of being able to detect responses from underlying signal generators with different orientations, which tend to cancel one another out when recording a summed response in the conventional VEP. Standardisation of recording guidelines for the mfVEP is yet to be agreed by ISCEV but does exist for the mfERG.

References

1. Odom J, Bach M, Barber C, Brigell M, Marmor M, Patrizia Tormene A, Holder G, Vaegan A (2004) Visual evoked potentials standard (2004). Doc Ophthalmol 108:115–123
2. Ducatti A, Fava E, Motti EDF (1988) Neuronal generators of the visual evoked potentials: intracerebral recordings in awake humans. Electroencephalogr Clin Neurophysiol 71: 89–99
3. Kraut MA, Arezzo JC, Vaughan HG (1985) Intracortical generators of the flash VEP in monkey. Electroencephalogr Clin Neurophysiol 62:300–312
4. Maier J, Dagrielle G, Spekreije H, van Duk BW (1987) Principal component analysis for source localisation of VEPs in man. Vis Res 27: 165–177
5. Blumhardt LD, Barett G, Halliday AM (1977) The asymmetrical visual evoked potential to pattern reversal in one half field and its significance for the analysis of visual field defects. Br J Ophthalmol 61:454–461
6. Barrett G, Blumhardt LD, Halliday AM, Halliday E, Kriss A (1976) A paradox in the lateralisation of the visual evoked response. Nature 261:253–255
7. You Y, Klistorner A, Thie J, Graham SL (2011) Latency delay of visual evoked potential is a real measurement of demyelination in a rat model of optic neuritis. Invest Ophthalmol Vis Sci 52:6911–6918
8. Riggs LA, Wooten BR (1972) Electrical measures and psychophysical data on human vision. In: Jameson D, Hurvich LM (eds) Handbook of sensory physiology. Springer-Verlag, New York, pp 690–731
9. Sutter EE, Tran D (1992) The field topography of ERG components in man—1. The photopic luminance response. Vision Res 32(3): 433–446
10. Goldberg I, Graham SL, Klistorner A (2002) Multifocal objective perimetry in the detection of glaucomatous field loss. Am J Ophthalmol 133:29–39
11. Graham SL, Klistorner A, Goldberg I (2005) Clinical application of the multifocal VEP in glaucoma. Arch Ophthalmol 123:729–773
12. Graham SL, Klistorner AI, Grigg JR, Billson FA (2000) Objective VEP perimetry in glaucoma: asymmetry analysis to identify early deficits. J Glaucoma 9:10–19
13. Klistorner AI, Graham SL, Grigg JR, Billson FA (1998) Multifocal topographic visual evoked potential: improving objective detection of local visual field defects. Invest Ophthalmol Vis Sci 39:937–950
14. Arvind H, Klistorner A, Graham S, Grigg J, Goldberg I, Billson FA (2007) Dichoptic stimulation improves detection of glaucoma with multifocal visual evoked potentials. Invest Ophthalmol Vis Sci 48:4590–4596
15. Klistorner A, Graham SL (2000) Objective perimetry in glaucoma. Ophthalmology 107:2283–2299
16. Odom JV (1991) Kernel analysis. In: Heckenlively JR, Arden GB (eds) Principles and practice of clinical electrophysiology of vision. Mosby-Year book, Ink., St. Louis, pp 254–259
17. Sutter E (1992) A deterministic approach to non-linear systems analysis. In: Pinter RB, Nabet B (eds) Nonlinear vision: Determination of neural receptive fields, functions and networks. CRC Press, Cleveland, pp 171–220
18. Hood DC, Greenstein VC, Hood DC, Greenstein VC (2003) Multifocal VEP and

ganglion cell damage: applications and limitations for the study of glaucoma. Prog Retin Eye Res 22:201–251
19. Hood DC (2000) Assessing retinal function with the multifocal technique. Prog Retin Eye Res 19:607–646
20. Klistorner A, Graham SL (2001) Electroencephalogram-based scaling of multifocal visual evoked potentials: effect on intersubject amplitude variability. Invest Ophthalmol Vis Sci 42:2145–2152
21. Howe JW, Mitchell KW (1986) Visual evoked potential changes in chronic glaucoma and ocular hypertension. Trans Ophthalmol Soc UK 105:457–462
22. Motolko M, Drance SM, Douglas GR (1982) The early psychophysical disturbances in chronic open-angle glaucoma. A study of visual functions with asymmetric disc cupping. Arch Ophthalmol 100:1632–1634
23. Sokol S (1990) The visually evoked cortical potential in optic nerve and visual pathway disorders. In: Fishman GA (ed) Electrophysiological testing in disorders of the retina, optic nerve and visual pathway. American Academy of Ophthalmology, San Fransisco, pp 105–141
24. Towle VL, Moskowitz A, Sokol S, Schwartz B (1983) The visual evoked potential in glaucoma and ocular hypertension: effects of check size, field size, and stimulation rate. Invest Ophthalmol Vis Sci 24:175–183
25. Parisi V, Miglior S, Manni G (2006) Clinical ability of pattern electroretinograms and visual evoked potentials in detecting visual dysfunction in ocular hypertension and glaucoma. Ophthalmology 113:216–228
26. Johnson MA, Drum BA, Quigley HA, Sanchez RM, Dunkelberger GR (1989) Pattern-evoked potentials and optic nerve fiber loss in monocular laser induced glaucoma. Invest Ophthalmol Vis Sci 30:897–907
27. Horton JC, Hoyt F (1991) The representation of the visual field in the human striate cortex. Arch Ophthalmol 109:816–824
28. Regan D (1972) Evoked potentials in psychology, sensory physiology and clinical medicine. Chapman & Hall, London
29. Weinstein GW, Odom JV, Cavender S (1991) Visually evoked potentials and electroretinography in neurologic evaluation. Review. Neurol Clin 9:225–242
30. Harter MR (1970) Evoked cortical responses to checkerboard patterns: effect of check-size as a function of retinal eccentricity. Vision Res 10:1365–1376
31. Graham SL, Fortune B (2009) Electrophysiology in glaucoma. In: Shaaraway T et al (eds) Glaucoma. Elsevier, London, pp 151–171
32. Fraser C, Klistorner A, Graham S, Garrick R, Billson F, Grigg J, Fraser C, Klistorner A, Graham S, Garrick R, Billson F, Grigg J (2006) Multifocal visual evoked potential latency analysis: predicting progression to multiple sclerosis. Arch Neurol 63:847–850
33. Leaney JC, Klistorner A, Arvind H, Graham SL (2010) Dichoptic suppression of mfVEP amplitude: effect of retinal eccentricity and simulated unilateral visual impairment. Invest Ophthalmol Vis Sci 51(12):6549–6555
34. James AC (2003) The pattern-pulse multifocal visual evoked potential. Invest Ophthalmol Vis Sci 44:879–890
35. Graham SL, Martins A, Klistorner A, Arvind H, Leaney JC, Grigg JR (2010) High resolution multi-focal VEP: towards better definition of scotomas. Invest Ophthalmol Vis Sci 51:E-Abstract 4332
36. Klistorner A, Graham SL, Martins A, Grigg JR, Arvind H, Kumar RS, James AC, Billson FA (2007) Multifocal blue-on-yellow visual evoked potentials in early glaucoma. Ophthalmology 114:1613–1621
37. Arvind H, Graham SL, Leaney J, Grigg JR, Goldberg I, Billson F, Klistorner A (2009) Identifying pre-perimetric functional loss in glaucoma: a blue-on-yellow multifocal visual evoked potentials study. Ophthalmology 116: 1134–1141
38. Fortune B, Demirel S, Bui BV, Fortune B, Demirel S, Bui BV (2009) Multifocal visual evoked potential responses to pattern-reversal, pattern-onset, pattern-offset, and sparse pulse stimuli. Vis Neurosci 26:227–235
39. Klistorner A, Arvind H, Grigg JR, Graham SL (2010) Selective stimulation of non redundant pathways enhances the detection of glaucoma using multifocal VEPs. Invest Ophthalmol Vis Sci 51:E ARVO Abstract 5489

Chapter 18

Improving Motor Activation Patterns After Stroke with Wii-based Movement Therapy

Angelica G. Thompson-Butel, Sarah E. Scheuer, and Penelope A. McNulty

Abstract

Stroke is one of the leading causes of acquired motor disability worldwide. With no cure, the only way to recover motor function is through rehabilitation. Recent trials of Wii-based Movement Therapy have demonstrated improved functional ability that has transferred to greater independence in activities of daily living. This brief but intense 2-week programme can be tailored to the deficits of individual patients to promote motor recovery. More recent studies have investigated changes in electromyography and joint movement *during* Wii-based Movement Therapy. Results from a heterogeneous group of stroke patients suggest that improved movement ability and motor control are promoted by functionally relevant changes in patterns of muscle activation and joint movement.

Key words: Motor control, Stroke rehabilitation, Hemiplegia, Upper limb function, Electromyography

1. Stroke and Motor Disability

Stroke is one of the largest contributors to the total burden of disease worldwide (1). The incidence of stroke is expected to rise with an ageing population and burgeoning epidemics of obesity, physical inactivity and diabetes (2). One of the most common deficits after a unilateral stroke is hemiparesis, a weakness on one side of the body contralateral to the lesion. Hemiparesis immediately post-stroke reflects reduced descending corticospinal drive from the lesioned cortex (3, 4). Secondary changes develop over many months and include a decrease in motor unit numbers (5–7), changes to motor unit firing rates (8, 9) and recruitment patterns (10), loss of co-ordination (11, 12) and physiological changes in muscle fibre size (13, 14) and phenotype (15). Although both the upper and lower extremities may be affected (16, 17), changes in

Paul M. Pilowsky et al. (eds.), *Stimulation and Inhibition of Neurons*, Neuromethods, vol. 78,
DOI 10.1007/978-1-62703-233-9_18, © Springer Science+Business Media, LLC 2013

the upper extremity recover less and have a larger impact on functional disability (18, 19).

Motor disability within the upper limbs is typically higher for more distal muscles (20, 21) which have greater corticospinal drive (22–24). Upper limb disability is exacerbated by learned non-use of the more-affected arm and hand. Non-use develops as stroke survivors find it easier, more successful and less frustrating to use the less-affected (ipsilateral) limb (25, 26). However, the ipsilateral side is not entirely unaffected (22, 27). Uncrossed (28) or bilateral (22) projections from the lesioned cortex or changes in cortical excitability (29) may contribute to this deficit compounding the post-stroke motor disability. For this reason, we refer to the contralesional side as the more-affected side and the ipsilesional side as the less-affected side.

2. Wii-Based Movement Therapy

Wii-based Movement Therapy was developed to target rehabilitation of the upper limb (30). It provides a form of massed practice and forced use that is the basis of more conventional approaches such as Constraint-induced Movement Therapy (31, 32) but overcomes some of the concerns relating to patient compliance, motivation and safety (33, 34). The Nintendo Wii provides a cheap, engaging basis for therapy that can be tailored for individual patients (35, 36). The internal scoring system assists patient motivation and provides a means of monitoring improvement in game performance. In this chapter we demonstrate the success of Wii-based Movement Therapy and provide early evidence for functionally relevant changes in muscle activation patterns and joint movement that promote improved motor control to restore independence in activities of daily living.

Wii-based Movement Therapy was designed as a rehabilitation tool to target upper limb hemiparesis in patients after a unilateral stroke. Patients with unstable blood pressure, co-morbidities significantly affecting sensorimotor function, e.g. diabetic neuropathy, or cognitive deficits that limit the ability to follow instructions and/or participate in self-guided practice as assessed by a Mini-Mental State Exam score of <24 (37) are excluded. Each patient must have a carer available throughout the study to ensure patient safety. The results presented in this chapter are for 19 patients, aged 22–83 years (61.7 ± 14.7 years, mean and standard deviation) and 1–91 months post-stroke (19.7 ± 22.6 months). All patients gave signed, informed consent. This study was approved by St Vincent's Hospital Human Research Ethics Committee.

The brief, but intense, 2-week protocol (Fig. 1) has several elements. On 10 consecutive weekdays, therapist-guided formal

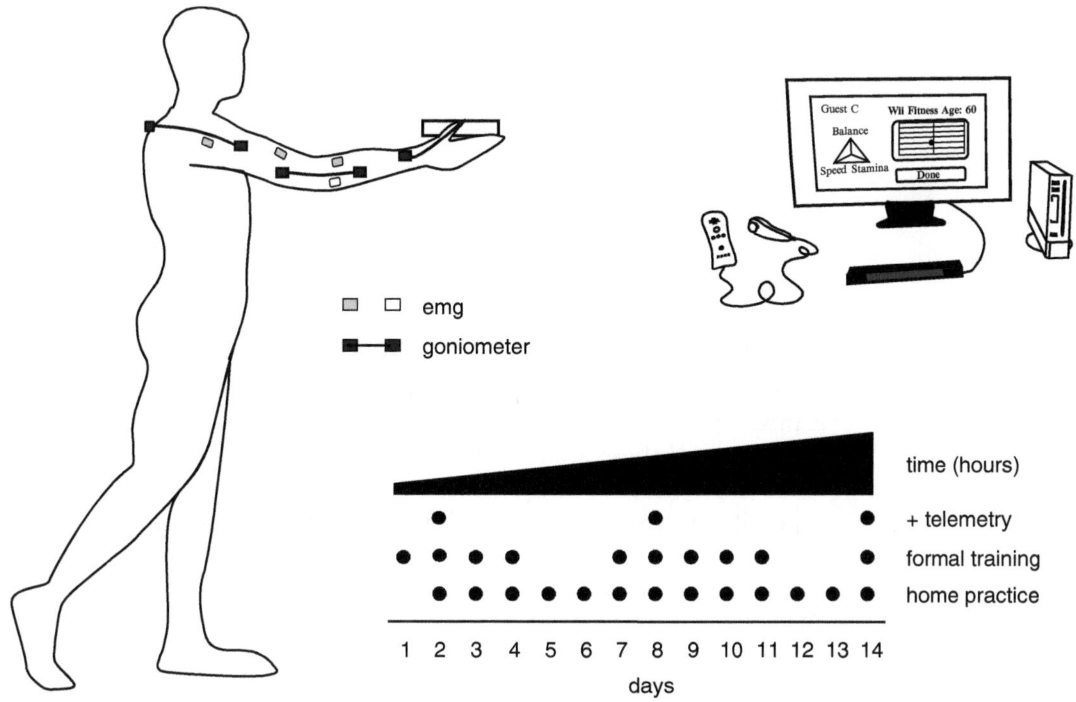

Fig. 1. Wii-based Movement Therapy protocol and schedule. The time indicates the total for formal sessions and home practice combined. All formal sessions were 1 h in duration. The home-practice sessions began with 15 min on day 2 and progressively increased towards 180 min by the end of the 2-week period. + telemetry indicates days on which neurophysiological recordings were made. The position of telemetry transducers is illustrated for the more-affected side only; FCR EMG position is indicated for the transducer position on the medial aspect of the forearm. ECR, FCR and biceps EMG were recorded on the more-affected side.

sessions last for 1 h with the time measured as time on task. Rest breaks are taken as necessary so that early sessions may take up to 2 h to complete. A second Wii system is installed in patients' homes on day 2, and daily home practice progressively builds from 15 to 180 min per day giving a combined therapy time of up to 4 h per day. Patients do not simply play Wii games; rather, the Nintendo Wii and Wii Sports (Nintendo, Japan) are used as a rehabilitation tool to train specific movement patterns. Wii games of tennis, golf, baseball, bowling and boxing are used to create a structured, targeted movement-based therapy. Tasks are introduced and varied according to individual patient's disability and progress. To prevent the potential for overuse injuries, formal therapy sessions and home-practice schedules require a change in activity at least every 10–20 min, and a variety of games and activities are used to target different joints and muscle groups. Home practice is monitored by the use of daily schedules, motor diaries and daily review of progress. When necessary, patient responses are verified with their carer.

The controller is only held in the more-affected hand. If necessary a self-adhesive wrap can be used to assist grip, but this support is progressively reduced over time. Whenever possible, training is undertaken with patients standing. Patients with unstable balance begin by holding on to a support with one hand and the level of support is progressively reduced. For patients with lower-limb disabilities, Wii therapy can be undertaken seated or in a wheelchair if necessary. All games can be effectively played in either position, with increased standing as locomotion, balance and endurance improve.

Each formal session concludes with patient-centred problem solving and goal setting. Goals are determined by the patient and range from functionally important tasks to game-based achievements, and these are reviewed daily. This aspect of therapy stresses the importance of using the more-affected hand and arm whenever possible to overcome learned non-use. In trials of constraint therapy, it has been suggested that this element of the protocol is fundamental to the success of therapy (38, 39).

3. Neurophysiological Assessments During Therapy

We have made recordings of muscle and joint activity during therapy using wireless telemetry (Myomonitor IV, Delsys, USA) (Fig. 1) in a subset of patients ($n=6$) aged 63–73 years (67.0 ± 3.6 years) and 7–91 months post-stroke (27.3 ± 32.0 months). Measurements occur in early, mid and late therapy typically during formal therapy sessions 2, 6 and 10. Surface electromyography (EMG) was recorded from the more-affected extensor carpi radialis (ECR), flexor carpi radialis (FCR), biceps brachii and deltoid and from the less-affected ECR, FCR and biceps (Fig. 1). The positioning of electrodes to record EMG in muscles affected by spasticity, atrophy, weakness and poor motor control is not simple. Initially placement is based on palpation of the muscle before a sequence of voluntary or passive movements is used to refine the optimal site. Possible anatomical and physiological changes to the muscle after stroke complicate the standardisation of EMG measurements over time. This difficulty is compounded when the goal of therapy is to alter the muscle and its activity. To maximise between session consistency of electrode placement, 9-megapixel digital images are taken at each session (Finepix S950, Fuji, Japan).

EMG signals are calibrated using a sequence of increasing load: with the limb supported at rest (supported), unsupported against gravity (unsupported) and resisting a 1-kg weight (weighted). If tonic activity is present, patients are encouraged to relax the muscle fully and the calibration procedure is repeated. Each condition is held for 5 s, and data are analysed from 1 s epochs of stable activity.

Although the EMG calibration methods employed in this study are appropriate in healthy subjects (40, 41), they are limited by paresis (9, 10), atrophy (42), hypertonia (43), spasticity (8), fatigue (41) and functional recovery in stroke patients. A better method may be to evoke the maximum muscle response either by supramaximal stimulation of the motor nerve or by voluntary efforts with calibrated force measurements.

Quantification of global EMG during Wii therapy, particularly with phasic activation during game play, was complicated by a number of factors. Single motor unit activity that was both spontaneous and task driven may have obscured the detection of systematic changes in EMG activity patterns. This was more noticeable for lower-functioning patients. During early sessions of Wii-based movement therapy, EMG signals were dominated by single motor unit activity, particularly in distal muscles. Such prolific activity may reflect the paucity of viable motor units in atrophied muscles post-stroke (44). Functional improvement coincided with the development of compound muscle activity. This change from single unit to compound muscle activity prevented meaningful measurement of changes in EMG during Wii Sports. However, we were able to measure changes in EMG during the calibration sequence for four patients. EMG data were amplified 1,000 times, filtered from 20 to 450 Hz, sampled at 1,000 Hz with EMGworks (Delsys, USA) and analysed using Spike2 software (CED, UK). All EMG signals were root mean square (RMS) processed using a sliding 250-ms window. High-definition video recordings enabled post hoc correlation between EMG data and therapy activities (30 frames/s at 1,280 × 720 9 Mb/s, VPC-CA9, Sanyo, Japan).

There was no difference in the pattern of activation between ECR and FCR, or between biceps brachii and deltoid. We therefore grouped the former into distal EMG and the latter into proximal EMG. The pattern of changes in EMG was different for proximal and distal muscles, and the amplitude of EMG activity in proximal muscles was approximately twice that of the distal muscles. EMG activity was graded across the calibration conditions, from supported to unsupported and weighted. There was a significant change in both proximal and distal muscle activities between conditions ($p < 0.001$) (Fig. 2a).

A decrease in distal muscle activity was observed, this reduction was most apparent during the weighted condition where mean values decreased between early and late therapy, from 10.2 ± 2.0 μV to 7.8 ± 2.1 μV (mean ± standard error of the mean (SEM)). Conversely, proximal muscle activity increased in both the unsupported and weighted condition where the mean value increased between early and late therapy, from 6.4 ± 0.9 μV to 10.1 ± 3.0 μV and 11.6 ± 2.3 μV to 17.8 ± 4.0 μV, respectively. The decrease in distal EMG was associated with a clear reduction in tonic muscle activity. Analysis of therapy videos suggests that

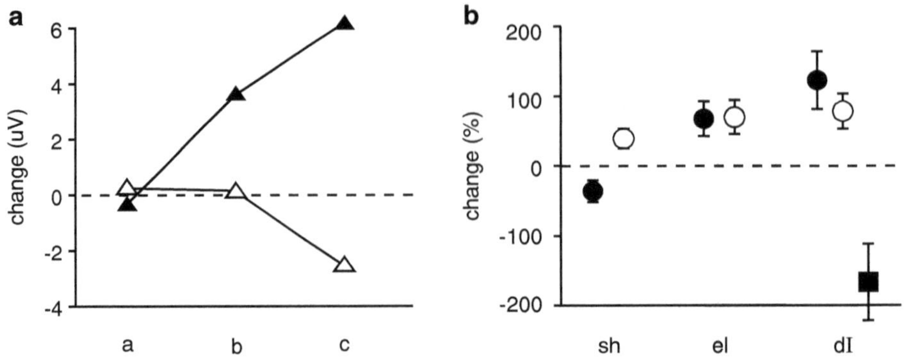

Fig. 2. Changes during therapy. (**a**) Mean change in RMS EMG from early to late therapy for proximal (*filled symbols*) and distal muscles (*open symbols*) during calibration sequences: *a* supported, *b* unsupported against gravity and *c* weighted with 1 kg ($n=4$). (**b**) Average change in maximum and minimum movement speed in the shoulder (*sh*), elbow (*el*) and thumb (*dI*) ($n=6$). Significant changes in shoulder maximum movement speed were specific to golf. When averaged across all sports, the change in shoulder maximum velocity was 2.2 ± 9.4%. The only significant change in joint excursion was in digit I (see text for explanation).

the reduced tonic activity is associated with more functionally appropriate movements.

Spontaneous activity of single motor units is evident under various physiological conditions including at rest, after voluntary contractions and with reflex muscle activation (7, 45). The origin of this drive is uncertain but may be due to tonic, low-level ionotropic inputs maintaining a proportion of the motoneurone pool closer to firing threshold and thus more readily activated (46). The onset of single motor unit discharges coincided with therapy activities, suggesting a response to descending motor commands. Some motor units clearly displayed both spontaneous and task-driven behaviour.

The development of single motor unit activity in otherwise inactive muscles, and the recruitment of additional motor units in muscles with minimal activity, presumably reflects neural plasticity in the cortex (3) driven by Wii Therapy, although this has not been directly assessed. Immediately post-stroke, there is an expanded area of recruitment during voluntary contraction within both the lesioned and contralesional hemispheres (47), which returns towards control values with motor recovery (48–50). Within the lesioned hemisphere, increased recruitment of secondary motor areas includes the premotor and supplementary motor areas of the cortex (51, 52). However, the level of motor recovery is limited as secondary motor regions are less efficient at generating motor output (3). We hypothesise that Wii-based Movement Therapy promotes improved movement ability through cortical reorganisation. The pattern of changes in motor function after Wii Therapy is similar to that of other protocols such as Constraint-induced

Movement Therapy for which brain imaging techniques have demonstrated cortical reorganisation (49, 53).

Joint range of motion of the more-affected shoulder, elbow and digit I interphalangeal joint was recorded during therapy with electrogoniometers positioned across the joints. Maximum joint excursion was measured, and movement speed was calculated from the first time derivative of the signal with maximum and minimum movement velocity recorded for each sport.

In contrast to active and passive measures of static range of motion (see below), there was no significant change in joint excursion measured during therapy except for a decrease of 166.7 ± 55.3% ($p = 0.002$) at the interphalangeal joint of digit I (Fig.2b). At this joint the maximum velocity of movement increased by 123.1 ± 40.7% ($p = 0.008$) which was accompanied by an increase in the velocity of slowing of 78.4 ± 25.2% ($p = 0.006$). A review of therapy videos suggests these changes corresponded to a smaller, more efficient button press by the thumb when playing golf. In early therapy there was an exaggerated movement involving much of the hand and wrist, but by late therapy this was more controlled and isolated to the thumb.

Maximum movement velocity significantly increased at the elbow by 67.8 ± 24.5% ($p = 0.008$), but there was no change in the minimum velocity. However, there was a baseball-specific increase in minimum movement velocity of 70.3 ± 24.9% ($p = 0.007$) (Fig. 2b). This more effective braking of movement speed reflected greater motor control particularly at end of range of motion. The 39.9 ± 14.4% increase in minimum velocity at the shoulder was significant ($p = 0.011$). Although the maximum velocity of shoulder movement did not change overall, being 2.2 ± 9.4%. Video analysis suggested there was a less jerky and more controlled movement that was particularly evident in golf putting. When analysed separately the change in the maximum velocity of shoulder movement during golf putting was −36.3 ± 16.0% ($p = 0.07$).

Joint range of motion is commonly restricted in the upper limb after stroke, particularly in the shoulder (54). This contributes to many movement deficits including abnormal postural adjustments (55) and movement synergies (20), incorrect timing within movement sequences (56) and loss of interjoint coordination (57). Assessment of joint movement during therapy should reflect changing motoneurone activation better than static range of motion measurements, whether active or passive. We were surprised not to see increased range of motion during therapy given the magnitude of the changes in active and passive measures of range of motion during functional assessments (see below). This may reflect the complex movements used during play or the difficulty of positioning uniaxial transducers to capture multidimensional movements. A potential solution may be afforded by accelerometry in multiple planes. Despite these limitations, a significant increase in minimum velocity

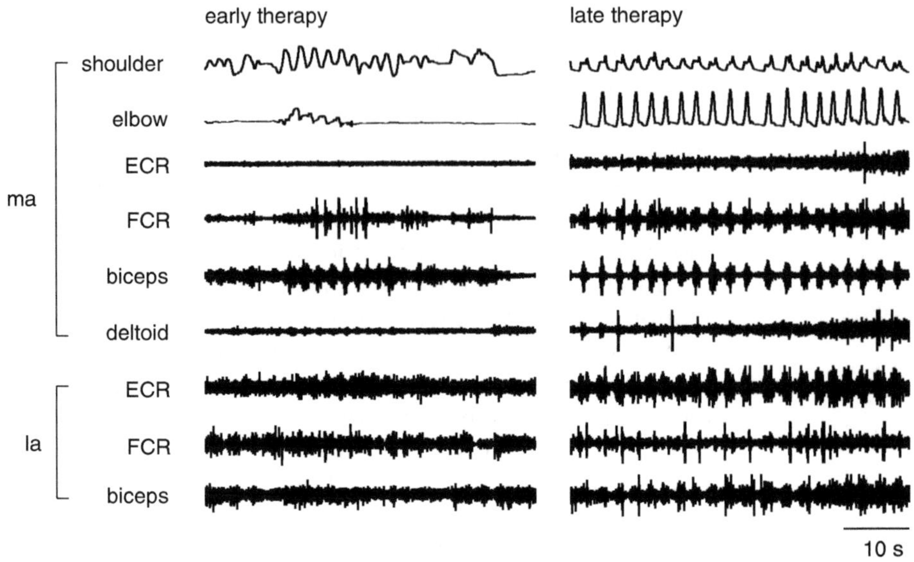

Fig. 3. Raw data during therapy for a 63-year-old male, 91 months post-stroke during Wii boxing. Top two traces are electrogoniometer recordings from the more-affected side; remaining traces are EMG from the more-affected side (*ma*) and less-affected side (*la*). Clearer more distinct bursts of EMG coinciding with crisper joint movements in late therapy (*right panel*) demonstrate increased levels of co-ordinated muscle activity.

at the shoulder and elbow and first digit was demonstrated. This is the functional equivalent of greater control in ending movements and corresponded qualitatively with the development of more purposeful and functional control in patients who initially presented with uncontrolled, jerky actions. These changes, particularly when game specific, suggest a re-establishment of the ability to update and modify motor commands (58). The perception that increased movement velocity is a functionally relevant goal of rehabilitation (e.g. (59)) may not always be advantageous or appropriate for individual patients.

Qualitative comparisons between early and late telemetry recordings revealed striking changes in muscle activation and joint movement. A more co-ordinated and coherent pattern of activity is clearly evident on both the more-affected and less-affected sides of the body (Fig. 3). This was seen on video analysis as smoother more purposeful movement. In this patient the wrist extensors and flexors were thought to be completely paralysed on the more-affected side before therapy, yet single motor unit activity can clearly be seen in the FCR trace in Fig. 3. By late therapy, this had developed into movement-related bursts of compound EMG activity, and single motor unit activity could be discerned in ECR where no activity was evident in early therapy.

4. Functional Assessments

Many functional assessments have been validated for use with post-stroke rehabilitation. Sensitivity may be limited in some tests by scoring over a small range and by floor and ceiling effects for low- and high-functioning patients, respectively. For this reason, we tested functional ability immediately before and after therapy using the Wolf Motor Function Test (WMFT) (60), the upper limb motor subscale of the Fugl–Meyer Assessment (54), the Quality of Movement subscale of the Motor Activity Log (MAL-QOM) (61, 62), and active and passive joint range of motion.

Both the WMFT and the FMA were specifically designed for use after stroke. The WMFT is a quantitative assessment of 15 timed tasks and two strength tasks (63). The tasks simulate movements and activities of daily life, such as grasping a can of drink and turning a key in a lock. The FMA uses a three-point scale to assess impairment of simple movements and more complex movement synergies (64). Although we do not have sufficient numbers for meaningful statistical comparisons yet, the WMFT appears to be more sensitive for high-functioning patients with a floor effect for low-functioning patients. Conversely, the FMA appears to be more sensitive for low-functioning patients with a ceiling effect for high-functioning patients.

In this cohort of patients, there was a significant improvement for the 15 timed tasks of the WMFT. Pre-therapy summed times 541.7 s (28.1–1697.9 s) (median and range) decreased to 464.6 s (18.7–1698.3 s) ($p<0.001$). The maximum time allowed per task is 120 s, with a total possible time of 1,800 s. The mean time for tasks that could be completed in the allotted 120 s significantly decreased from 57.7 ± 9.7 s to 38.4 ± 5.3 ($p<0.001$) (Fig. 4a). The ability to complete tasks within the 120 s also increased on average by one task per patient (range 0–3). For the strength-based tasks, the amount of weight that could be lifted improved from 8.5 ± 1.7 lb to 10.6 ± 2.0 lb ($p=0.009$); while handgrip strength increased but not significantly, 15.3 ± 2.5 kg pre-therapy, 16.2 ± 2.8 kg post-therapy. FMA scores increased from a mean of 41.8 ± 3.8 to 46.6 ± 3.8 after therapy. The maximum possible score for the FMA upper limb motor subscale is 66. This 13% change was significant ($p<0.001$) with scores improving over a range from 0, for high-functioning patients who scored 66 points pre-therapy (ceiling effect), to 20 points for low-functioning patients.

Range of motion was measured using a hand-held goniometer for both passive and active shoulder flexion, extension and lateral abduction and elbow flexion. After 2 weeks of Wii-based movement therapy, both passive and active range of motion improved for all patients who did not have full range of motion. The mean improvement for shoulder and elbow passive range of motion was

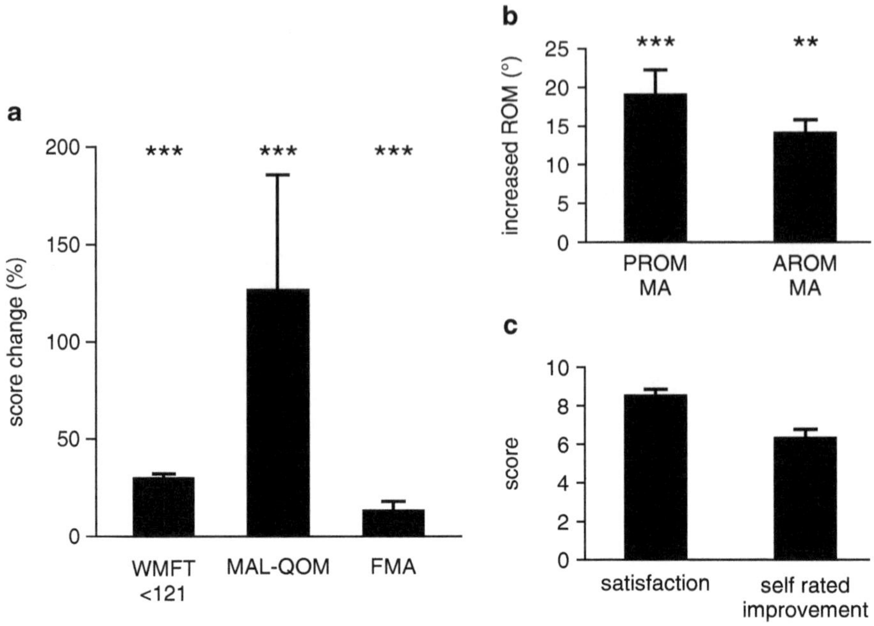

Fig. 4. Functional assessment results. (**a**) Improvement in functional ability assessed using the Wolf Motor Function Test (WMFT) (30% improvement), Fugl-Meyer Assessment (FMA) (13% improvement) and Motor Activity Log Quality of Movement (MAL-QOM) scale (128%). The change score for WMFT timed tasks is for tasks completed in the allocated 120 s, and this measure highlights improvements in existing movement rather than acquisition of new movement. (**b**) Improved active (AROM) and passive (PROM) range-of-motion score combined for shoulder flexion, extension and lateral abduction and elbow flexion. (**c**) Patient satisfaction and self-perceived improvement rated using a 10-point visual analogue scale (***$p<0.001$, **$p<0.01$).

19±3° ($p<0.001$) and 14±2° ($p=0.004$) for active range of motion (Fig. 4b).

The MAL-QOM provides an index of the transfer of therapy-induced gains in motor ability to activities of daily living (62). Thirty everyday activities are scored using a six-point scale to assess use of the more-affected upper limb. Scores range from 0 where the more-affected hand and arm are not used at all to 5 in which performance with the more-affected hand is equivalent to pre-stroke ability. A score of 3 represents independent use of the hand. The averaged sum of the scores for all 30 activities significantly improved from 65.6±10.8 to 91.1±10.4 ($p<0.001$) (Fig. 4a). The mean score for all 30 items increased after therapy from 2.2±0.4 to 3±0.3 ($p<0.001$). The number of tasks that were scored >3 improved for all patients by an average of 5.3 tasks post-therapy (range 0–23). It is generally accepted that the MAL-QOM reflects 'how much' as well 'how well' the more-affected hand and arm are used, and although the Motor Activity Log also has an amount of use subscale, the two measures are highly correlated and convergent (65). Thus, the simpler MAL-QOM is increasingly being administered alone.

During the post-therapy assessments, patients were asked to rate their satisfaction with therapy and the extent of improvement using a 10-point visual analogue scale. Patients were highly satisfied with mean scores of 8.5 ± 0.3 (Fig. 4c). Self-perceived improvement was measured on the same scale, averaging 6.3 ± 0.5 (Fig. 4c). Many patients commented that the hard work of therapy was no longer considered a chore, therapy was fun; they were now able to undertake simple everyday tasks they never thought possible; they felt more confident and now had hope for the future.

Wii-based Movement Therapy uses the Nintendo Wii and Wii Sports as a rehabilitation tool to improve patterns of muscle activation after stroke to promote the recovery of functional movement. All patients successfully completed an intense 2-week therapy programme which resulted in increased proximal muscle activation, decreased tonic activity of distal muscles, more controlled movement speed and greater range of motion in the shoulder and elbow joints. Qualitatively, by the end of therapy, there was a more coherent, co-ordinated pattern of activation clearly evident in both EMG and joint movement. These changes were associated with significant improvements in functional motor ability and independence in activities of daily living assessed using the WMFT and MAL-QOM, respectively. Wii-based Movement Therapy provides a form of massed practice, invoking forced use of the more-affected hand but without the use of restraints such as those of Constraint-induced Movement Therapy (32, 66). Not only do patients improve their functional ability and range of motion but these data suggest a physiological basis for such improvements via changes in the activation of motoneurones. These results suggest that Wii-based Movement Therapy is effective as an upper limb post-stroke rehabilitation strategy.

Acknowledgments

This work was supported by the National Health and Medical Research Council of Australia and the NSW Office of Science and Medical Research (Spinal Injury and other Neurological Diseases Research Program).

References

1. Yach D, Hawkes C, Gould CL, Hofman KJ (2004) The global burden of chronic diseases: overcoming impediments to prevention and control. JAMA 291(21):2616–2622
2. Senes S (2006) How we manage stroke in Australia. Canberra AIHW.
3. Ward NS, Newton JM, Swayne OB, Lee L, Thompson AJ, Greenwood RJ, Rothwell JC, Frackowiak RS (2006) Motor system activation after subcortical stroke depends on corticospinal system integrity. Brain 129(Pt 3): 809–819

4. Lemon RN (2008) Descending pathways in motor control. Annu Rev Neurosci 31:195–218
5. Hara Y, Akaboshi K, Masakado Y, Chino N (2000) Physiologic decrease of single thenar motor units in the F-response in stroke patients. Arch Phys Med Rehabil 81(4):418–423
6. Hara Y, Masakado Y, Chino N (2004) The physiological functional loss of single thenar motor units in the stroke patients: when does it occur? Does it progress? Clin Neurophysiol 115(1):97–103
7. Lukacs M (2005) Electrophysiological signs of changes in motor units after ischaemic stroke. Clin Neurophysiol 116(7):1566–1570
8. Rosenfalck A, Andreassen S (1980) Impaired regulation of force and firing pattern of single motor units in patients with spasticity. J Neurol Neurosurg Psychiatry 43(10):907–916
9. Tang A, Rymer WZ (1981) Abnormal force—EMG relations in paretic limbs of hemiparetic human subjects. J Neurol Neurosurg Psychiatry 44(8):690–698
10. Gemperline JJ, Allen S, Walk D, Rymer WZ (1995) Characteristics of motor unit discharge in subjects with hemiparesis. Muscle Nerve 18(10):1101–1114
11. Levin MF, Selles RW, Verheul MH, Meijer OG (2000) Deficits in the coordination of agonist and antagonist muscles in stroke patients: implications for normal motor control. Brain Res 853(2):352–369
12. Neckel N, Pelliccio M, Nichols D, Hidler J (2006) Quantification of functional weakness and abnormal synergy patterns in the lower limb of individuals with chronic stroke. J Neuroeng Rehabil 3:17
13. Scelsi R, Lotta S, Lommi G, Poggi P, Marchetti C (1984) Hemiplegic atrophy. Morphological findings in the anterior tibial muscle of patients with cerebral vascular accidents. Acta Neuropathol 62(4):324–331
14. Slager UT, Hsu JD, Jordan C (1985) Histochemical and Morphometric Changes in Muscles of Stroke Patients. Clin Orthop Relat Res 199:159–168
15. Dattola R, Girlanda P, Vita G, Santoro M, Roberto ML, Toscano A, Venuto C, Baradello A, Messina C (1993) Muscle rearrangement in patients with hemiparesis after stroke: an electrophysiological and morphological study. Eur Neurol 33(2):109–114
16. Duncan PW, Goldstein LB, Horner RD, Landsman PB, Samsa GP, Matchar DB (1994) Similar motor recovery of upper and lower extremities after stroke. Stroke 25(6):1181–1188
17. Desrosiers J, Malouin F, Richards C, Bourbonnais D, Rochette A, Bravo G (2003) Comparison of changes in upper and lower extremity impairments and disabilities after stroke. Int J Rehabil Res 26(2):109–116
18. Kalra L, Dale P, Crome P (1993) Improving stroke rehabilitation. A controlled study. Stroke 24(10):1462–1467
19. Feys HM, De Weerdt WJ, Selz BE, Cox Steck GA, Spichiger R, Vereeck LE, Putman KD, Van Hoydonck GA (1998) Effect of a therapeutic intervention for the hemiplegic upper limb in the acute phase after stroke: a single-blind, randomized, controlled multicenter trial. Stroke 29(4):785–792
20. Twitchell TE (1951) The restoration of motor function following hemiplegia in man. Brain 74(4):443–480
21. Beebe JA, Lang CE (2009) Active range of motion predicts upper extremity function 3 months after stroke. Stroke 40(5):1772–1779
22. Colebatch JG, Gandevia SC (1989) The distribution of muscular weakness in upper motor neuron lesions affecting the arm. Brain 112(Pt 3):749–763
23. Palmer E, Ashby P (1992) Corticospinal projections to upper limb motoneurones in humans. J Physiol 448:397–412
24. Turton A, Lemon RN (1999) The contribution of fast corticospinal input to the voluntary activation of proximal muscles in normal subjects and in stroke patients. Exp Brain Res 129(4):559–572
25. Taub E (1980) Somatosensory deafferentation research with monkeys: implications for rehabilitation medicine. Behav Psychol Rehabil 30:1303–1313
26. Taub E, Uswatte G, Mark VW, Morris DM (2006) The learned nonuse phenomenon: implications for rehabilitation. Eura Medicophys 42(3):241–256
27. Brodal A (1973) Self-observations and neuro-anatomical considerations after a stroke. Brain 96(4):675–694
28. Nyberg-Hansen R (1963) Some comments on the pyramidal tract, with special reference to its individual variation in man. Acta Neurol Scand 39:1–30
29. Hwang IS, Abraham LD (2001) Quantitative EMG analysis to investigate synergistic coactivation of ankle and knee muscles during isokinetic ankle movement. Part 1: time amplitude analysis. J Electromyogr Kinesiol 11(5):319–325

30. Mouawad MR, Doust CG, Max MD, McNulty PA (2011) Wii-based movement therapy to promote improved upper extremity function post-stroke: a pilot study. J Rehabil Med 43(6):527–533
31. Mark VW, Taub E (2004) Constraint-induced movement therapy for chronic stroke hemiparesis and other disabilities. Restor Neurol Neurosci 22(3–5):317–336
32. Wolf SL, Winstein CJ, Miller JP, Taub E, Uswatte G, Morris D, Giuliani C, Light KE, Nichols-Larsen D (2006) Effect of constraint-induced movement therapy on upper extremity function 3 to 9 months after stroke: the EXCITE randomized clinical trial. JAMA 296(17):2095–2104
33. Page SJ, Levine P, Sisto S, Bond Q, Johnston MV (2002) Stroke patients' and therapists' opinions of constraint-induced movement therapy. Clin Rehabil 16(1):55–60
34. Sterr A, Szameitat A, Shen S, Freivogel S (2006) Application of the CIT concept in the clinical environment: hurdles, practicalities, and clinical benefits. Cogn Behav Neurol 19(1):48–54
35. Saposnik G, Mamdani M, Bayley M, Thorpe KE, Hall J, Cohen LG, Teasell R (2010) Effectiveness of Virtual Reality Exercises in STroke Rehabilitation (EVREST): rationale, design, and protocol of a pilot randomized clinical trial assessing the Wii gaming system. Int J Stroke 5(1):47–51
36. Yong Joo L, Soon Yin T, Xu D, Thia E, Pei Fen C, Kuah CW, Kong KH (2010) A feasibility study using interactive commercial off-the-shelf computer gaming in upper limb rehabilitation in patients after stroke. J Rehabil Med 42(5):437–441
37. Folstein MF, Folstein SE, McHugh PR (1975) Mini-mental state. A practical method for grading the cognitive state of patients for the clinician. J Psychiatr Res 12(3):189–198
38. Taub E, Uswatt G (2006) Constraint-Induced Movement therapy: answers and questions after two decades of research. NeuroRehabilitation 21(2):93–95
39. Gauthier LV, Taub E (2009) CI therapy: a method for harnessing neuroplastic changes to improve rehabilitation after damage to the brain. In: Schmorrow DD, Estabrooke IV, Grootjen M (eds) Foundations of augmented cognition neuroergonomics and operational neuroscience. Springer, Berlin, pp 792–799
40. Mirka GA (1991) The quantification of EMG normalization error. Ergonomics 34(3): 343–352
41. Oberg T (1995) Muscle fatigue and calibration of EMG measurements. J Electromyogr Kinesiol 5(4):239–243
42. Toffola ED, Sparpaglione D, Pistorio A, Buonocore M (2001) Myoelectric manifestations of muscle changes in stroke patients. Arch Phys Med Rehabil 82(5):661–665
43. Shaw JBJ, Deuvorst N, Macfie C, Brouwer B (1999) Clinical and physiological measures of tone in chronic stroke. J Neurol Phys Ther 23:19–24
44. English C, McLennan H, Thoirs K, Coates A, Bernhardt J (2010) Loss of skeletal muscle mass after stroke: a systematic review. Int J Stroke 5(5):395–402
45. Mottram CJ, Suresh NL, Heckman CJ, Gorassini MA, Rymer WZ (2009) Origins of abnormal excitability in biceps brachii motoneurons of spastic-paretic stroke survivors. J Neurophysiol 102(4):2026–2038
46. Mottram CJ, Wallace CL, Chikando CN, Rymer WZ (2010) Origins of spontaneous firing of motor units in the spastic-paretic biceps brachii muscle of stroke survivors. J Neurophysiol 104(6):3168–3179
47. Butefisch CM (2004) Plasticity in the human cerebral cortex: lessons from the normal brain and from stroke. Neuroscientist 10(2):163–173
48. Johansen-Berg H, Dawes H, Guy C, Smith SM, Wade DT, Matthews PM (2002) Correlation between motor improvements and altered fMRI activity after rehabilitative therapy. Brain 125(Pt 12):2731–2742
49. Liepert J, Hamzei F, Weiller C (2004) Lesion-induced and training-induced brain reorganization. Restor Neurol Neurosci 22(3–5):269–277
50. Nelles G (2004) Cortical reorganization—effects of intensive therapy. Restor Neurol Neurosci 22(3–5):239–244
51. Ward NS, Brown MM, Thompson AJ, Frackowiak RS (2003) Neural correlates of outcome after stroke: a cross-sectional fMRI study. Brain 126(Pt 6):1430–1448
52. Ward NS, Brown MM, Thompson AJ, Frackowiak RS (2003) Neural correlates of motor recovery after stroke: a longitudinal fMRI study. Brain 126(Pt 11):2476–2496
53. Kim YH, Park JW, Ko MH, Jang SH, Lee PK (2004) Plastic changes of motor network after constraint-induced movement therapy. Yonsei Med J 45(2):241–246
54. Fugl-Meyer AR, Jaasko L, Leyman I, Olsson S, Steglind S (1975) The post-stroke hemiplegic patient. 1. A method for evaluation of physical

performance. Scand J Rehabil Med 7(1):13–31
55. Di Fabio RP, Badke MB, Duncan PW (1986) Adapting human postural reflexes following localized cerebrovascular lesion: analysis of bilateral long latency responses. Brain Res 363(2):257–264
56. Archambault P, Pigeon P, Feldman AG, Levin MF (1999) Recruitment and sequencing of different degrees of freedom during pointing movements involving the trunk in healthy and hemiparetic subjects. Exp Brain Res 126(1):55–67
57. Levin MF (1996) Interjoint coordination during pointing movements is disrupted in spastic hemiparesis. Brain 119(Pt 1):281–293
58. Shadmehr R, Smith MA, Krakauer JW (2010) Error correction, sensory prediction, and adaptation in motor control. Annu Rev Neurosci 33:89–108
59. Van Vliet PM, Kerwin D, Sheridan M, Fentem P (1995) The influence of functional goals on the kinematics of reaching following stroke. J Neurol Phys Ther 19:11–16
60. Wolf SL, Thompson PA, Morris DM, Rose DK, Winstein CJ, Taub E, Giuliani C, Pearson SL (2005) The EXCITE trial: attributes of the Wolf Motor Function Test in patients with subacute stroke. Neurorehabil Neural Repair 19(3):194–205
61. Taub E, Miller NE, Novack TA, Cook EW, Fleming WC, Nepomuceno CS, Connell JS, Crago JE (1993) Technique to improve chronic motor deficit after stroke. Arch Phys Med Rehab 74(4):347–354
62. Uswatte G, Taub E, Morris D, Vignolo M, McCulloch K (2005) Reliability and validity of the upper-extremity Motor Activity Log-14 for measuring real-world arm use. Stroke 36(11):2493–2496
63. Wolf SL, Lecraw DE, Barton LA, Jann BB (1989) Forced use of hemiplegic upper extremities to reverse the effect of learned nonuse among chronic stroke and head-injured patients. Exp Neurol 104(2):125–132
64. Gladstone DJ, Danells CJ, Black SE (2002) The fugl-meyer assessment of motor recovery after stroke: a critical review of its measurement properties. Neurorehabil Neural Repair 16(3):232–240
65. Uswatte G, Taub E, Morris D, Light K, Thompson PA (2006) The Motor Activity Log-28: assessing daily use of the hemiparetic arm after stroke. Neurology 67(7):1189–1194
66. Taub E, Uswatte G, King DK, Morris D, Crago JE, Chatterjee A (2006) A placebo-controlled trial of constraint-induced movement therapy for upper extremity after stroke. Stroke 37(4):1045–1049

INDEX

Paul M. Pilowsky et al. (eds.), *Stimulation and Inhibition of Neurons*, Neuromethods, vol. 78,
DOI 10.1007/978-1-62703-233-9, © Springer Science+Business Media, LLC 2013

MIX
Papier aus verantwortungsvollen Quellen
Paper from responsible sources
FSC® C105338

If you have any concerns about our products,
you can contact us on
ProductSafety@springernature.com

In case Publisher is established outside the EU,
the EU authorized representative is:
Springer Nature Customer Service Center GmbH
Europaplatz 3, 69115 Heidelberg, Germany

Printed by Libri Plureos GmbH
in Hamburg, Germany